CISM COURSES AND LECTURES

The series presents lecture notes, monographs, edited works and proceedings in the field of Mechanics, Engineering, Computer Science and Applied Mathematics.
Purpose of the series is to make known in the international scientific and technical community results obtained in some of the activities organized by CISM, the International Centre for Mechanical Sciences.

INTERNATIONAL CENTRE FOR MECHANICAL SCIENCES

COURSES AND LECTURES - No. 367

ARCH DAMS

DESIGNING AND MONITORING FOR SAFETY

EDITED BY

JOSÉ O. PEDRO
NATIONAL LABORATORY OF CIVIL ENGINEERING

Springer-Verlag Wien GmbH

This volume contains 153 illustrations

Originally published by Springer-Verlag Wien New York in 1999
SPIN 10730576

In order to make this volume available as economically and as rapidly as possible the authors' typescripts have been reproduced in their original forms. This method unfortunately has its typographical limitations but it is hoped that they in no way distract the reader.

ISBN 978-3-211-83149-6 ISBN 978-3-7091-2488-8 (eBook)
DOI 10.1007/978-3-7091-2488-8

PREFACE

Arch dams are in general large structures integrated in hydraulic projects of social and economical significance and impact on the environment, their safety and performance being therefore of the utmost importance. The assessment of safety and performance of arch dams, either for design of new dams or for monitoring of the existing ones, requires the utilisation of appropriate criteria, models and methods of analysis.

A short course on Mathematical modelling for arch dam design and safety evaluation, was held in Udine (Italy) in the week 4-8 of September, 1995. This course was organised by the International Centre for Mechanical Sciences (Udine) in co-operation with the National Laboratory of Civil Engineering (Lisbon, Portugal), and provided a comprehensive understanding of the modern criteria, models and mathematical methods of analysis of arch dams. It was developed in thirty lectures dealing with the following five themes: models and methods of analysis, scenarios for the construction phase, scenarios for the operation conditions, dynamic behaviour, and safety and performance of arch dams. These themes were developed respectively by Professors M. Fanelli (National Power Board, Italy), W. Wittke (Technical University of Aachen, Germany), G. Heigerth (University of Technology of Graz, Austria), R.T. Severn (University of Bristol, United Kingdom), and myself. The course was attended by twenty four post-graduate students and professionals from Universities, Research Institutions and Consultant Firms of the following countries: Austria, Bulgaria, Canada, Germany, India, Iran, Italy, Poland, Portugal, Spain, Sweden, Switzerland and United Kingdom.

Since the publication of the lectures of the course was considered of interest, the present volume was prepared. It is entitled Arch dams: designing and monitoring for safety, and is broken down into three parts: Part 1 - Safety and performance of arch dams; Part 2 - Models and melhods of analysis; Part 3 - Dynamic behaviour of arch dams. Unfortunately, it has not been possible to obtain the texts of the lectures concerning the other two themes.

Part 1 presents a brief characterisation of arch dams in the world, and a historical background of these structures since the Roman times up to our days. The basic concepts of safety and performance, and the criteria, models and methods of analysis used for their evaluation are discussed and illustrated using examples. Reference is made to the legal aspects conceming the safety of dams.

Part 2 develops a detailed and thorough explanation of the models and mathematical methods of analysis of arch dams. The simple approaches for preliminary designs, as well as the state-of-the-art finite element approaches are discussed, both for static and dynamic loads, and are illustrated by examples.

Part 3 deals with the dynamic behaviour of arch dams, especially under seismic actions. A comprehensive presentation is made concerning the definition of the design seismic actions, the dam-reservoir interaction, the model analysis, and the full scale dynamic tests. Examples are given for dams all over the world.

Thanks are due to the International Centre for Mechanical Sciences and the National Laboratory of Civil Engineering for their support to the course on arch dams and to the present publication. Special acknowledgements are due to Professors M. Fanelli and R.T. Severn for having participated in the course and in the preparation of this volume, as well as to Professors W. Wittke and G. Heigerth for their participation in the course. Nuno Grossmann, Research Officer of the National Laboratory of Civil Engineering, deserves also our gratitude for his contribution in the preparation of Part 1 of the present volume.

José Oliveira Pedro

CONTENTS

PART 1

SAFETY AND PERFORMANCE OF ARCH DAMS

by

J. O Pedro

National Laboratory of Civil Engineering, LNEC

Portugal

1
INTRODUCTION

1 - The supply of *fresh water* is a matter of great concern for a farseeing water management because fresh water is a finite and limited resource (it corresponds to only 2% of all the water existing in the world). The demand for it increases rapidly, in accordance with the increase of the world population. An adequate management of the fresh water resources is therefore necessary, and the management of the river water resources, which form only a small part of the total fresh water resources, is particularly important for a *sustained development*.

In general, the river water management projects (whenever possible multi-purpose projects for *water supply*, *irrigation*, *power production*, *navigation*, *flood control* and *recreation*) require the construction of *dams*. A dam project includes several components, such as:

- the *dam body* and its *foundation*;

- the *appurtenant works*, which are necessary either for the operation (intake towers, channels, power station and associated hydraulic circuits, penstocks, tunnels and other underground works, protection devices and restitution circuits, navigation and fish locks, hydraulic and electromechanical equipment, logways) or for the safety (spillways, surface and bottom outlets, energy dissipaters - stilling basins, roller buckets or plunge pools - hydraulic and electromechanical equipment);

- the *reservoir* and the *slopes downstream of the dam*.

The problems of the projects for large dam have been considered within the frame of the *International Commission on Large Dams* (ICOLD). According to ICOLD, *large dams* are those which are:

- *above 15 m in height* (measured from the lowest point of the foundation to the top);

or

- *between 10 m and 15 m in height and meet at least one of the following conditions*: 1) the crest length is not less than 500 m; 2) the capacity of the reservoir is not less than 10^6 m^3; 3) the maximum flood discharge is not less than 2,000 m^3/s; 4) it has specially difficult foundations; 5) it is of an unusual design.

Dams may be built with different materials and may have different characteristics. They are usually classified as follows:

- *concrete* and *masonry* dams (*gravity*, *buttress*, *arch* and *multiple arch* types);

- *earthfill* and *rockfill* structures.

The present course deals with arch dams only; we shall also refer to multiple arch dams, particularly to large-span multiple arch dams.

2 - An *arch dam* is a concrete or masonry structure curved in plan, so as to transmit the major part of the water load to the abutments.

Arch dams may be classified into the two following groups:

- *arch-gravity* dams, in general, thick vaults, with a single curvature in plan;

- *double curvature arch dams*, in general, thin shells or domes, with a vertical and horizontal curvature.

As mentioned above, we shall also consider the single or double curvature arches included in multiple arch dams.

The main terminology used for arch dams is summarised in Fig.1.

The *choice of an arch structure* as a suitable design solution for a dam project may be affected by different factors, particularly by *geotechnique*, *topography*, *hydrology* and *climate*.

The existence of *rock abutments* is a condition of vital importance. Both U or V shaped valleys may be appropriate for arch dams, provided the span/height ratio is not very large.

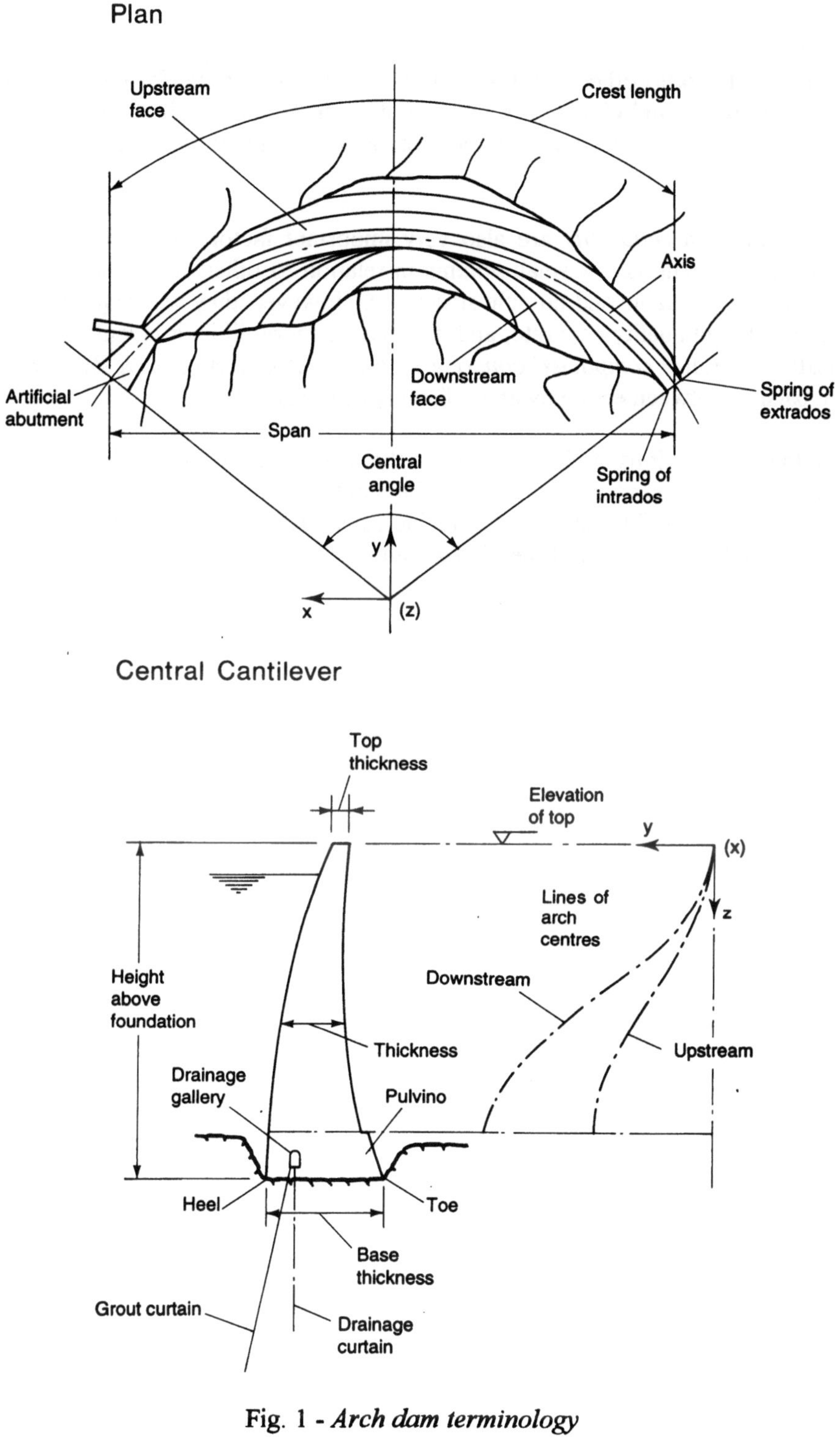

Fig. 1 - *Arch dam terminology*

Arch dams have been built in valleys with a span/height ratio up to 5, or even more, but for large values of this ratio other solutions may be more appropriate, for instance, multiple arch dams. The hydrology and climate are not conditioning factors; arch dams have been built on great rivers and on sites with large temperature variations, aggressive waters, and freezing and thawing.

Economy and *reliability* are also very important factors in the choice of a dam. In general, arch dams will have a small volume in relation to other dams, particularly to gravity dams, and are very reliable and durable structures. However, they will require more expensive formworks and high quality concrete, and may have a more difficult layout. It is expected that the utilisation of *roller compacted concrete in the construction of arch dams* will lead to an optimum balance between economy and reliability (Geringer 1994).

In the next sections, a classification of the existing arch and multiple arch dams, and a brief historical background are presented. The safety and performance as well as the major deterioration scenarios for these dams are indicated. Reference is made to the models and methods used in the evaluation of their behaviour.

2
ARCH DAMS
HISTORICAL BACKGROUND

1 - *Existing arch dams*

1.1 - According to the World Register of Dams (ICOLD 1984/88), 36 235 large dams of all types were operating in the world in 1986(not considering 328 other large dams situated in non-member countries of the ICOLD). *Among the 36 235 dams, only 1 592 were arch dams, and 141 were multiple arch dams* (Table 1).

The number of dams of all types under construction in 1986 was 1010; among these, 110 (11%) were arch and multiple arch dams (ICOLD 1984/88). Data of recent inquiries (ICOLD 1992/93/94) indicate that the number of dams of all types under construction in recent years remains about the same (1 100 to 1 200), as well as the proportion of arch and multiple arch dams, among the dams of all types. These inquiries indicate also that, in average, 300 dams of all types (and therefore about *30 arch and multiple arch dams*) are being *started* and *commissioned every year*.

In order to define the general characteristics of the currently operating dams in the world, the last available data (end of 1986) are used. These data are still considered as appropriate for that purpose, since the increase from 1986 to 1995 in the number of dams in operation is only 15%.

1.2 - From the existing 36 235 dams, 84% (30 498) are situated in the *USA*, *Europe*, *China*, and *Japan*. In a similar way, 1 384 arch dams (87% of the total) and 106 multiple arch dams (75% of the total) are situated in those regions (Table 1). Table 1 also shows that, among the existing dams, only 4.8% are *arch or multiple arch* dams (3.5% *in USA*, 11.6% *in Europe*, 4.2% *in China*, and 2.3% *in Japan*). increases with the height class, from 3.1% (for the dams

Considering the *European countries* only, Table 2 shows that a large number of the existing arch or multiple arch dams are built in the *Alps* and in the *Iberian peninsula,* namely in *Austria, France, Italy, Portugal, Spain, and Switzerland.* In these countries, with the exception of Spain, the ratio of the arch and multiple arch dams in relation to the dams of all

Country	*Number of dams*			*(Arch and multiple arch) / All types (%)*
	All types	*Arch*	*Multiple arch*	
USA	5 459	165	27	3.5
Europe (1)	3 991	418	43	11.6
China	18 820	753	32	4.2
Japan	2 228	48	4	2.3
SUBTOTAL (USA, Europe, China, Japan)	30 498	1 384	106	4.9
Other ICOLD member countries	5 737	208	35	4.2
TOTAL	36 235	1 592	141	4.8

(1) *Albania, Austria, Belgium, ex-Czechoslovakia, Denmark, Finland, France, Netherlands, Norway, Poland, Portugal, Romania, Spain, Sweden, Switzerland, Turkey, ex-USSR, ex-Yugoslavia*

Table 1 - *Dams in the world*

types is higher than the same ratio for all the above mentioned 26 European countries (Table 2). The ratio is particularly high in Switzerland, where more than one third of the existing dams are arch dams.

On the other hand, as may also be seen in Table 2, about 50% of the dams of all types and 75% of the arch and multiple arch dams, built in the 26 European countries, are situated in the above indicated 6 countries.

1.3 - The characterisation of the arch and multiple *arch dams by height classes* is presented in Table 3. This table shows that the ratio between the arch and multiple arch dams and the dams of all types increases with the height class, from 3.1% (for the dams with heights from 15 m to 30 m) to 47.8% (for the dams above 150 m in height). Therefore, *a large number of the exis-*

Country	*Number of dams*			*(Arch and multiple arch) / all types %*
	All types	*Arch*	*Multiple arch*	
Austria	123	17	-	13.8
France	468	85	19	22.2
Italy	440	97	10	24.3
Portugal	81	17	2	23.5
Spain	737	46	1	6.4
Switzerland	144	51	-	35.4
SUBTOTAL (*Austria, France, Italy, Portugal, Spain, Switzerland*)	1 993	313	32	17.3
Other European countries	1 998	105	11	5.8
TOTAL	3 991	418	43	11.6

Table 2 - *Dams in Europe*

ting higher dams are arched structures (about *one third* of the *dams above 100 m*, and *one half* of the *dams above 150 m*).

Dam type	*Number of dams by height class*				
	15-30 (m)	*30-60 (m)*	*60-100 (m)*	*100-150 (m)*	*Above 150 (m)*
Arch (1)	793	456	214	87	42
Multiple arch (2)	78	48	13	1	1
All types (3)	28 546	6 031	1 247	321	90
$\frac{(1)+(2)}{(3)}$ (%)	3.1	8.4	18.2	27.4	47.8

Table 3 - *Dams in the world by height classes*

The arch and multiple arch dams completed in the different *countries* and different *construction periods* are indicated in Table 4.

Construction period	Dam type	Number of dams				
		USA	Europe	China	Japan	Other countries
Up to 1950	Arch Mult. arch	119 25	110 23	- -	1 2	66 12
1950 - 1986	Arch Mult. arch	46 2	308 20	753 20	47 2	142 23

Table 4 - *Arch and multiple arch dams by construction period, in different countries*

Table 4 shows that the construction of arch and multiple arch dams has been particularly important to *Europe*. In *China* and *Japan* it occurred mainly in the second half of the 20th century, whereas in the *USA* arch and multiple arch dams were mainly built in the first half of this century.

2 - *First arch dams*

2.1 - Arch dams have been built since ancient times, mainly for irrigation and water supply. The *Romans* were great builders of arch structures and constructed many arch dams in different regions of the Roman Empire.

The *Roman arch dams* were arch-gravity structures, circular or polygonal in plan, some including wing walls and buttresses. Multiple arch dams were also built. In general, masonry was the used material, in some cases together with earth or concrete cores. The design was based on *empirical criteria*.

Major *examples* of Roman arch dams are (Quintela 1987, Schnitter 1994):

- The *Vallon de Baume arch dam*, built for supplying water to Saint-Rémy de Provence (France). At the end of 19th century (1891), the site of this dam was almost entirely covered by a modern dam, but a plan of the foundation excavations is still available (Fig. 2). The dam was about 12 m high, the crest length 18 m, and the thickness 3.9 m.

- The *Esparragalejo multiple arch dam* (5.6 m high), situated near Merida (Spain), and still in a good condition; and the *Monte Novo arch dam* (5.7 m high, a crest length of 52 m, including the wing walls, and 6.5 m of base thickness), situated near Evora (Portugal) (Fig 2).

- The *Derb arch dam* (10 m high), situated near Kasserine (Tunisia); and the *Dara Est arch dam* (5 m high), situated near Mardim (Turkey).

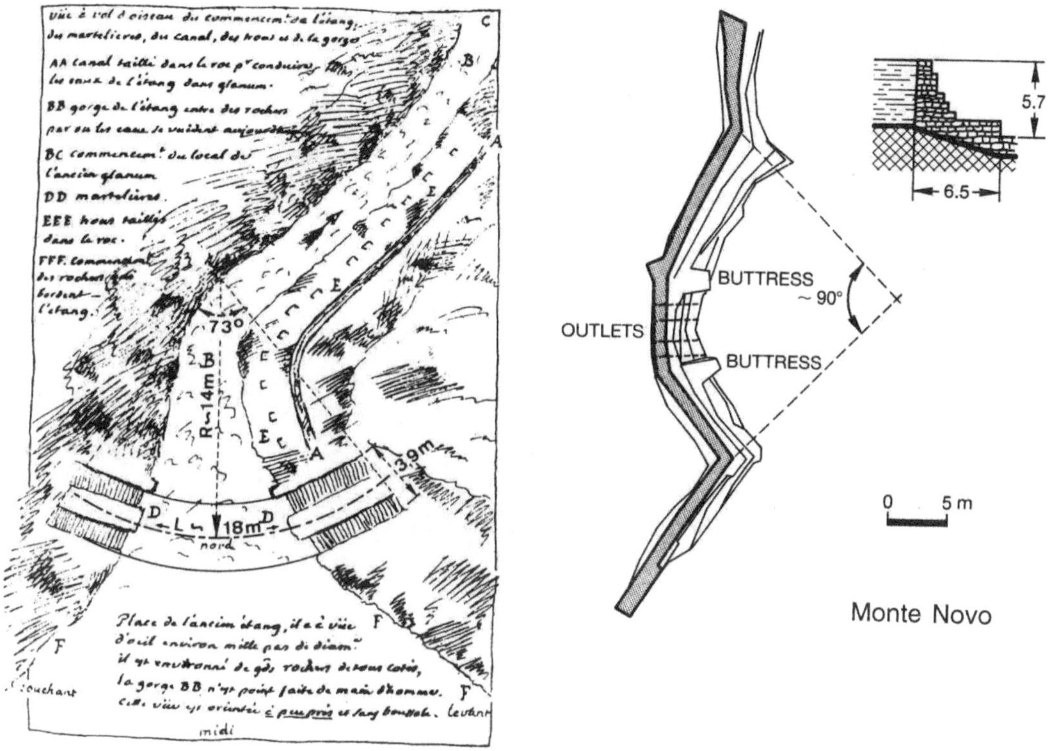

Vallon de Baume

Fig. 2 - *Vallon de Baume and Monte Novo arch dams* (Quintela 1987, Schnitter 1994)

2.2 - In the *14th century*, remarkable masonry arch-gravity dams were built by the *Mongols in Iran*, for irrigation and water supply.

Examples of these structures are (Schnitter 1994):

- The *Kebar arch dam*, situated near Kashen (26 m high, a crest length of 55 m, including the wing walls, and a thickness between 6 m and 9 m) (Fig. 3).

- The *Kurit arch dam* (60 m high, increased by 4 m in 1850) (Fig. 3); and *the Abbas arch dam* (20 m high), both situated near Tabas.

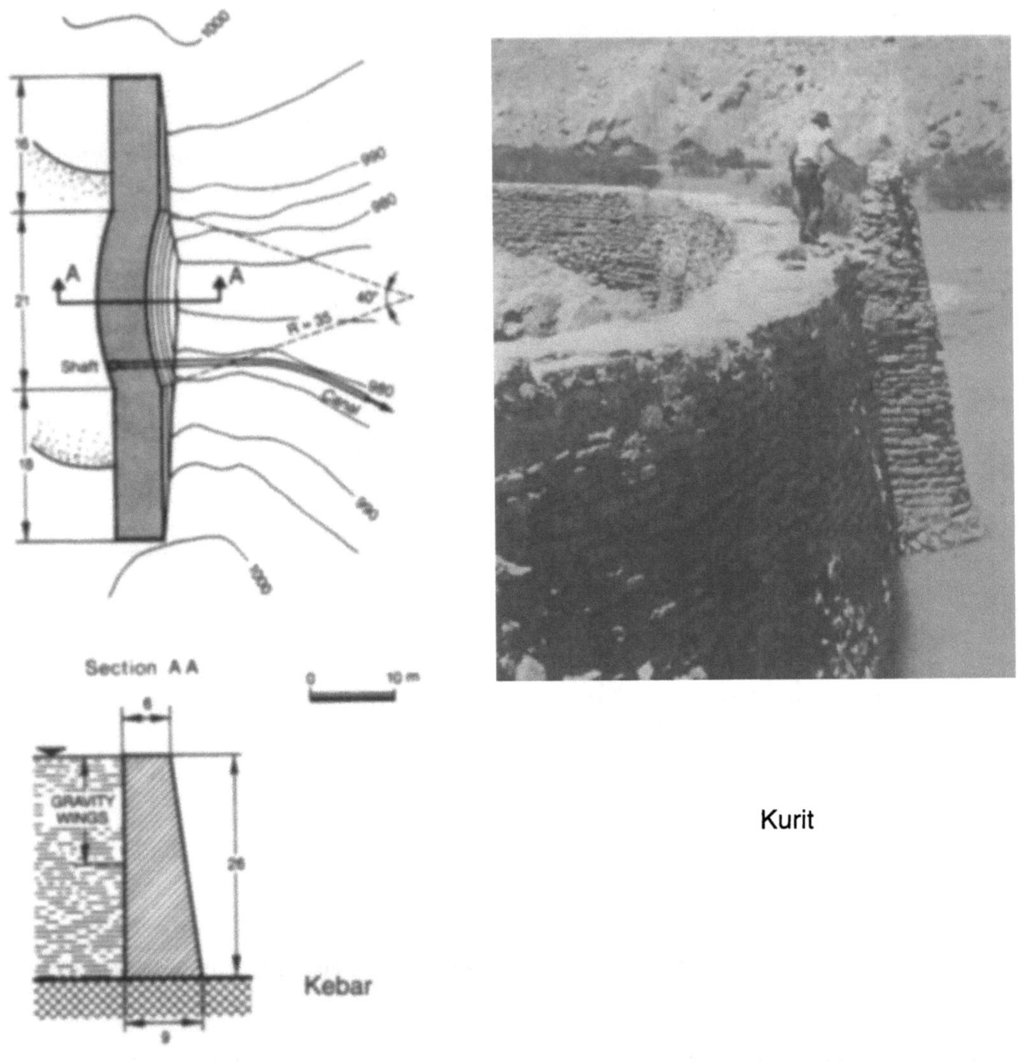

Fig. 3 - *Kebar and Kurit arch dams* (Schnitter 1994)

In *Europe*, several masonry arch and multiple arch dams were built *between the 14th and the 19th centuries*.

Major *examples* of those dams are (Smith 1992, Schnitter 1994):

- The arch dams built near Valência (Spain) for irrigation, namely the *Almansa dam*, with a first stage completed in 1384 and a second stage completed in 1586; the *Tibi dam*, completed in 1594 (46 m high, a crest length of 65 m, and a thickness varying between 20 m and 34 m); the *Elche dam*, completed in 1640 (23 m high, a crest length of 95 m, and a thickness varying between 9 m and 13.5 m); and the *Relleu dam*, with a first stage completed in 1776 (29 m high), and a second stage completed in 1879 (33 m high).

- The small *multiple arch dams* built by Villareal de Berriz (1670-1740) in the North of Spain, for power production, among others, the *Guizaburuaga dam*, near Bilbau (2 m high, a crest length of 40 m and a thickness of the arches varying between 1 m and 2 m).

- *The Pontalto arch dam*, situated near Trento (Italy), built in stages between 1611 and 1887, up to 40 m high (a crest length of 22 m, and thickness of 4 m); and the *Joux Verte arch dam*, situated near Lausanne (Switzerland), built for fluming (8 m high, a crest length of 30 m, and a thickness of 3.5 m).

The Almansa dam has been supplying water for irrigation for about 600 years. Its plan and section are presented in Fig. 4. In the first stage, the Almansa dam was an arch-gravity structure with a height of about 17 m and a thickness varying from 9 m (at the top) to 16 m (at the base). In the second stage, the dam was heightened to 33 m by means of a wall, which is polygonal in plan, and about 3 m thick at the top. Later on, in 1921, an outlet tower was built.

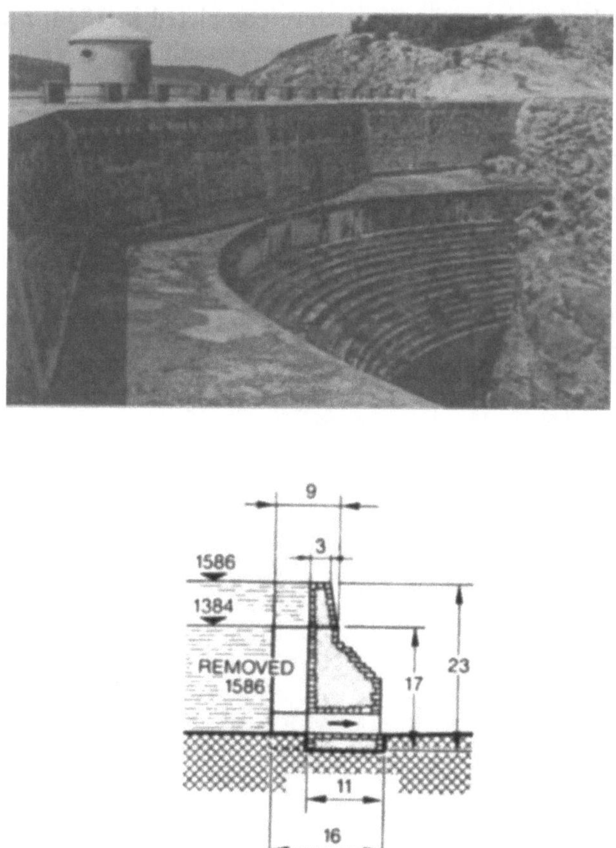

Fig. 4 - *Almansa arch dam* (Smith 1992, Schnitter 1994)

3 - *Arch dams designed according to Solid and Fluid Mechanics*

3.1 - *In the second half of the 19th century*, important developments were made in the design of arch dams, with the application of *Solid and Fluid Mechanics*, particularly the *Strength of Materials*. The arch dams began to be modelled as independent arches, which were analysed using the cylinder formula as linear structures (initially, as circular arches on rigid abutments).

According to these methods, several arch dams were designed and built until the beginning of the 20th century, mainly for water supplying, especially in *Europe, the USA, and Australia*. The construction material for most of these structures was still masonry. Some dams were built in valleys with very large span- height ratios.

Examples of those arch dams are (BNCOLD 1983, ICOLD 1984.88, FNCOLD 1985 USCOLD 1988, Schnitter 1994):

- The *Zola arch dam*, a masonry structure completed in 1854, near Aix-en-Provence (France) (36 m high, a central angle of 77°, and a crest length of 70 m (Fig. 5); the *Abbeystead*

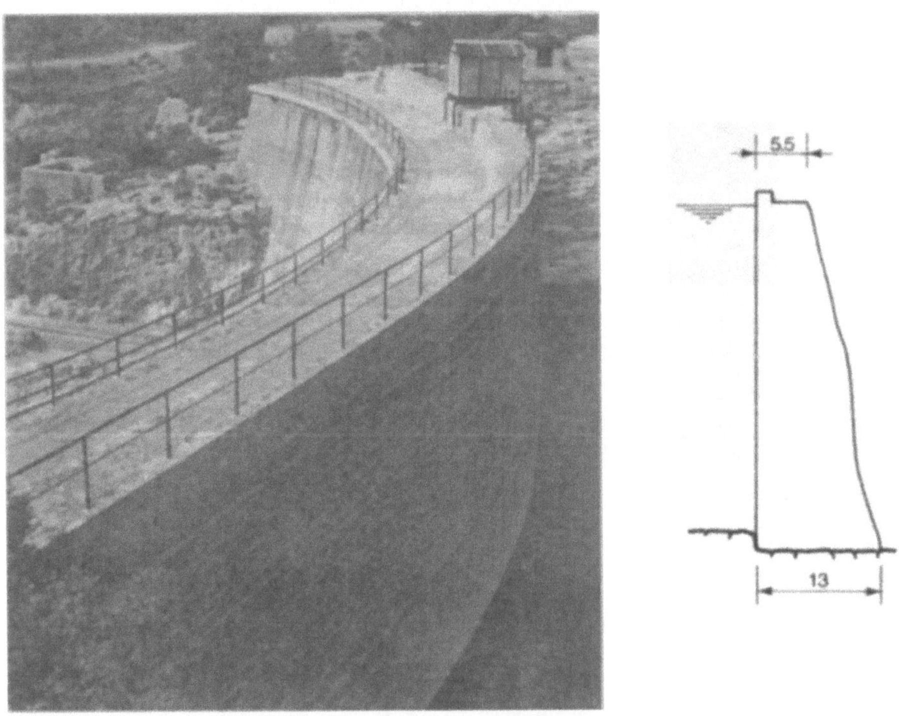

Fig. 5 - *Zola arch dam* (FNCOLD 1985, Schnitter 1994)

arch dam, a masonry structure completed in 1881, near Lancaster (Great Britain) (20 m high, a crest length of 133 m); and the *Pontebba arch dam*, a concrete structure completed in 1902, near Udine (Italy) (24 m high, a crest length of 19 m).

- The *Bear-Valley arch dam*, a masonry structure completed in 1884 near Los Angeles (USA) (18.5 m high, a crest length of 70 m, and a thickness varying from 1 m to 6 m); and the *Sweetwater arch gravity dam*, a concrete structure completed in 1888 near Chula Vista (USA) (25 m high, a crest length of 120 m) and heightened in 1911 (33 m high, a crest length of 213 m).

- The *Parramatta arch dam*, a masonry structure completed between 1856 and 1898, near Sydney (Australia) (16 m high, a crest length of 70 m, 4.5 m thick).

3.2 - The scientific development in the design and construction of arch dams, initiated in the second half of the 19th century, continued and led to important achievements during the *first half of the 20th century.*

The arch-gravity masonry structures were developed to *shell double curvature concrete dams*. These dams were built in U or V shaped valleys, and, for the latter, the radius of the lower arches was reduced in order to maintain the curvature, leading to the so-called *constant-angle arch dams* (USCOLD 1988). The thickness of the arches increased near the abutments, by means of fillets and socles. *Multiple arch dams* were often used in large valleys. These achievements were made possible by great improvements in the mass concrete technology.

The modelling of arch dams as *independent arches* continually improved (*elastically supported arches*, *active arches*, and *inclined arches*). Since the beginning of the 20th century however, this model was more and more replaced by the *arch-cantilever model*. The *arch-cantilever model* consists of a system of arches and cantilevers with displacements adjusted at the intersection nodes, which became well known world-wide as the *"Trial-Load method"*, from the trial method used to the adjustment of the displacements. Initially, only the adjustment of the *radial displacements* of the *arches and the central cantilever* were made; later the adjustment of the *radial displacements* for a system of *several arches and cantilevers*, and finally, the adjustment of the *radial*, *tangential* and *twist* displacements for the *arch-cantilever system* (USBR 1938, 1956).

Efforts were also made to develop the *shell* and the *three- dimensional elasticity models*, but the practical application of these models, by means of mathematical methods, remained very difficult.

The study of the three-dimensional models was thus essentially made by means of *scaled models*, among others (US Engineering Foundation 1928/31/33, Rocha et al 1960, LNEC 1963): 1) the *celluloid and concrete models* developed at the Universities of Princeton and Boulder (USA), for the studies of the Stenvenson Creek dam (an arch structure built for study purposes); 2) the *plaster-diatomite models* developed at LNEC (Lisbon); and 3) the *pumice-concrete models* developed at ISMES (Italy).

Monitoring of the dam behaviour was also developed. By the fifties, devices and techniques were available to monitor different control variables (namely displacements, strains, temperatures and joint movements) (LNEC 1955, Rocha et al 1958).

Many *arch dams of great dimensions* were built in this period (by 1960, *44 arch dams higher than 100 m* had been commissioned). Among these dams, a special reference is made to the *24 arch dams with more than 120 m in height*, built: 5 in the USA; 17 in Europe (7 in Italy, 4 in Switzerland, 3 in France, and 1 in Austria, Portugal and Spain); and 2 in Africa (Morocco and Zimbabwe-Zambia).

It is also referred to the *10 multiple arch dams higher than 60 m*, built: 4 in the USA; 2 in Europe (France and Italy); 2 in China; and 1 in Japan and Africa (Tunisia).

Examples of the above mentioned dams are (ICOLD 1984/88, PNCOLD 1982, FNCOLD 1985, SWCOLD 1985, USCOLD 1988):

- The *Hoover (Boulder)* arch-gravity dam, completed in 1936 near Boulder (USA), 221 m high, a crest length of 379 m; the *Ross arch dam*, completed in 1949 near Newhalem (USA), 165 m high, a crest length of 396 m; and the *Hungry Horse arch-gravity dam*, completed in 1953 near Kalis Pell (USA), 172 m high, a crest length of 645 m.

- The *Santa Giustina arch dam*, completed in 1950 near Bolzano (Italy), 153 m high, a crest length of 124 m; *the Tignes arch dam*, completed in 1952 near Albertville (France), 180 m high, a crest length of 375 m; the *Cabril arch dam*, completed in 1954 near Coimbra (Portugal), 136 m high, a crest length of 290 m; the *Mauvoisin arch dam*, completed in 1957 near Fionnay (Switzerland), 237 m high, a crest length of 520 m; the *Zeuzier arch dam*, completed in 1957 near Sion (Switzerland), 156 m high, a crest length of 256 m; the *Zervreila arch dam*, completed in 1957 near Vals (Switzerland), 151 m high, a crest length of 504 m; the *Frera arch-gravity dam*, completed in 1959 near Sondrio (Italy), 138 m high, a crest length of 316 m; and the *Grandval multiple arch dam*, completed in 1959 near St. Flour (France), 88 m high, a crest length of 400 m.

The cross-section of some large arch dams built in the USA and Europe are presented in Fig. 6 and 7.

4 - *Modern arch dams*

4.1 - During the *sixties and seventies* the trend for the construction of double curvature arch dams, particularly defined by *three-centred circular, parabolic and elliptic arches*, became more intensive. Those arches made it possible to decrease the curvature towards the abutments, reducing the central angle and therefore improving the support conditions (Rocha 1965).

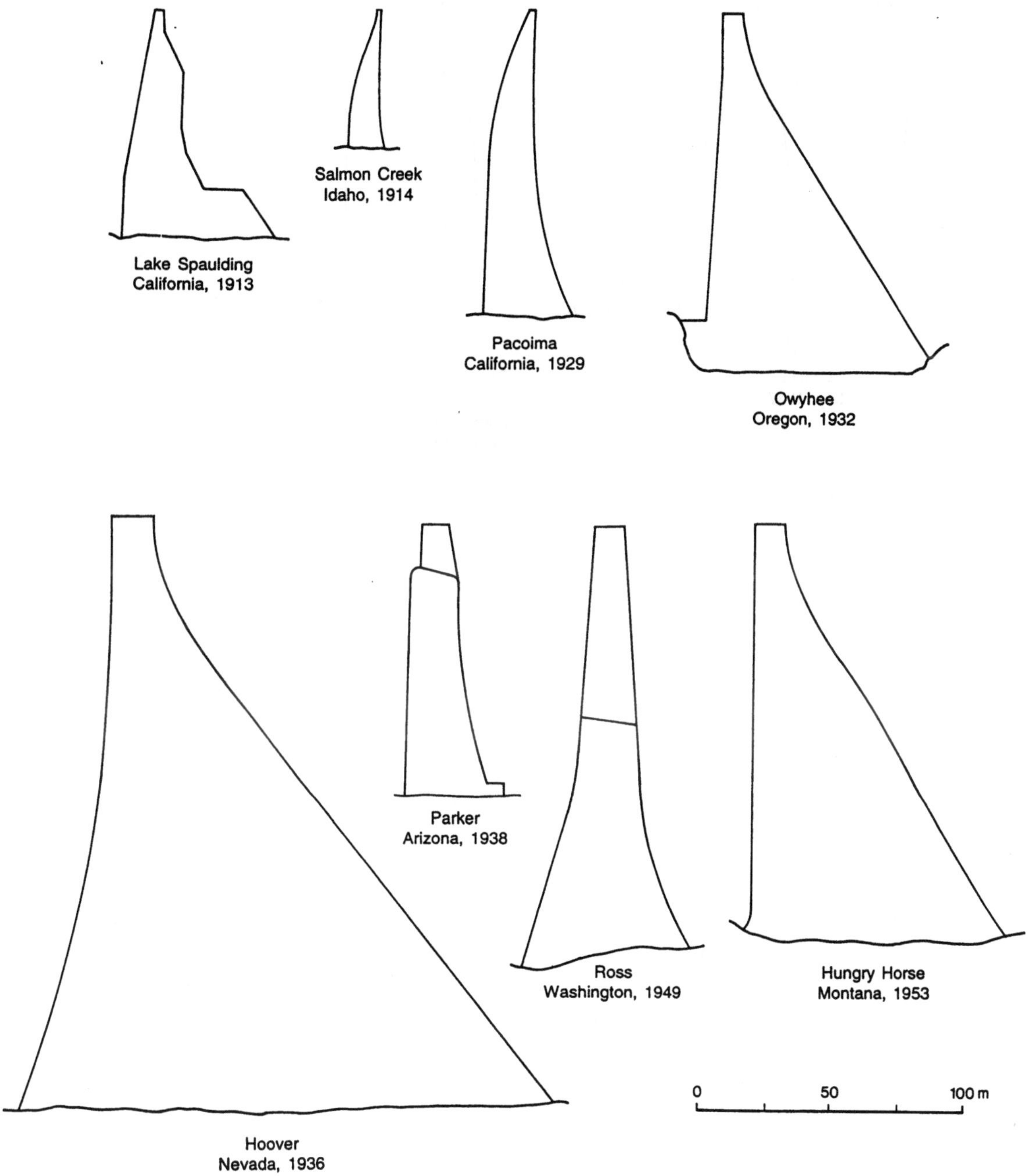

Fig. 6 - *Large arch dams built in the USA up to 1960* (USCOLD 1988)

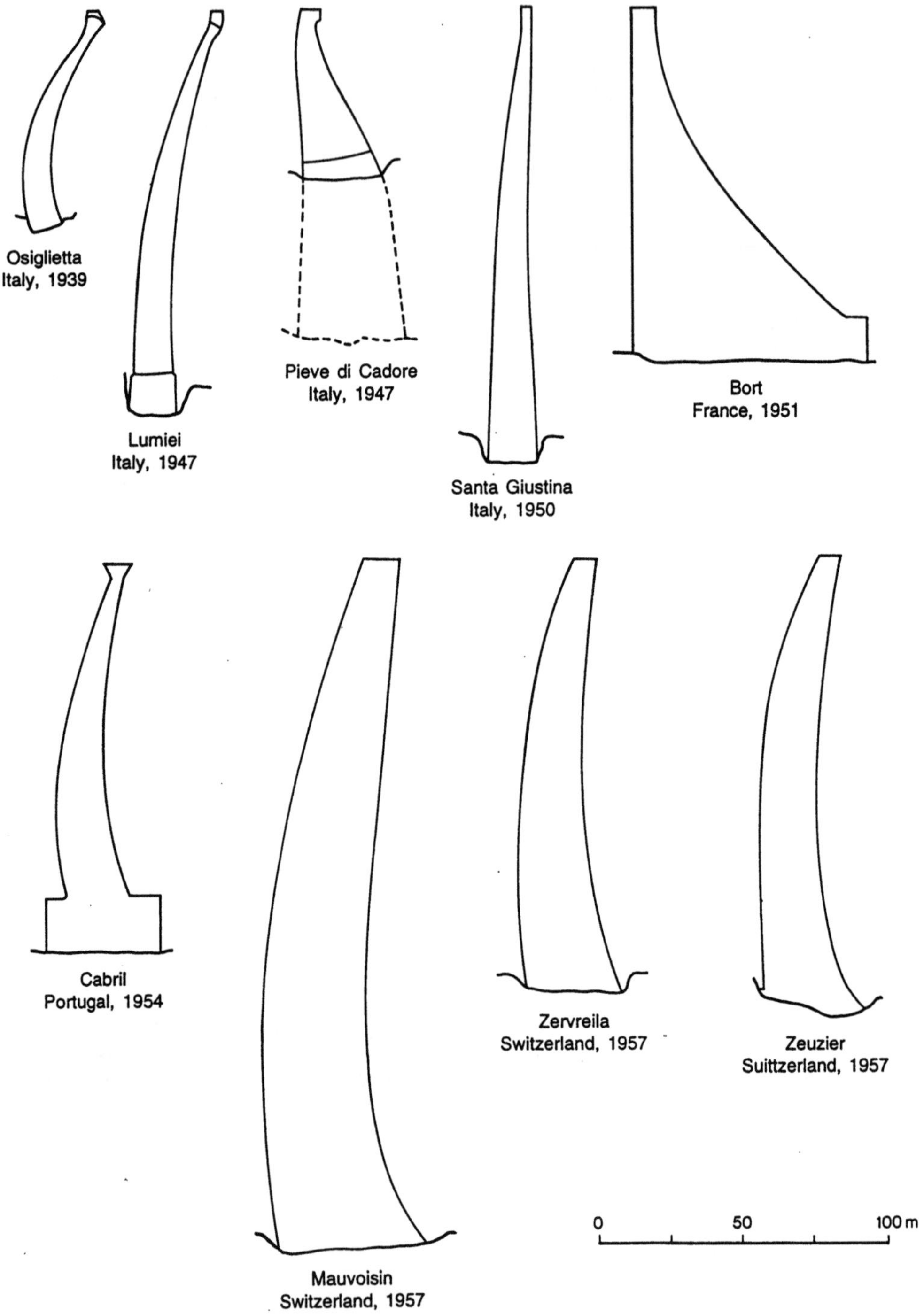

Fig. 7 - *Large arch dams built in Europe up to 1960* (FNCOLD 1985, SWCOLD 1985, PNCOLD 1992)

For large valleys, *large-span multiple arch dams* were developed, as well as the use of *pre-stressed arch dams* by means of tendons and jacks.

A *reduction in the number of contraction joints*, particularly at the higher elevations, was also made in thin arch dams (Serafim 1973).

The construction of large arch dams in sites with difficult foundations called the attention to the important problems of the *foundation stability and watertightness*. The importance of these problems was dramatically illustrated by the accidents of the *Malpasset dam* (1958), and the *Vajont reservoir* (1962). Therefore, a great effort was made in the development of the appropriate techniques and materials for *the rock mass characterisation and treatment* (consolidation, watertightness and drainage), (Rocha 1964, 1973, 1974, Stagg and Zienkiewicz 1968, Serafim 1973).

Important developments concerning the *spillways* included in the arch dam projects were also made, which allowed the *incorporation of spillways and outlets into the dam body*, even for very large flows (Fernandes and Molico 1970, Serafim 1973).

A great improvement was made in the *models and methods of analysis of arch dams*, particularly due to the development of the *computer technology*. The computers enabled the practical application of the *shell and three-dimensional models*, for static and dynamic actions, by numerical methods. In spite of some late developments, the "Trial-Load method" was replaced by the *finite element method*, both in studies of the construction sequence, and in studies for the different actions, under normal operation conditions (water, temperature, earthquake, etc.) (Rydzewski 1965, Zienkiewicz and Holister 1965, Severn and Taylor, CIRIA 1968, Pedro 1973, 1976, ICOLD 1978/82/87).

In the case of arch dams studies for normal operation conditions, the *scaled models lost much of their importance*, in spite of some late developments. However, they enabled important developments in studies of the *rupture conditions* of arch dams (both for static and dynamic loads) and *stability analysis of their rock mass foundations* (Rocha et al 1964, 1965, 1965a, 1969, Stagg and Zienkiewicz 1968, LNEC 1974, Silveira et al 1979).

As for as *monitoring* is concerned, the major developments were made in the *interpretation of results* by means of computers, although some improvements in the techniques and devices were achieved (LNEC 1962, Silveira and Pedro 1964).

A large number of *arch dams of great dimensions* were built in the sixties and seventies (by 1980, *70 arch dams higher than 100 m* had been commissioned - about twice the number of similar dams commissioned until 1960). Among these dams, a special reference is made to the *55 arch dams with more than 120 m in height* (*9 of them higher than 200 m*), built: 6 in the USA; 35 in Europe (11 in Switzerland, 9 in Spain, 4 in Italy, 3 in Austria and France, 2 in ex- Yugoslavia and 1 in Romania, Turkey and ex-USSR); 6 in Japan; and 8 in other countries (3 in Iran, and 1 in Argentina, India, Taiwan and Thailand).

Reference is made also to *3 multiple arch dams higher than 60 m*, which were built in Algeria, Canada and France.

Examples of these dams (those with more than 200 m) are (SPCOLD 1976, ICOLD 1984/88, SWCOLD 1985, USCOLD 1988, AUCOLD 1991):

- The *Glen Canyon arch dam*, completed in 1966 near Lees Ferry (USA), 216 m high, a crest length of 475 m.

- The *Vajont arch dam*, completed in 1961 near Belluno (Italy), 262 m high, a crest length of 190 m; the *Luzzone arch dam*, completed in 1963 near Olivone, (Switzerland), 208 m high, a crest length of 530 m; the *Contra arch dam* completed in 1965 near Locarno (Switzerland), 220 m high, a crest length of 380 m; the *Almendra arch dam*, completed in 1970 near Almendra (Spain), 202 m high, a crest length of 567 m; the *Mratinje arch dam*, completed in 1976 near Foca (ex-Yugoslavia), 220 m high, a crest length of 268 m; and the *Kölnbrein arch dam*, completed in 1977 near Gmünd (Austria), 200 m high, a crest length of 626 m.

- The *Dez arch dam*, completed in 1962 near Desful (Iran), 213 m high, a crest length of 212 m of; and the *Daniel Johnson multiple arch dam*, completed in 1968 near Baie Comeau (Canada), 214 m high, a crest length of 1,314 m.

The cross-section of some large arch dams built in the USA and Europe in the sixties and seventies are presented in Fig. 8. The plan and a cross-section of the Daniel Johnson multiple arch dam are presented in Fig. 9.

4.2 - During the *eighties and nineties* the construction of arch dams *decreased in the USA and Europe*, and *increased* in other countries (for instance, in *China* and *Japan*). Arch dams of double curvature were built, but with a tendency to use *simple and thicker shapes*. The *roller compacted concrete* was developed, initially for gravity dams and afterwards for arch-gravity dams (Hollingworth et al 1989, Geringer 1994).

A great progress was made in *mathematical methods* associated with the rapid improvement of computers. *Three-dimensional finite element models of the whole dam-foundation-reservoir*, for static and dynamic loads, and for linear and non-linear behaviour, became current use (Severn et al 1980, Pedro et al 1985, 1986, 1987a, 1988, Pina and Câmara 1987, ICOLD 1978/82/87, Wittke 1984).

Thermal models (particularly for the dam body) and *hydraulic models* (particularly for the foundation) were also developed. *Coupled analysis*, particularly of hydraulic and structural models, was used in some cases.

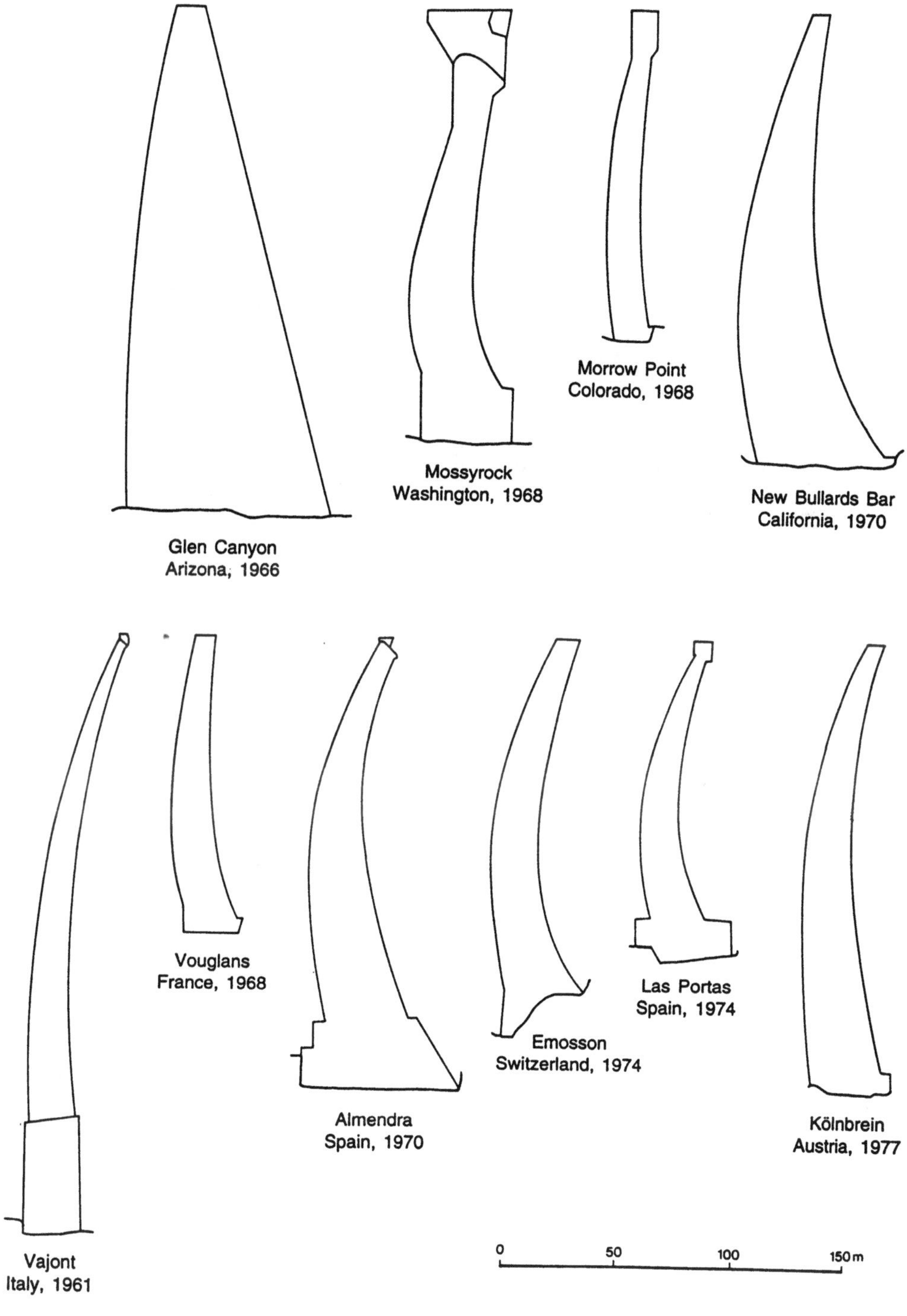

Fig. 8 - *Large arch dams built in the USA and Europe, during the sixties and seventies* (SPCOLD 1976, FNCOLD 1985, SWCOLD 1985, USCOLD 1988, AUCOLD 1991)

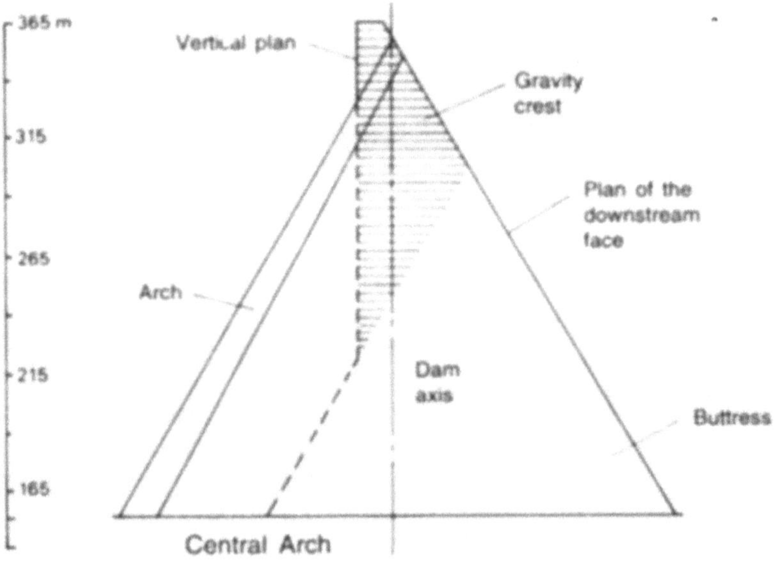

Fig. 9 - *Daniel Johnson multiple arch dam* (Bulota et al 1991)

Modelling the *time effects*, associated with the *construction sequence* and with the *ageing phenomena*, as well as the structural analysis for *hazard scenarios* related to exceptional events (floods, earthquakes, large movements of the rock mass foundation, etc.), became also available by means of mathematical methods (Pedro et al 1989, 1989a, Pina et al 1994).

Scaled models were applied rarely; however, they remained an important tool, both for the structural analysis of *unusual hazard scenarios* (Gilg et al 1987, Pedro and Azevedo 1988), and for studies of the hydraulic behaviour of *appurtenant works* (Pedro et al 1992).

Geomechanical scaled models were also developed for the study of discontinuous media, such as rock mass foundations of arch dams (Ferreira 1983). The use of geomechanical models, however, remained very limited, because of difficulties arising from their construction and high cost. Moreover, comprehensive *mathematical methods* (*finite discrete methods*) were developed for these problems (Lemos 1987, Oliveira et al 1992, Pina and Costa 1993).

As far as the *monitoring* is concerned, great improvements were made in the *monitoring of the foundation*, in the *interpretation of the results* and in the *automation* of the monitoring tasks. New *techniques and devices* were developed for the monitoring of the foundation. For the interpretation of the results methods of *quantitative analyses* and of *back-analysis* were developed (Gomes 1985, Ramos and Pinho 1986), which enabled a fundamental validation of the models and methods of analysis. Finally, the application of the new technologies enabled the *automation* of the *collection, processing, validation* and *storage* of the information, and the *control* of the structures in *real time* (ICOLD 1982, 1989, Fanelli 1990).

General concerns about the *environmental problems* gave rise to develop of the studies on the impacts of the construction and operation of large dams, particularly studies on safety and performance conditions. The *deterioration and ageing* of dams became fundamental subjects of research, as well as methods for *repair*, *rehabilitation* and *upgrading* of dams (ICOLD 1974/81, 1984, 1987, 1988, 1988a, 1994, 1994a, Pedro et al 1990, 1991, 1993, 1993b, 1993c, Heigerth et al 1994).

The construction of *arch dams of great dimensions* continued (by 1986, *23 arch dams higher than 120 m* had been completed since 1980 or were still under construction). These arch dams were situated, 11 in Europe (3 in ex-USSR, 2 in Turkey and Spain, and 1 in Austria, France, Norway and Romania); 7 in China and 1 in Japan; and 4 in other countries (Equator, Honduras, Iran, Taiwan).

Two multiple arch dams higher than 60 m have also been completed in Portugal and Sri Lanka.

Examples of the dams built since 1980 are (JPCOLD 1979-84, ICOLD 1984/88, CNCOLD 1987, Technopromexport 1988, AUCOLD 1991, PNCOLD 1992, SPCOLD 1993):

- The *Inguri arch dam* completed in 1980 near Zugdidi, (Georgia, ex-USSR), 272 m high, a crest length of 680 m; the *Aguieira multiple arch dam* completed in 1981 near Coimbra (Portugal), 89 m high, a crest length of 400 m; the *Oymapinar arch dam* completed in 1984 near Antalya (Turkey), 185 m high, a crest length of 360 m; the *Beznar arch dam* completed in 1986 near Beznar (Spain), 134 m high, a crest length of 408 m; and the *Zillergründl arch dam* completed in 1986 near Mayrhoffen (Austria), 186 m high, a crest length of 506 m.

- The *Kawaji arch dam* completed in 1981 near Imaichi (Japan), 140 m high, a crest length of 360 m; the *El Cajon arch dam* completed in 1984 near San Pedro (Honduras), 234 m high, a crest length of 382 m; the *Victoria multiple arch dam* completed in 1984 near Mahaweli (Sri Lanka), 122 m high, a crest length of 520 m; and the *Baishan arch dam* completed in 1986 near Huadian (China), 146 m high, a crest length of 664 m.

The plan and cross-section of some dams mentioned above are presented in Figs. 10 and 11.

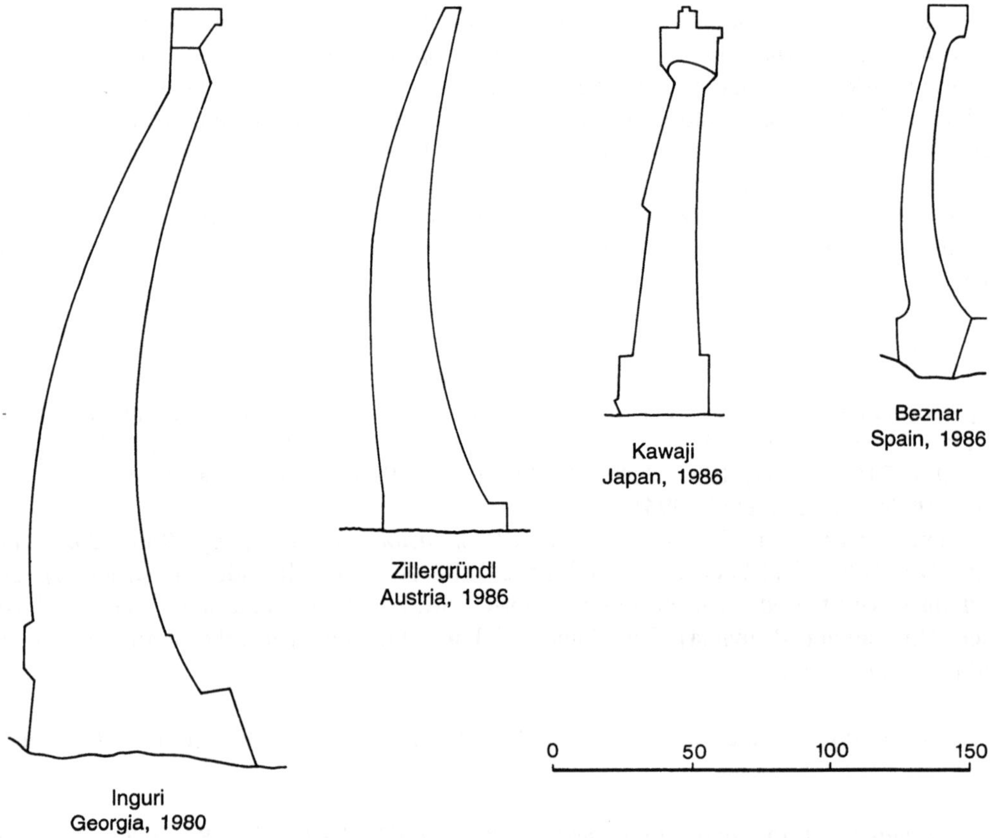

Fig. 10 - *Large arch dams completed after 1980* (JPCOLD 1979-1994, Technopromexport 1988, AUCOLD 1991, SPCOLD 1993)

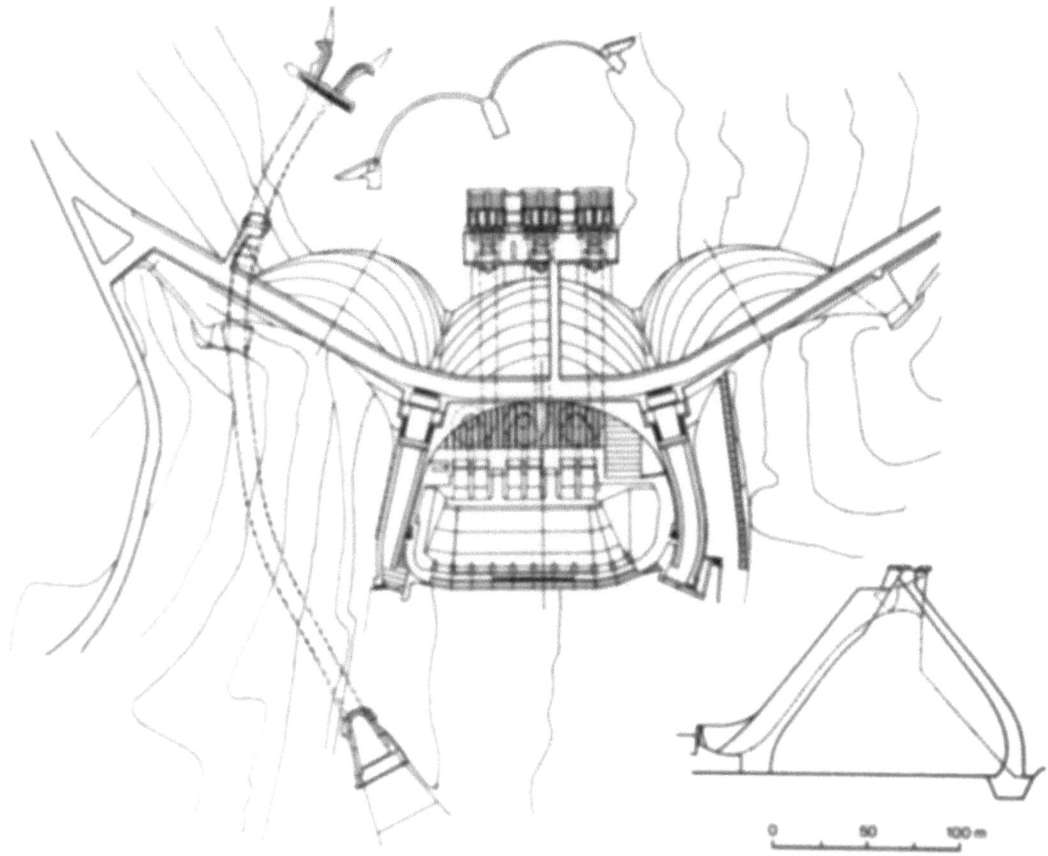

Fig. 11 - *Aguieira multiple arch dam* (Heitor and Pedro 1977, PNCOLD 1992)

3
SAFETY AND PERFORMANCE OF ARCH DAMS

1 - *Basic concepts*

1.1 - In order to fulfil the *behaviour requirements* during their whole *lifetime*, arch dams, like other structures, should be appropriately designed, constructed, and operated.

The *lifetime* of a dam begins with its construction and lasts as long as the existence of the dam represents a hazard to people and property.

In general, the following phases are considered during the lifetime of a dam: the *construction*, the *first filling of the reservoir*, the *operation*, and the *abandonment and demolition*. Although the control of the dam's safety and performance is required during its whole lifetime, a *special care* is needed during the *first filling of the reservoir* and the *first five years* of operation.

The *fundamental behaviour requirement* for any important structure, like a dam, is its *reliability*. Other requirements should, however, also be considered, such as the *lowest total cost* (construction, operation, and maintenance) and the *smallest negative impact on the environment*.

Dams should be *reliable* during their whole lifetime to *avoid any kind of deterioration*, i.e., both *accidents* (*safety requirement*), and *incidents* (*performance requirement*).

Both safety and performance of dams are affected by different factors, namely:

- *structural factors*, associated with the mechanical properties of the structures;

- *hydraulic factors*, particularly related to the hydraulic behaviour of appurtenant works, and to sealing and drainage systems;

- *operational factors*, associated with the operation of the equipment;

- *environmental factors*, associated with land and water, fauna and flora, population and production systems.

Safety and performance conditions can never be absolutely ensured so that a minute *small probability of the occurrence of deterioration* must be accepted. Basically, any deterioration occurring during the construction, the first filling of the reservoir and the first five years of operation should be attributed to *inadequacies of the design, construction, or operation*. Any deterioration occurring after the five initial years of the operation should be attributed to *ageing*. Furthermore, there may also occur a deterioration due to *exceptional events*.

A summary of the above concepts related to safety and performance is presented in Fig 12.

1.2 - As referred to above, a minute probability of occurrence of deterioration in the course of the dam's lifetime must be accepted. In fact, some *scenarios* (time related cause-effect-consequence chains of events) may lead to incidents (*current or utilisation scenarios*) or accidents (*hazard scenarios*) (ICOLD 1988, 1994a, Pedro 1990, 1993, 1993c).

For each dam, the major deterioration scenarios should be *foreseen and thus prevented*, by means of appropriate *design, construction, and operation* measures. However, the only relative efficiency of these measures and the possible *occurrence of unforeseen scenarios* call for a permanent *monitoring* throughout the whole lifetime.

Monitoring allows a continuous updating of the knowledge about the dam's condition throughout its lifetime, and therefore provides a great improvement in the *timely detection and correction of* eventual deterioration scenarios. When a careful monitoring is being carried out, the probability of the occurrence of an accident (due to natural hazards, etc.) becomes very small, although never zero. *Civil defence measures* should therefore be foreseen to be launched in the case of an accident especially to protect the population.

The *efficiency of the preventive and mitigation measures* taken in the design, construction, operation, monitoring, and civil defence, depends very much on its correct implementation. In fact, many dam accidents happened due to *human errors*. So, the implementation of adequate *quality assurance procedures* is required for all measures mentioned above.

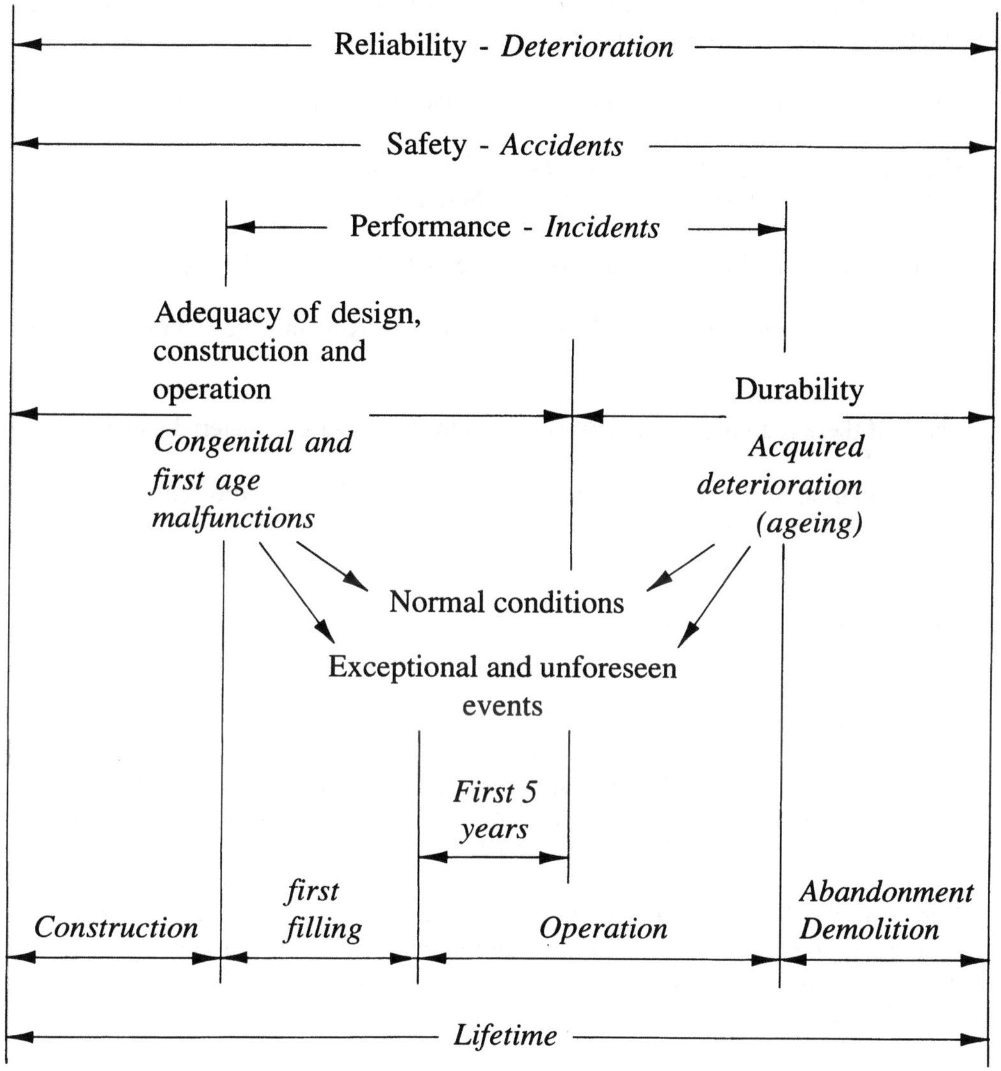

Fig. 12 - *Safety and performance along the lifetime of dams* (Pedro et al 1993, 1993c, ICOLD 1994a)

The evaluation of the *risks associated with a given deterioration scenario*, should consider:

- the *cost of the consequences of the deterioration* (incident or accident), regardless of its probability of occurrence (the *potential risk*);

- the *product of the potential risk by its probability of occurrence* (the *effective risk*).

The following three *levels of potential risk* have been considered for dam structures (USCOLD 1982, NWREA 1986, NBWE 1987, CSOPT 1990):

- *low*, if there is no loss of human lives, and only little material damage;

- *significant*, if there is a loss of only a few human lives, in spite of important material damage;

- *high*, if there is an important loss of human lives and serious material damage.

The evaluation of the risk associated with a deterioration scenario requires appropriate *criteria, models and methods* of analysis.

2 - *Major deterioration scenarios*

2.1 - The deterioration of dams has been discussed in many papers, including the *ICOLD Congress* and *other ICOLD papers*, therefore a comprehensive information has been accumulated about this subject. The analysis of this information allowed several *studies on the deterioration scenarios* for the different types of dams (Navarro 1945, ASCE 1956, ICOLD 1974/81, 1984, 1994, 1994a, ASCE/USCOLD 1975/88, Silveira 1990).

Data on *accidents and incidents which occurred in concrete and masonry dams* (gravity, buttress, arch, and multiple arch) are presented in Table 5. These data correspond to the *accidents and incidents reported up to the end of 1975* for a sample of *5,150 dams*, situated *in 33 different countries* (no Chinese dams are included, because China only joined ICOLD after that time).

The sample concerns *a large number of dams*, built over a period of about one century. It is, however, *very heterogeneous*, because the data were reported by *different authors* (with their own criteria), the respective *dams were built in different periods and regions*, and they were built with *different dimensions, purposes, and monitoring systems*. In spite of the deficiencies of the sampling for exact *statistical interpretations*, the following conclusions may be drawn:

- The *frequency of accidents* for arch dams (0.5%) *is smaller than that for other concrete dams* and it represents, in average, one accident per year for 200 arch dams.

Dam type	*Number of dams* (1)	*Number of cases* (1)	
		Deterioration (2) (% *of the number of dams)*	*Accident* (3) (% *of the number of dams)*
Arch	758	103 (13.6%)	4 (0.5%)
Multiple arch	142	26 (18.4%)	1 (0.7%)
Gravity	3,966	268 (6.8%)	27 (0.7%)
Buttress	284	44 (15.5%)	5 (1.8%)

(1) *At the end of 1975*
(2) *Including accidents (global failure, affecting the safety) and incidents (punctual failure, affecting the performance)*
(3) *Not including failures due to study purposes, bombing, etc.*

Table 5 - *Deterioration in concrete and masonry dams by the end of 1975* (ICOLD 1984, 1994)

- The *frequency of any kind of deterioration* (incidents and accidents) for arch and multiple arch dams is 26 times the respective frequency of accidents. This ratio is nearly three times bigger than that for gravity and buttress dams, and may be interpreted that *only one, among 26 cases of deterioration, was not timely detected and controlled*, and led to an accident.

The above indicated data are a fundamental source of information to identify the *major deterioration scenarios* in dams, appurtenant works, and reservoirs (ICOLD 1984, 1994, Pedro 1990, 1993c, Combelles 1991). For concrete and masonry dams, the major deterioration scenarios are (Pedro et al 1991, 1993a, 1993b):

- *Loss of strength and/or watertightness of the rock mass foundation*, due to cracking, shearing, sliding, erosion, or solution (in the rock or along the joints), caused by several phenomena, such as: 1) *stresses and hydraulic gradients* in the foundation, as well as *differential displacements at the dam-foundation interface*; 2) *physic-chemical phenomena*, including *solution and erosion*; 3) *degradation of the foundation treatment* (stresses and deformations, erosion, solution and chemical attacks of the grout, clogging of the drains, etc.).

- *Loss of strength and/or watertightness of the dam body*, due to overturning, cracking, shearing, sliding or solution of concrete or mortar, caused by several phenomena, such as: 1) *stresses* developed under normal operation conditions (particularly from *actions of the water and temperature*); 2) *freezing and thawing* for dams in cold regions; 3) *chemical reactions between the material compounds, resulting in swelling* (alkali-aggregate reactions, sulphate reactions); 4) *shrinkage, creep, dissipation of the hydration heat, and chemical reactions leading to contraction* (mainly for young concrete); 5) *chemical reactions of the material compounds with elements from the environment* (particularly aggressive waters from the reservoir or underground); 6) *degradation of the structural elements of the dam* (joints, upstream facings, prestressing devices).

Among the *case histories* on deterioration of concrete and masonry dams, particularly on arch and multiple arch dams, a large number concerns *dams built in the USA and Europe*. This situation results from the fact that, not considering the Chinese dams, about 70% of all arch and multiple arch dams are situated in these regions. On the other hand, the systematic reporting and discussion on incidents and accidents, in order to take from them all important lessons, is very common in the USA and Europe.

2.2 - The *major case histories on deterioration* concerning *the foundations of arch or multiple arch dams* are listed below.

The reported *accidents* were (ASCE 1956, ICOLD 1974/81, USCOLD 1988):

- the *Gleno multiple arch dam*, a concrete structure (35 m high, a crest length of 224 m) completed in 1923, near Bergamo (Italy), which failed in the same year;

- the *Moyie (Eileen) arch dam*, a concrete structure (16 m high, a crest length of 47 m) completed in 1924, in Idaho (USA), which failed in 1926;

- the *Lake Lanier arch dam*, a concrete structure (19 m high) completed in 1925, in the Vaughn Creek river (North Carolina, USA), which failed in 1926;

- the *Malpasset arch dam*, a concrete structure (66 m high, a crest length of 222 m) completed in 1954, near Nice (France), which failed in 1959;

- the *Idbar arch dam*, a concrete structure (38 m high, a crest length of 108 m) completed in 1959, in ex-Yugoslavia, which failed in 1960.

The accidents referred to above occurred *until 1960* (before the development of Rock Mechanics) and *during the initial years of the operation* of the works. They have been attributed mainly to deficiencies in the design, construction or operation.

The reported *incidents* concerning the foundation of arch and multiple arch dams, which occurred *after the first five years of operation* (and which are, therefore, associated with a lack of durability), were (ICOLD 1994a):

- *Loss of strength*: the *Gerlos* (Austria), *Avon*, *Nepean* and *Umberumberka* (Australia), *Pieve di Cadore* and *Comelico* (Italy), *Varosa* (Portugal), *Santa Maria*, *Punt del Gall*, and *Zöt* (Switzerland) arch dams; the *La Girotte* (France), *Santa Chiara* and *Combamala* (Italy) multiple arch dams.

- *Solution*: the *Venda Nova* (Portugal) and *Roode Elsberg* (South Africa) arch dams.

- *Degradation of the foundation treatment*: the *Place Molin* (Italy), *Venda Nova* and *Bouçã* (Portugal), *Isola* and *Les Toules* (Switzerland), and *Kariba* (Zimbabwe-Zambia) arch dams; the *Ville Hatte* (France) multiple arch dam.

This list shows that the majority of the incidents in the foundation of arch or multiple arch dams are caused by a *loss of strength*.

Among the reported incidents concerning the foundation of arch dams, which occurred in the *first period of the dam operation*, were the important cases of the *Kölnbrein* and *Schlegeis* (Austria), and *Zeuzier* (Switzerland) arch dams (Tischler and Schlosser 1985, Demmer and Ludescher 1985, Gicot 1981, Müller and Pougatsch 1985).

2.3 - The *major case histories on deterioration* concerning the *dam body of arch or multiple arch dams* are listed below.

An important *accident* caused by the degradation of the dam body occurred with the *Leguaseca (Fonsagrada) multiple arch dam* (20 m high, a crest length of 70 m), which was completed in 1958, near Lugo (Spain), and failed in 1987 (Guerreiro et al 1991).

Another reported accident concerns the *Meihus arch dam* (22 m high, a crest length of 64 m), a masonry structure completed in China, in 1981. This dam failed in the year of completion, due to unforeseen actions (ICOLD 1994).

Furthermore, reference is also made to the failures of the *Eder arch dam* (48 m high, a crest length of 400 m) and *Mohne arch dam* (40 m high, a crest length of 650 m) in the Ruhr basin (Germany), which were completed in 1914 and 1918, respectively, and destroyed by bombing in 1943.

The reported *incidents* concerning the dam body of arch or multiple arch dams, which occurred *after the first five years of operation* (and which are, therefore, associated with a lack of durability) were (ICOLD 1994a):

- *Loss of strength*: the *Le Gaje* (France), *Corfino*, *Montagna Spaccata*, *Muro Lucano*, *Gurzia*, *Rimasco*, and *Ponte Pia* (Italy), *Alto Ceira* and *Cabril* (Portugal) arch dams; the *Daniel Johnson* (Canada), *Grandval* (France), *Venina*, *Molato*, *Pian Sapeio*, *Ozola*, and *Santa Chiara* (Italy), *Lake Hodges* and *Murray* (USA) multiple arch dams.

- *Freezing and thawing*: the *Bort* (France), *Gurzia* and *Rimasco* (Italy), and *Lake Spaulding* (USA) arch dams; the *La Girotte* (France), *Molato, Fontamallucia, Combamala, Pian Sapeio,* and *Ozola* (Italy), and *Florence Lake* (USA) multiple arch dams.

- *Swelling*: the *Bimont* (France), *Meantwrog* (Great Britain), *Corfino* (Italy), *Alto Ceira* (Portugal), *Poortjies-Kloof, Keerom, Pietersfontein,* and *Roode Elsberg* (South Africa), *Cooper Basin, Gene Wash, Gibraltar, Matilija,* and *Stewart Mountain* (USA) arch dams; the *Churchil* and *Stompdrift* (South Africa) multiple arch dams.

- *Shrinkage and creep*: the *Laouzas* (France), *Corfino, Montagna Spacata,* and *Nuraghe Arrubio* (Italy), and *Zervreila* (Switzerland) arch dams; the *Venina* (Italy) and *Churchil* (South Africa) multiple arch dams.

- *Aggressive waters*: the *Avon* and *Victoria* (Australia), *Rasgão* (Brazil), *Lanau* and *Monceaux-la-Virole* (France), *Meantwrog* and *Dinas* (Great Briatain), *Muro Lucano, Gurzia, Spigno Monferrato, Rimasco, Cancano, Orichella,* and *Comelico* (Italy), *Rochon* (Seychelles), *Roode Elsberg, Groendal, Duivenhoks,* and *Korente Vette* (South Africa), *Victoria* (Sri Lanka), *Stewart Mountain, Salmon Creek, Glenn Canyon,* and *Lake Spaulding* (USA) arch dams; the *La Girotte* and *Migoëlou* (France), *Venina, Molato, Fontanallucia, Combamala, Pian Sapeio,* and *Ozola* (Italy), and *Florence Lake* (USA) multiple arch dams.

- *Degradation of structural elements*: the *Monceaux-la-Virole* and *Rassisse* (France), *Muro Lucano* and *Spigno Monferrato* (Italy), and *Les Toules* (Switzerland) arch dams; the *Neuber* (Tunisia) and *Florence Lake* (USA) multiple arch dams.

The above list of incidents shows that the most frequent causes of deterioration concerning the dam body of arch or multiple arch dams are, on one side, *environmental actions* (particularly aggressive waters, freezing and thawing) and *structural deficiencies* (lack of strength and swelling) on the other side.

The list above shows also that, as far as the *deterioration of the dam body* is concerned, arch dams are more reliable than *multiple arch dams*. Indeed, the number of incidents reported for arch dams (61) is about twice the one for multiple arch dams (31), whereas the number of arch dams is about seven to eight times higher than the one for multiple arch dams.

Some incidents concerning the dam body of arch dams have occurred already *during the first five years of operation*. Important incidents occurred at the *El Atazar* (Spain), *Chambon* and *Voughans* (France) arch dams (Urbistondo and Yges 1985, Bister et al 1991). In this context it should be mentioned that, in general, arch dams are *large structures* with *appropriate monitoring systems*. Therefore eventual deficiencies in design, construction and operation can usually be rapidly detected.

3 - *Illustration of accidents and incidents*

3.1 - The case histories of the *Moyie, Lake Lanier and Malpasset* dams are well known illustrations of accidents concerning the foundation of arch dams.

The *Moyie dam* (presently named *Eileen dam*) was built on the Moyie river, near Bonner's Ferry (Idaho, USA). It is a 16 m high cylindrical dam with an upstream face radius of about 200 m, a crest length of 47 m and a thickness varying from 1.6 m at the base to 0.6 m at the top. The thin arch was reinforced with a mesh of railway rails, both along the arches and the cantilevers, near the upstream face. A timber-lined spillway was placed on the left abutment on soft and stratified rock.

In 1925, an exceptional flood destroyed the spillway canal, and the water cut a deep by-pass through the soft and stratified rock, and carried away the left abutment (Fig. 13). *The dam, however, did not fail* (and is still standing), perhaps on account of its steel reinforcement (ASCE 1956, ICOLD 1974/81, USCOLD 1988).

The *Lake Lanier dam* was built on the Vaughn Creek river (North Carolina, USA). It is a 19 m high cylindrical dam with an upstream face radius of 45 m, a crest length of 72 m, and a thickness of 3.8 m at the base and 0.3 m at the top. Steel reinforcements were placed in

Dam after the failure (1924)

Dam (1986)

Fig. 13 - *Moyie dam after the abutment failure* (USCOLD 1988)

the upper 4 m of the dam and in the cut-off wall. The major part of the dam crest served as overall spillway.

Shortly after the first filling of the reservoir (January 1926), the spillway flow undermined the left bank, where the dam rested on a large boulder instead of sound bed-rock. A considerable flow developed through that bank and caused a *settlement and overturning of the abutment*. The *dam was repaired* by means of a new gravity abutment, designed as a spillway, *and since then has been performing well* (Fig. 14) (ASCE 1956, ICOLD 1974/81, USCOLD 1988).

Dam after the failure (1926)

Reconstructed abutment

Fig. 14 - *Lake Lanier dam* (ICOLD 1974/81, USCOLD 1988)

The *Malpasset dam* was a thin, 66 m high, double curvature arch dam, built on the Reyran river, near Nice (France). A gravity abutment, protected by a wing wall, was built on the left bank. The upstream face radius of the crest arch was about 100 m, and the central angle 122°.

The failure occurred during the *first filling of the reservoir* (December 1959), when the water level approached the elevation of the crest spillway, about 2.5 m below the dam crest. Successive cracks and a violent wind were reported by the site resident (ICOLD 1974/81).

The remainder of the dam (a little part of the wall on the right bank and only the abutment on the left bank) is represented in Fig. 15. An approximately *30 m deep and 40 m wide wedge was excavated* on the left bank which was defined by a downstream running fault, near the dam base, and the foliation (which is approximately perpendicular to this fault).

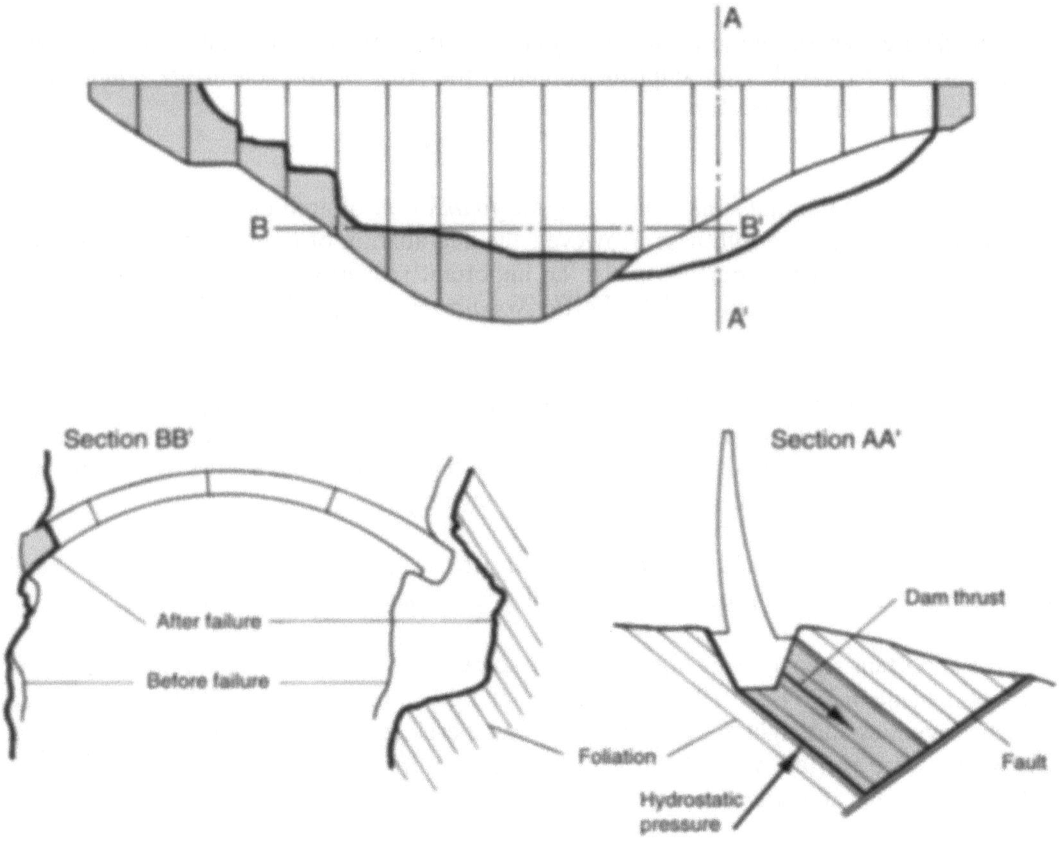

Fig. 15 - *Malpasset dam after the failure* (ICOLD 1974/81)

The failure of the *Malpasset dam* was attributed to the *sliding of that rock wedge*, along the face corresponding to the fault, under the dam thrust and the water action. The foliation of the gneiss rock led to a concentration of the dam thrust in a small zone of the rock mass (underneath the base of the blocks), which became very impermeable as compared with the upstream zone (a ratio of the respective permeability of about 1 to 100 was estimated); as a result, the full hydrostatic pressure developed upstream of that impermeable zone, causing the sliding along the fault.

The case history of the *Leguaseca (Fonsagrada) dam* is the sole reported illustration of a failure concerning the body of a multiple arch dam (Guerreiro et al 1991).

The 20 m high *Leguaseca multiple arch dam*, with concrete arches and masonry buttresses, was completed near Vigo (Spain) in 1958. *After 30 years of operation* (March 1987), *an arch of the dam failed*, as a result of a large opening (with an area of about 7 m2) which had developed in it (Fig. 16).

The failure was attributed to the *poor quality of the concrete*, particularly to withstand the *aggressive water* (acid, high sulphate content), the *frost action*, the *large temperature variations* and the *thermal gradients*.

3.2 - The majority of the *incidents concerning the foundation* of arch dams were caused by a loss of strength and/or watertightness. However, the degradation of the rock mass, as a result of *erosion and solution* phenomena, is also an important scenario, as illustrated by the case history concerning the foundation of the *Venda Nova arch dam* (PNCOLD 1985, Pedro et al 1993a, 1993b, ICOLD 1994a).

The *Venda Nova arch-gravity dam* is a concrete structure (97 m high, a crest length of 294 m), completed near Guimarães (Portugal) in 1951 (Fig. 17). The *foundation rock mass* is mainly formed by a granite, with schistose rocks at the higher elevations on the right bank. It is crossed by many joints and faults, filled with clay or mylonite.

The *treatment of the foundation* consisted in consolidation, grouting, and waterproofing. A 50 m deep grout curtain as well as a drainage system were installed.

An important seepage was observed after some years of operation, and, in spite of the repair works made in 1964 and 1968, the seepage and the uplift pressure continued to increase during the seventies and eighties. The reservoir was therefore emptied in 1985, and important repair works were carried out, which included the *consolidation and waterproofing* of the foundation by means of an *acrylic resin*. Since then, the dam has been performing well.

The comprehensive studies undertaken at the Venda Nova dam showed that the *opening of the joints and faults*, namely in the left bank and in the bottom of the valley, allowed the *washing out and dissolution of their infillings*. This was confirmed by the X-ray

Fig. 16 - *Leguaseca dam after the failure* (Guerreiro et al 1991)

diffraction identification of the materials washed out and deposited at the outlets of the drains as well as by chemical analyses of the water. These analyses indicated that the content of dissolved salts in the seepage water was much higher than in the reservoir water (22 mg/l) (Fig. 17).

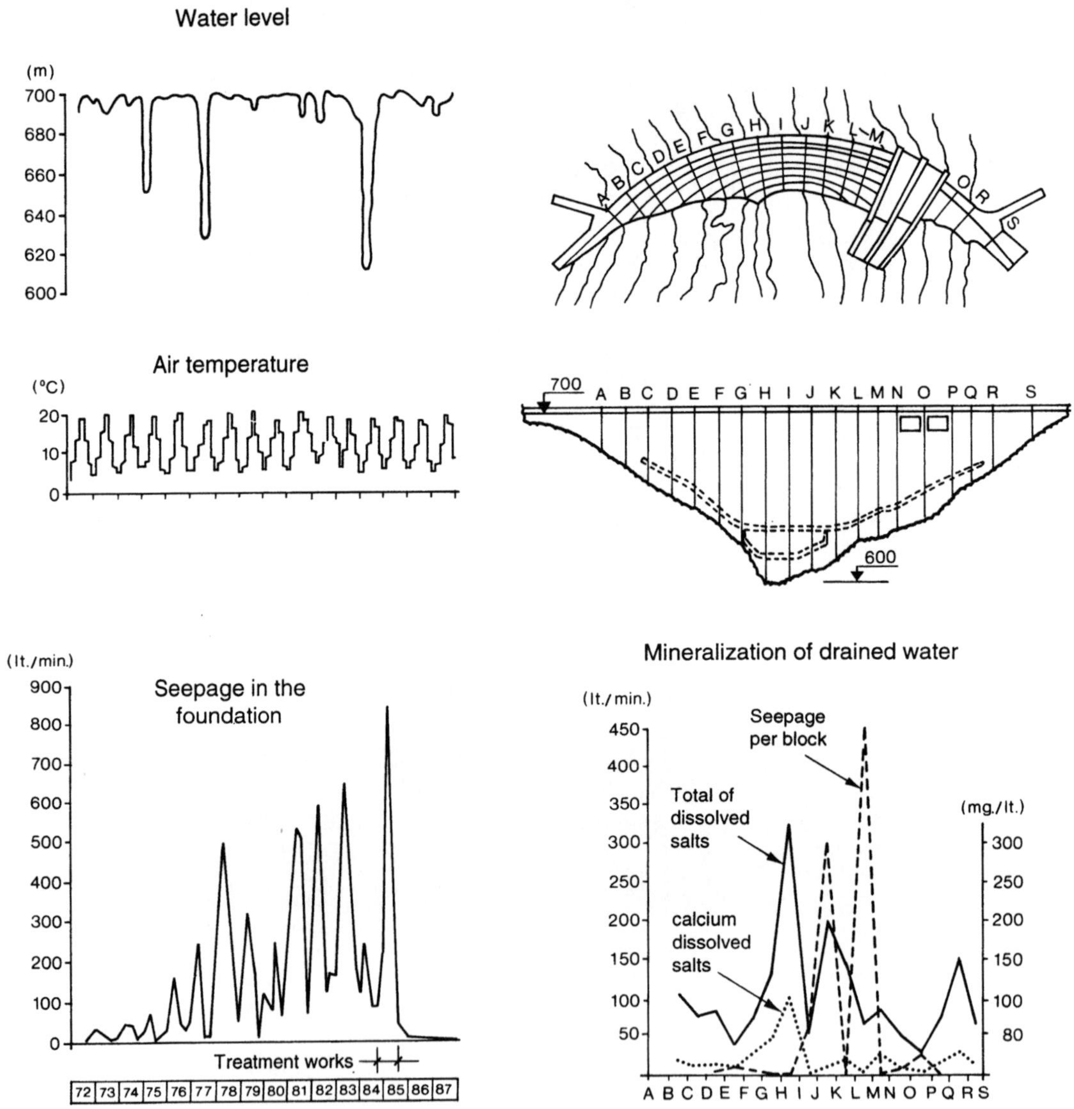

Fig. 17 - *Venda Nova dam. Some monitoring results* (ICOLD 1994a)

Among the *incidents concerning the dam body* of arched structures, the case history of the *Daniel Johnson multiple arch dam* is an important and typical one (Bulota et al 1991, ICOLD 1994a).

The *Daniel Johnson dam is one of the world largest dams.* It was completed on the Manicouagan river (Canada) in 1968 (Fig. 9).

During the 10 years of the reservoir's impounding, an *intensive cracking* developed in the dam (Fig. 18). There were:

- *Oblique cracks,* on the downstream face of the arches, following the direction of the principal compressive stresses, which were *caused* by the combined effect of the water pressure with the severe *climatic conditions* at the site (the annual temperature extremes are -45° C and +35° C, and 50 freezing - thawing cycles develop in an average year).

- *Plunging cracks,* in large zones near the arch heels, nearly all of which penetrate the entire thickness of the arches and cause leakage. These cracks are *attributed to the water loads,* combined with the *lack of vertical curvature in the arches,* near the foundation.

- *Opening of the rock-concrete contact surface,* near the heel of the arches, due to the same causes of the plunging cracks.

- *Cracks, parallel to the arch faces,* at the bottom of some arches, which were attributed to the *construction process.*

The comprehensive investigation carried out on the dam's condition and behaviour, showed that the *safety conditions were ensured,* but that *remedial measures should be taken.* The most necessary ones are:

- *grouting* of the plunging cracks and of the concrete-rock contact;

- *grouting and bolting* of the parallel cracks;

- installation of a *permanent thermal shelter* at the base of the main arches, to control the oblique cracks.

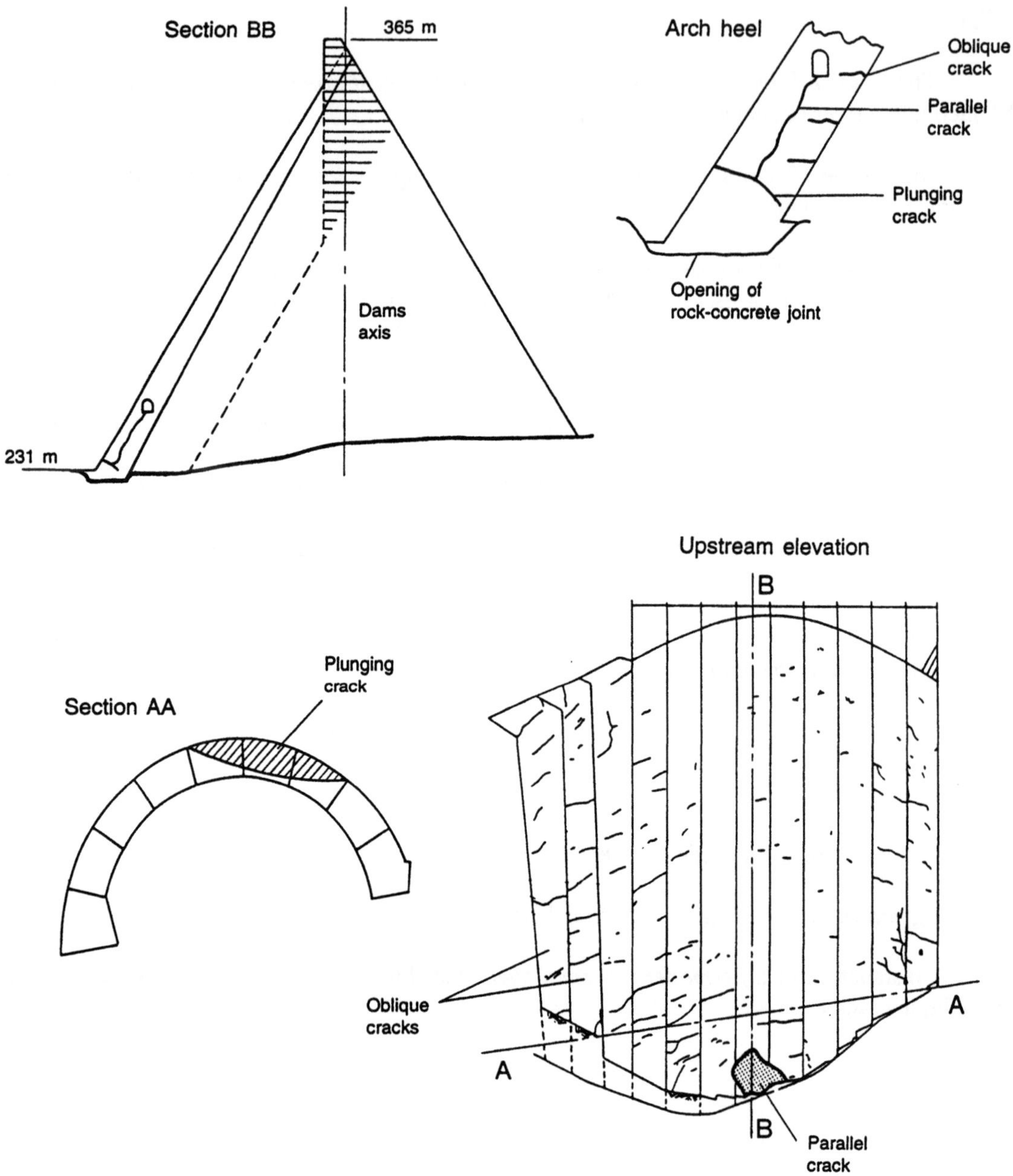

Fig. 18 - *Daniel Johnson dam. Cracking in the arch contiguous to the central arch, in the left bank* (Bulota et al 1991)

4
EVALUATION OF SAFETY AND PERFORMANCE OF ARCH DAMS

1 - *General*

1.1 - The construction of dams should be seen in the *frame of Man's intervention in the Environment*, which calls for the following typical activities (Oliveira and Pedro 1993):

- *monitoring* of the whole dam and its behaviour (beginning with site investigation) to assemble necessary data;

- *modelling* of the behaviour of the whole dam to allow simulation studies and the assessment of its safety and performance;

- *design* (defining the appropriate measures), either for construction, repair, rehabilitation or upgrading, or for operation;

- *execution* of the above mentioned measures.

The *evaluation of the safety and performance* of a dam is therefore not only necessary *for the design* but also *for the control of the behaviour* throughout its entire lifetime, especially (ICOLD 1994a):

- During the *construction phase*, to ensure the safety conditions of the building process and the quality control of the construction.

- During the *first filling of the reservoir*, to control one of the most important real tests in the course of the dam's lifetime.

- At the *end of the 5 initial years of operation*, to characterise the real behaviour of the dam under normal operation conditions. This characterisation will usually allow a safety and performance control by means of simpler procedures, up to the detection of an abnormal behaviour.

- Immediately after the *occurrence of any exceptional event* (flood, earthquake, etc.), or the *detection of an abnormal behaviour*.

- For any *repair, rehabilitation, or upgrading works*, to assist in the choice of the appropriate measures, in the control of their implementation, and in the characterisation of the new dam behaviour, after these works.

- In the case of *abandonment or demolition*, to support the definition of the appropriate measures to ensure that the safety conditions remain satisfactory.

1.2 - *Modelling* the dam behaviour is a fundamental step in the evaluation of its safety and performance. As stated above, several different kinds of models have been developed for arch and multiple arch dams. They all satisfy the *general pattern* indicated in Fig. 19.

According to this pattern, two types of models can be defined:

- The *models of the actions*, which define the relationship between the *actions* (water, temperature, etc.), the *structural properties* (hydraulic, thermal, etc.), and the *corresponding effects* (forces, expansion, etc.) and eventual *consequences* (change of the structural properties or a global failure).

- The *structural models*, which define the relationship between the *mechanical effects* derived from the models of the actions (forces, strains, etc.), the *mechanical properties* (deformability, strength, etc.), and the corresponding *structural effects* (displacements, stresses, etc.) and eventual *consequences* (punctual failures and change of the structural properties or a global failure).

For instance, in the case of the action of the *water*, which is usually defined by the *water levels* of the reservoir and immediately downstream of the dam, the above models are:

- The *seepage model* (model of the water action), which allows to determine *water velocities and pressures* in the dam and its foundation; and derived from these, the mechanical effects *forces and strains* acting on the structure, as well as direct consequences (*erosion, solution*, etc.).

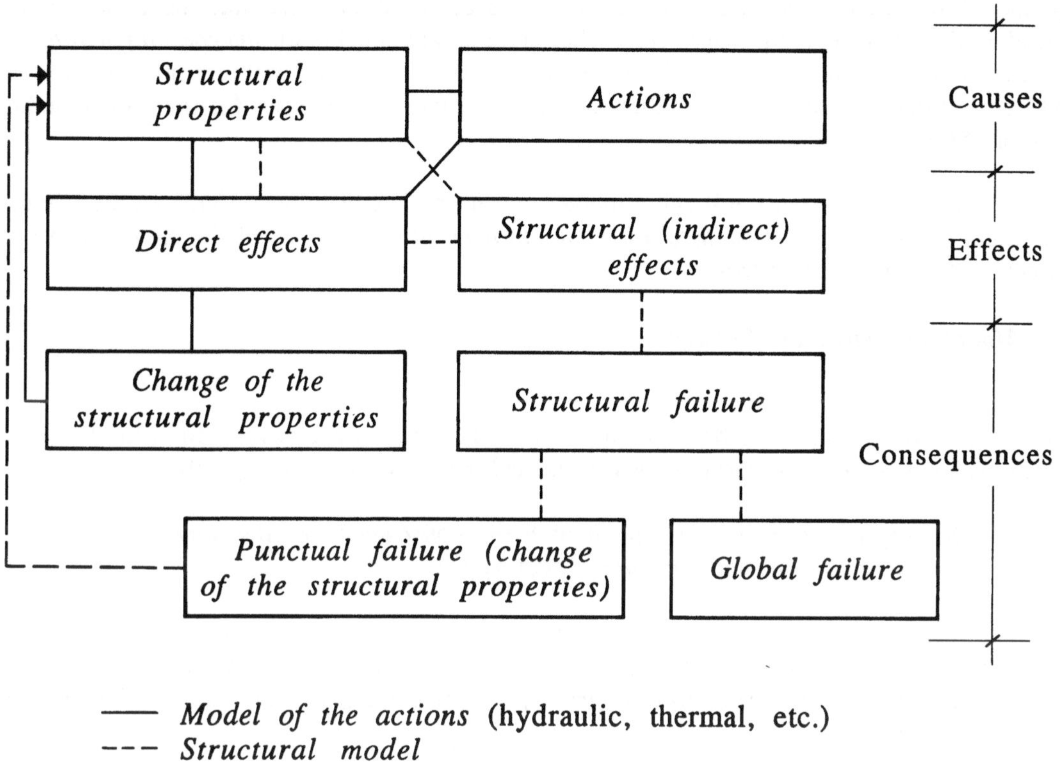

Fig. 19 - *General pattern of the dam models* (Oliveira and Pedro 1986, Pedro et al 1987, 1987a, 1990, 1991, 1993, 1993c, ICOLD 1994a).

- The *structural model*, defined by the structural properties and the applied forces and strains (derived from the seepage model), which allows to determine *displacements, strains, and stresses* in the dam and its foundation; and derived from these, to evaluate punctual failures (*cracking*) leading to a change of the structural properties (mechanical, hydraulic, etc.), or an eventual *global failure*.

- The *change* which may occur *in the structural properties*, requires a *re-analysis of the seepage model*, the results of which, in turn, require a *re-analysis of the structural model*. This process continues until no more significant changes occur.

This example shows that *a coupled analysis of the seepage and structural models* is required. However, the simplified approach which considers the mechanical effects of the water action represented by the *hydrostatic pressure*, is often used.

Some models simulate *histories of the actions and structural properties* in a certain period of the lifetime (such as the period of the construction sequence, the period of the development of an ageing process, etc.). These models are suitable for *interpreting monitoring results*, because in those cases the histories of the actions and structural responses are known. The aim of the analysis is the identification of the structural properties for the corresponding period of time.

For *design purposes*, instead of models which simulate histories throughout the lifetime, the usual approach consists in *modelling the unfavourable situations* of the dam, which may occur *in the course of its lifetime*.

2 - *Models and methods of analysis*

2.1 - As shown before the basic variables of a dam model are associated with *the causes* (actions and structural properties), *the effects*, and *the consequences* (Fig. 19).

The major *actions* are associated with the *construction of the dam*, the *filling and operation of the reservoir*, and *exceptional occurrences*.

The *construction actions* are the *dead weight* of the materials and equipment; the *thermo-hygrometric* and *autogeneous volume variations of the concrete*, particularly for the young concrete; *pre-stressing forces*, *grout pressures*, etc.; and the *external temperature variations* (daily and annual waves).

The *actions due to the filling of the reservoir* are the *water pressure* on the dam and in the foundation; the *temperature variations* at the dam faces, associated with the rise of the water level and with temperature waves in the air and water.

The *actions due to the operation and to the environment* are the *variations of the water pressure and temperature*; the *dynamic actions* (natural and induced earthquakes, water pressure in the spillways and stilling basins, operation of the power house, etc.); and the *pressure of the sediments* in the reservoir, *of the ice*, etc.

The *exceptional occurrences* in the course of the dam's lifetime may be *large floods*, *strong earthquakes*, *differential movements of the foundation rock mass*, etc.

In general, both for design and monitoring studies, appropriate *combinations of the above actions* are considered.

For the *design studies*, the following combinations of actions are often considered (CSOPT 1990):

1) the *construction actions* (for different stages, and specially for the end of the construction);

2) the combination 1) for the end of the construction, together with the *actions due to the filling of the reservoir* (for successive stages of the reservoir filling);

3) the combination 2) for the end of the filling, together with the actions associated with the *changes of water level and temperature waves*, and *others*, deemed to represent the most unfavourable situations foreseen to occur in the course of the lifetime;

4) the appropriate combination 3) with the actions due to *each of the considered exceptional occurrences* (the possibility of having two simultaneous exceptional occurrences is usually not considered).

The *shape and dimensions* of an arch or multiple arch dam and its foundation can be accurately represented by *three-dimensional models*. However, the simplified approaches of *independent arches and cantilevers* (one-dimensional models), and the *shell approach* (two-dimensional models) are often used.

The characteristics of the materials (rock mass, concrete, and masonry), i.e. their *discontinuity, heterogeneity, and anisotropy*, as well as their hydraulic, thermal, and mechanical properties (*permeability, diffusibility, thermal expansion, deformability* and *strength*), associated both to volume and joint elements, can be easily considered with modern mathematical approaches (*finite elements, discrete elements*).

Models which include several zones of different continuous media, linked by joints, and with general laws for the *water flow*, *heat transfer*, and *mechanical behaviour* (*elastic, elastoplastic,* and *visco-elastoplastic*), are nowadays of widespread use.

The representation of the *contraction joints*, as well as of the properties of the young concrete (*hydration heat, thermal properties, shrinkage, creep, and relaxation*), are important to be considered in the models for the *construction sequence*.

The *discontinuities and heterogeneity in the foundation rock mass* may have a fundamental role in the safety conditions of arch dams (Rocha 1965, ICOLD 1974/81, Silveira et al 1991). An appropriate modelling of the foundation, taking into account the treatment and other measures to improve the behaviour (*aprons, grout curtains, cut-off walls, drainage systems*), is therefore important.

In the *modelling of a dam*, an important problem is the definition of the *extent of the foundation to be incorporated in the model*, and of the appropriate *boundary conditions*. In general, this definition will be different for hydraulic, thermal or mechanical problems, and for static or dynamic loads.

Simple methods have been used to take into account the foundation deformability, such as among others, the well known *Vogt's coefficients* (USBR 1938, 1956, Pedro 1973, 1976). Nowadays, however, the models include a *large extent of the foundation,* allowing a detailed analysis of the zone in the vicinity of the dam base, with an appropriate representation of the deformability. In the case of dynamic problems, boundary conditions avoiding the wave reflection for the foundation and the reservoir, are often used (Pedro and Câmara, 1986).

Boundary conditions may also be used to limit the volume of the model to be analysed (*symmetry or anti-symmetry conditions*), and to perform a *detailed study* of a specific part of the structure by means of successive analyses (Pedro et al 1973, 1976, 1987a).

For the hydraulic models, the main *direct effects* are the *discharge flow* and *water pressure* in the appurtenant works, the *seepage* and *leakage* through the dam and the foundation, the *pore pressure* and *uplift*, and the *turbidity* and *mineralization* of the seepage water.

For the thermal models, the main direct effects are the *temperature distribution* in the dam body, and the corresponding *volume variations*.

The consequences of the direct effects, particularly of the hydraulic effects, may be the *overtopping*, the *erosion* of the appurtenant works and foundation (*local scour*, *abrasion* and *cavitation*), and the *internal erosion* in the foundation (*piping*), particularly along joints and faults of the rock mass.

The presence of water in the dam body and foundation promotes the development of *physic-chemical phenomena*, which lead to a degradation of the concrete, masonry, and rock mass.

The *structural effects* are the *displacements* (horizontal, vertical, angular), *joint and crack movements, strains, and stresses.*

The *consequences of the structural effects* may be a *global failure* (overturning, settlement, heaving, or sliding of a large volume of the dam and the foundation) or *punctual failures* (cracking, shearing, and crushing).

2.2 - The *analysis* of a model is usually made by defining as *input* the *actions and structural properties*, and by obtaining as *output* the *effects* (direct and structural) *and the corresponding consequences* (Fig. 19).

It is, however, also possible to define as *input* the *actions, effects, and consequences* observed in a certain period of the lifetime, and to obtain as *output* the *structural properties* for that period. This is called *back-analysis, quantitative analysis, or identification of parameters* (Rocha et al 1958, Silveira and Pedro 1964, ICOLD 1982, Gomes 1985, Ramos and al 1986, 1992).

In the case of *coupled problems*, a joint analysis of the structural, hydraulic, and thermal models has to be made.

The *physical models*, based on analogies defined by the similitude conditions for different problems (stress analysis, heat transfer, seepage and other hydraulic problems, and the selection of the best arch dam shape) (Zienkiewicz and Holister 1965, LNEC 1957 to 1977, Rocha et al 1960, 1964a, 1964b, 1965b, 1969, Serafim 1990) are being replaced by the more general, faster and cheaper numerical approaches. Physical models are, however, still used in studies of the *hydraulic behaviour of the appurtenant works*, as well as in studies of *hazard scenarios* for arch dams.

Currently, the models are mostly analysed by *mathematical methods*, even for the interpretation of the results of physical models (Severn et al 1980, Pedro and Azevedo 1988, Pina et al 1993, 1994).

Numerical approaches used today are the *finite element, finite difference, and boundary element methods* (Rydzewski 1965, CIRIA 1968, Pedro et al 1976, 1985, 1986, 1987a, 1988, 1989, 1989a, ICOLD 1978/82/87). Recently, the *discrete element method* (initially developed for studies of rock masses), has also been used for dam studies (Lemos 1987, Pina et al 1993).

3 - *Reliability factors*

3.1 - To *assess the reliability* of a dam and its foundation, for a specific *current or hazard scenario*, the occurrence of this scenario has to be simulated in one of the following ways:

- an *increase* of the values of the variables related to *the actions* (water level, ground acceleration, relative displacements in the foundation, etc.);

- a *decrease* of the values of the variables related to *the structural properties* (usually the strength, i.e. cohesion and internal friction of the materials);

- an *appropriate combination* of an *increase of the actions* and a *decrease of the structural properties*.

The *probability of non-failure* is obtained by integrating the products of the probability density associated with each value of the actions, by the probability associated with the occurrence of structural properties with values larger than the respective value of the actions. This last probability is the integral of the probability densities associated with all values of the structural properties larger than the considered value of the actions (Borges and Castanheta 1983).

However, the *current and hazard scenarios for dams are complex phenomena*, which require a large number of variables for an appropriate definition. Furthermore, in general, the *probability densities* for these variables *are not available*. The practical application of the above probabilistic method becomes, therefore, very difficult, and *the usual method, based on the reliability factors*, remains the common practice, in spite of the attempts made to develop *half probabilistic approaches (method of the partial coefficients)* (ICOLD 1988).

The behaviour of the existing dams (which have been designed with *the application of the usual reliability factors*) in the course of the last century, remains the fundamental source of data for assessing the design criteria of dams. These data should be taken into account in the *validation of any method* to evaluate the safety and performance of dams.

The usually made *recommendations* for the *design of concrete dams*, in particular for arch and multiple arch dams, are listed below (Rocha 1974, 1978, ICOLD 1988, 1994, 1994a, CSOPT 1990, Serafim and Clough 1990).

For the *current scenarios*, the dam and its foundation should have a *reversible behaviour allowing the normal operation conditions*. The *permanent deformations* accumulated along the loading-unloading cycles, should therefore be *very small*, which will be the case if the *cracking is limited to small volumes* (punctual failures), but the majority of the dam and its foundation is far from the failure conditions.

Usually, the above objectives are ensured if the following conditions are verified:

- the *displacements are of the same magnitude as* those which are estimated when assuming an *elastic behaviour* of the materials;

- the *displacements*, both in the zone of the foundation near to the dam base and in the dam body, are *small enough* to avoid disturbances in the rock mass and in the watertightning and drainage devices, as well as in the operation of the equipment;

- the *stresses in the majority of the volume elements* of the dam and its foundation satisfy the *Mohr-Coulomb criteria* with *reliability factors not less than 4 to 5 as for as the cohesion is concerned*, and *1.5 to 2 as with regard to the internal friction* (in general, the cohesion is the most important parameter for current scenarios);

- the *seepage and leakage*, particularly in the foundation, do not exceed the flow corresponding to an *average permeability of 1 Lugeon*;

- the *uplift at the dam base*, in the zone of the drainage curtain, does not exceed *one third of the water pressure* at the dam heel.

For *hazard scenarios* (exceptional conditions), although many punctual failures may develop in the dam and its foundation, a *global failure* of large volumes *should not occur*.

Usually, the above objective is ensured if the following conditions are verified:

- the *displacements of the cracked blocks*, particularly near the top of the dam, satisfy the *non-overturning condition*;

- the *stresses in the majority of the volume elements* of the dam and its foundation satisfy the *Mohr-Coulomb criteria* with *reliability factors not less than 3 as far as the cohesion is concerned*, and *1.2 to 1.5 with regard to the internal friction*;

- the *leakage in the foundation* does not exceed the flow corresponding to an *average permeability of 5 Lugeon*.

3.2 - As referred to above, the experience already assembled on the evaluation of the safety and performance of dams, allows an *assessment of the used criteria, models and methods* of analysis. This assessment may be made at the following three different levels:

- at the *level of the methods of analysis*, by checking the results of different methods for the same model assumptions (results of finite elements, discrete elements, scaled models, etc.);

- at the *level of the models*, by checking the results of the model (which incorporate the approximations of the analysis and the data on the structural properties) with the corresponding results of the monitoring for a given period of the dam's lifetime (by means of quantitative analysis or back-analysis);

- at the *level of the criteria*, by checking the occurrence of the hazard and current scenarios, in the course of the dam's lifetime, with the evaluations of the safety and performance criteria (which incorporate the approximations of the models, the methods of analysis and the data on structural properties).

The reliability of dams, particularly of arch dams, has been much discussed over the recent years. In fact, in spite of the great improvements made in the models and methods of analysis, in the methods of reconnaissance, testing, and treatment of the rock mass foundations, and in the composition and placing of the mass concrete, the *used reliability criteria* are basically those which were developed for the models and methods of the *Strength of Materials* in the second half of the last century.

However, special care is required when *changing the reliability criteria*, because an assessment of these changes can only be made after their *application to a large number of works, during a long period of time*.

As far as the reliability level defined by the usual criteria is concerned, the above presented results of the *studies on the deterioration of concrete dams* show that incidents were reported for 13.6% of the arch dams, and accidents for 0.5%. These figures decrease for the *dams built after the sixties*, due to the improvement in the dam technology, but it is difficult to quantify this decrease. For the time being, it seems appropriate *to maintain the reliability level defined by the usual reliability factors*.

4 - *Selection of the best solutions for a dam project*

4.1 - In the selection of the best solutions for the project of a large structure, the *minimum generalised cost* (sum of the initial cost of the project with the effective risk of deterioration) has been proposed as the major criterion (Borges and Castanheta 1985).

The application of this criterion to a *dam project* is not easy, because a dam project includes several and complex structures (the dam and its foundation, the appurtenant works, and the reservoir). Furthermore, in order to evaluate the effective risk, hydraulic, operational, and environmental factors have also to be considered besides the usual structural factors (ICOLD 1987, CSOPT 1990). The solutions for a dam project are therefore, in general, selected in favour of the *minimum cost of construction and operation*, among those solutions which satisfy the reliability criteria for the major deterioration scenarios.

The search of the best solutions is usually carried out with the aid of the *trial and error* method.

In the case of arch or multiple arch dam projects, the definition of the *best shape and dimensions* for the dam itself may be an important part in the selection of the best dam project.

Different studies have been made to define the *best shape* of an arch or multiple arch dam *for a specific valley* by means of *membrane tests* (LNEC 1957 to 1977, Rocha 1964b), and also by *mathematical procedures* (Serafim and Clough 1990).

The *whole project* and, in particular (together with the topographical factors considered in the methods above), *the foundation rock mass*, the *environment*, the *purpose* of the project and the *methods of its construction and operation* should be taken into account when defining the shape and dimensions of a dam.

4.2 - As far as the shape of an arch dam is concerned the following *general recommendations* are given:

- the arches should have a *small central angle* (80° to 100°), a *decreasing curvature* and *increasing thickness towards the abutments*; three-centred circular, elliptic, parabolic, and logarithmic spiral arches are appropriate solutions;

- the dam should have a *vertical curvature* compatible with the construction techniques (a downstream overhang at the higher elevations and an upstream one near the foundation);

- *the arches* should be symmetric whenever possible (by means of artificial abutments and the pulvino);

- the *crest* should be *simple and slender* (even if it supports a roadway);

- the possibility of having the *spillway and bottom outlets* (and, in some cases, even the *power house*) incorporated into the dam body, should be considered;

- *drainage and visiting galleries*, large enough to allow the execution of eventual repair works, should exist in the dam body.

The construction by *independent cantilevers*, separated by contraction joints, is a common technique. In some projects, the spacing of the contraction joints is smaller near the foundation than at higher elevations. *Prestressing* of the arches by tendons and flat jacks has also been used (Serafim 1973, Serafim and Clough 1990).

For the upcoming *roller compacted concrete arch dams*, new construction techniques, as well as appropriate dam shapes, will certainly be required.

5 - *Examples*

5.1 - The evaluation of the reliability of arch or multiple arch dams for some typical scenarios is illustrated by a few *examples, selected from studies carried out at LNEC.*

The studies undertaken at LNEC have used mathematical and scaled models and concerned a large number of concrete dams (PNCOLD 1992, Pedro et al 1992). About fifty of these dams (approximately one third of the total amount) are situated in Portugal and in the majority of them, particularly in the *arch and multiple arch dams, monitoring systems have been installed and are in operation* since many years.

On the other hand, according to the Portuguese legislation, the *LNEC is involved in the safety control of the Portuguese dams* (CSOPT 1990). Therefore many of the model studies have been carried out aiming at the evaluation of the monitoring results of Portuguese dams.

A general conclusion of the above mentioned studies is that *adequate evaluations are made for current scenarios, with the usual models*. (Pedro et al 1976, 1982, 1984, 1986, 1987a, 1993b, among other papers).

However, for *hazard scenarios, the adequacy* of the safety evaluations has been *more difficult to test*. This difficulty has justified the utilisation of both *scaled models* (Rocha et al 1960, 1965a, 1969, 1988) and *mathematical methods* (Pedro et al 1985, 1987a, 1988, 1989, 1989a, Pina et al 1987, 1993, 1994).

The *examples* presented below are mainly concerned with the behaviour of arch dams for *hazard scenarios*, including strong earthquakes. A reference is also made to studies on the selection of the *best design solutions* for arch and multiple arch dams.

5.2 - The *failure of an abutment* of an arch dam is an important scenario, as illustrated above with the case histories of the *Moyie and Lake Lanier dams*.

An example of a study on the *simulation of an abutment failure* is the case of a double curvature arch dam (68 m high, a crest length of 300 m), built on a rock mass foundation formed by a *granite and metamorphosed clastic formations, crossed by several joint sets*, one being approximately horizontal.

The heightening of this dam up to 87.5 m has been envisaged. Therefore, a study was made to evaluate the risk associated with an *eventual sliding* along horizontal joints on the right bank due to an increase of water pressure. This study included analyses of a *geomechanical model*, as well as of *finite element models* (Ferreira 1983, Azevedo et al 1988, Pina et al 1991, Pina and Costa 1993) (Fig. 20).

It was concluded that the risk of a sliding in the foundation *is small*, since a sliding would only occur for extremely unfavourable values of the *angle of internal friction* and of the *uplift*. It was also concluded that such a sliding would only cause the *failure of a narrow zone of the dam near the artificial abutment* (like the one in the cases of the *Moyie* and the *Lake Lanier* dams), while the stresses in the rest of the dam remain within values which do not cause concrete failures (Fig. 21).

Geomechanical model

Finite element model

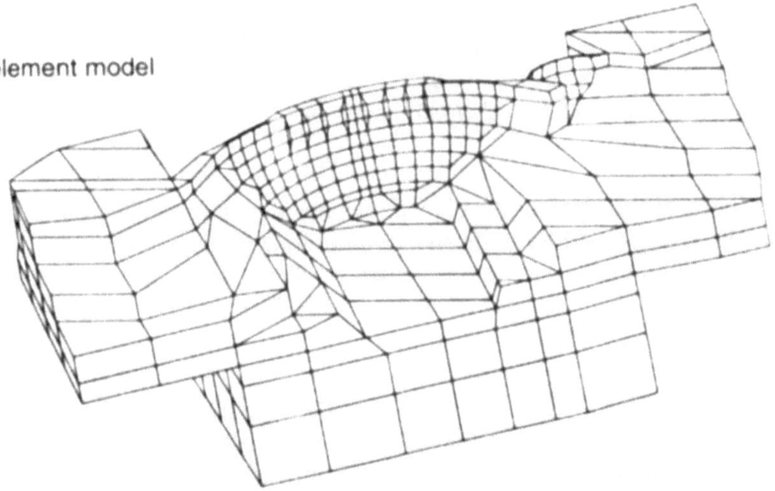

Fig. 20 - *Geomechanical and finite elements models of an arch dam, developed at LNEC* (Ferreira 1983, Pina and Costa 1993)

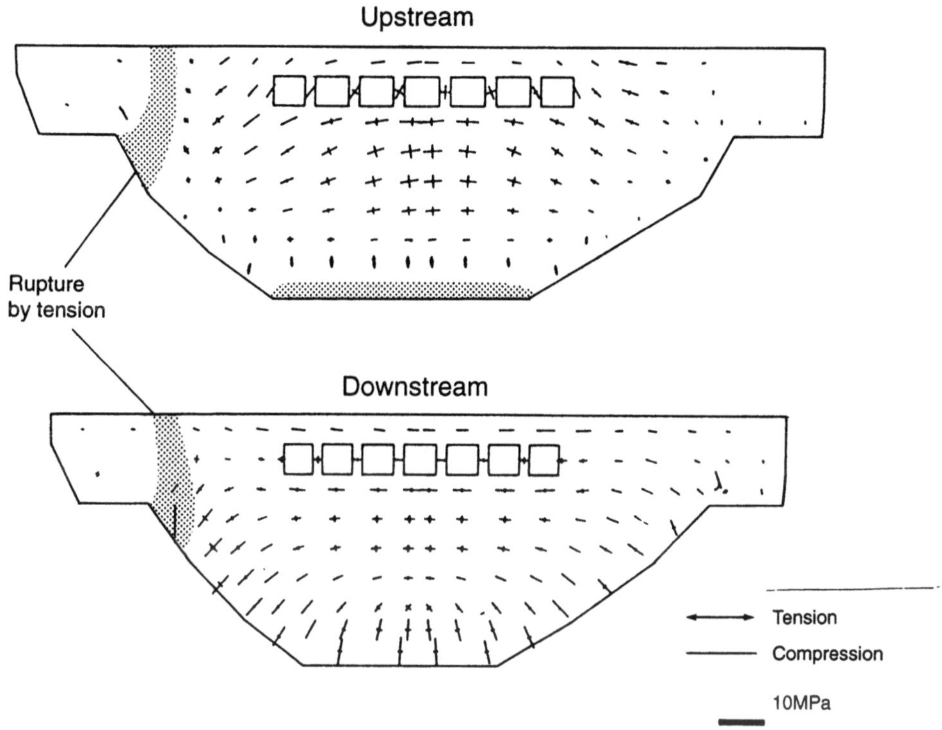

Fig. 21 - *Stresses on the dam faces near the global failure, caused by the abutment slide* (Pina and Costa 1993)

5.3 - Another important scenario of deterioration (ageing) of arch dams is the *loss of strength of the dam body*, caused by different phenomena (permanent and repeated actions, freezing and thawing, chemical reactions resulting in swelling, aggressive waters, etc.), under normal operation conditions.

The studies for this scenario have been carried out at LNEC by *simulating a decrease of the concrete cohesion* and have included *arch dams of different dimensions and types, built in several countries.*

Among other dams, the studies developed at LNEC concerned (Rocha et al 1960, 1969, Azevedo et al 1979, Pedro et al 1985, 1989, 1989a, 1990, 1991, 1993, Silveira et el 1991, Pina et al 1993): the *Alto Ceira dam* (Portugal), completed in 1949 (36 m high, a crest length of 85 m); the *Cabril dam* (Portugal), completed in 1954 (131 m high, a crest length of 290 m); the *Bouçã dam* (Portugal), completed in 1955 (65 m high, a crest length of 175 m); the *Slettedalen dam* (Norway), completed in 1959 (72 m high, a crest length of 170 m); the *Cachi dam* (Costa Rica), completed in 1966 (80 m high, a crest length of 70 m); the *Repulse dam* (Australia), completed in 1968 (42 m high, a crest length of 433 m); the *Murray II dam* (Australia), completed in 1968 (43 m high a crest length of 131 m); the *Mossyrock dam* (USA), completed in 1968 (185 m high, a crest length of 502 m); the *Salto Funil dam* (Brazil), completed in 1969 (85 m high, a crest length of 385 m); the *Cahora Bassa dam* (Mozambique), completed in 1974 (171 m high, a crest length of 321 m); the *Karoon dam* (Iran), completed in 1975 (200 m high, a crest length of 380 m); and the *Funcho dam* (Portugal), completed in 1989 (49 m high, a crest length of 165 m).

The above studies have led to different values of the reliability factors, namely to *factors between 5 and 16 with regard to the safety, and to factors between 3 and 10 with regard to the performance*. The reliability depends on the *ratio between the compressive strength of the concrete and the maximum compressive stress in the arches*, and on the *dam and foundation characteristics*, particularly:

- the *crest length/height ratio*;

- the *deformability and heterogeneity of the foundation*;

- the *strength and plasticity of the concrete*;

- the *existence of outlets and other holes* in the dam body (although, in the cases studied, no relevant influence of these holes has been found).

Some results of the *Alto Ceira dam* are presented as an example of the above studies (Pedro et al 1991, Ramos et al 1992).

The *Alto Ceira dam* is situated on a left bank tributary of the river Mondego. For many years, it has been subject to a process of *concrete swelling*, due to an *alkali-aggregate reaction*.

This concrete swelling caused a *permanent displacement* and *compression in the arch*. After 40 years, at the crown of the crest arch, the vertical and horizontal permanent displacements have amounted to about 15 mm upwards and 40 mm in the upstream direction (about 10 times the downstream displacements due to water pressure), respectively.

Dividing the dam into different zones and considering the displacements measured at each zone in the period 1950-90, an estimate of the swelling strains which would cause those displacements, was made by *back-analysis*. These strains are similar to those of a permanent temperature rise, which, besides moving the dam upwards and in the upstream direction, and inducing large compressive stresses in the arches, cause *shear cracks in the concrete*, which are approximately parallel to the slope. This crack pattern shows a *general agreement with that observed in the dam*.

The *cracking of the concrete and the increase of the compression stresses in the arches reduce the reliability* of the structure in the course of time. So, studies were made to evaluate the decrease in the reliability of the dam, including *tests on the dam and foundation materials, and analyses by mathematical models*.

For the concrete strengths of 30 MPa (one-axial compression) and 3 MPa (one-axial tension) and for the *initial conditions* (without the stresses induced by swelling), the results of the analyses showed that *the safety factor of the dam was about 20* (Fig. 22); and that, *after 40 years of operation* (with stresses in the upper arches of about 2 MPa), *the safety factor decreased to about 8*, a value still above the acceptable minimum.

5.4 - The available data on the *behaviour of arch dams under strong earthquakes* point to a *remarkable strength of these dams*. In fact, several dams have been affected by strong earthquakes in the last decades, among others (Glover 1957, Hatano and Takahashi 1957, Takahashi et al 1959, Okamoto et al 1964, JPCOLD 1973, Serafim and Oliveira 1987, Câmara 1989):

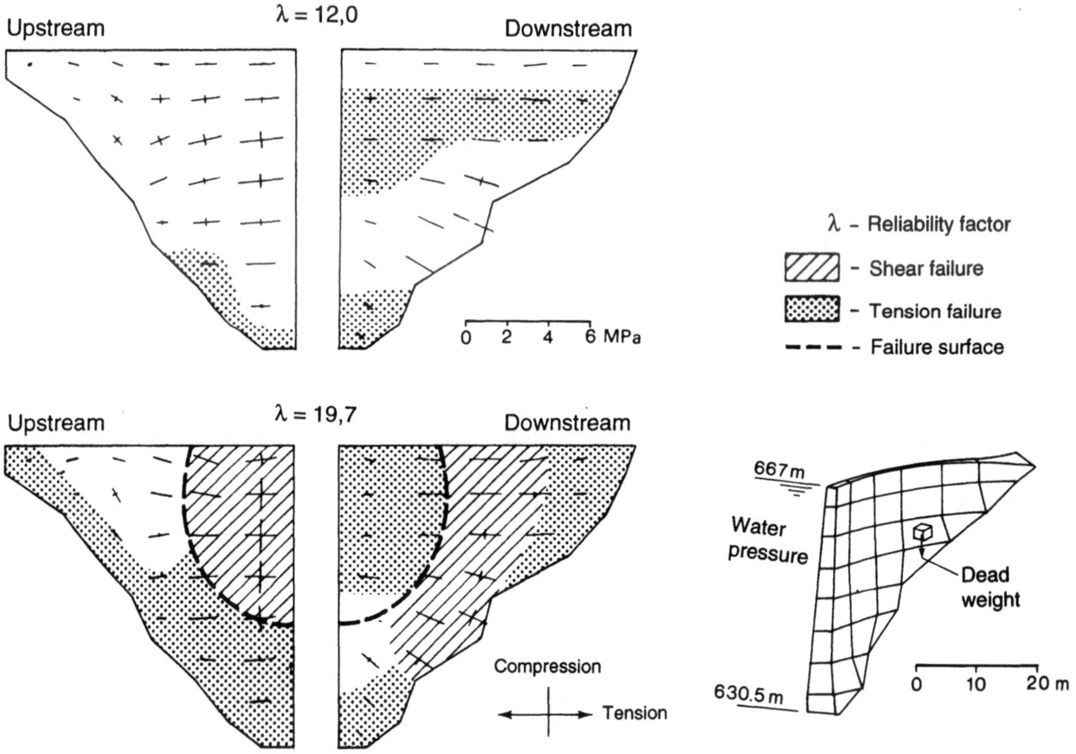

Fig. 22 - *Alto Ceira dam.* Reliability studies (Pedro et al 1991)

- the *Barossa* (Australia), *Rapel* (Chile), *Monteynard* (France), *Ambiesta, Barcis, Corfino, Main di Sauris,* and *Turrite di Gallicano* (Italy), *Ayakita, Kamishiiba, Kurobe, Nagawado, Naramata,* and *Tonoyama* (Japan), *Odeáxere* (Portugal), *Paltinul Doftana and Vidraru--Anges* (Romania), *Susqueda* (Spain), *Ceres* (South Africa), *Big Santa Anita, Big Tjunga, Brush Creek, Hoover,* and *Pacoima* (USA), and *Kariba* (Zimbabwe-Zambia) arch dams;
- the *Grandaval* (France) and *Honen-Ike* (Japan) multiple arch dams.

No accident and only small incidents have been reported for arch dams affected by strong earthquakes. Serafim and Oliveira (1987) state that, for a sample of 41 arch dams affected by earthquakes, damage was reported for only 7 of them.

An important case history concerns the behaviour of the *Pacoima arch dam* during the *San Fernando earthquake* (February, 1971).

The *Pacoima dam* is a very large structure (128 m high, a crest length of 195 m), completed in 1929, and is situated about 6 km from San Fernando (USA). During the 1971 earthquake, this dam was shaken by very strong vibrations, *ground movements of 1.25 and 0.7 times the gravitational acceleration* having been recorded for the horizontal and vertical directions in the left bank, respectively. *Only small disturbances were reported*, namely the opening of the vertical joint between the arch and the left abutment, and a crack in the thrust block (Swanson and Sharma 1979).

The good performance of arch dams under earthquakes has been also pointed out by the *analyses of mathematical and scaled models* made at LNEC. These studies concerned, among others: the *Cachi* (Costa Rica), *Kalaritikos* and *Steno* (Greece), *Karoon* and *Tang-e-Soleyman* (Iran), *Cahora Bassa* (Mozambique), *Alto Lindoso* and *Cabril* (Portugal), and *Mossyrock* (USA) arch dams; and the large-span *Aguieira* multiple arch dam (Portugal) (Pedro 1977, Pedro and Pereira 1981, Câmara 1989, Gilg et al 1987).

The study of the dynamic behaviour of the *Cabril arch dam* (Fig. 7) was made by means of different mathematical models (Pedro and Câmara 1986, Pina et al 1990). The results obtained were checked with *in-situ tests of small forced vibrations*, applied by means of an eccentric mass vibrator (Pedro et al 1986). The evaluation of the *dam behaviour under strong earthquakes* was subsequently made by *step-by-step analyses* of the so calibrated models *The seismic action was defined by accelerograms*, derived from the power spectra adopted in the Portuguese Regulations (RSA 1984). The studies included the static actions of the *dead weight* and the *water pressure*.

It was assumed that the concrete has a *visco-elastoplastic behaviour* (Young's modulus 30 GPa; Poisson's ratio 0.2; compressive strength 30 MPa; tensile strength 3MPa; damping ratio 5%) and that it will *lose its cohesion when the peak compressive strain exceeds 2.5‰.*

The results obtained for vibrations with *peak accelerations of 1.5 times the gravitational acceleration*, show that compressive strains over 2.5 ‰ *only occur at the dam faces in limited zones near the dam crest* (Fig. 23).

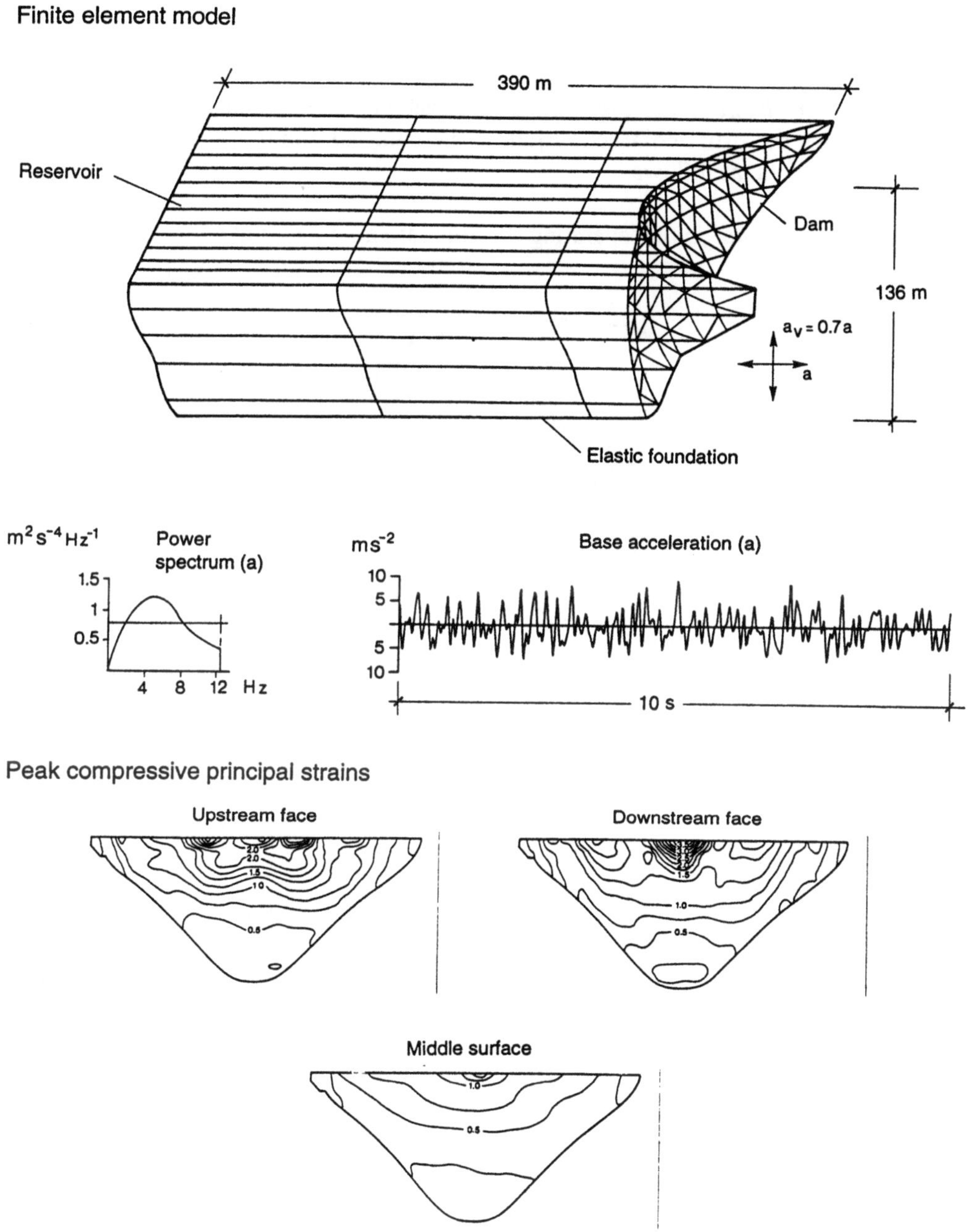

Fig. 23 - *Cabril dam. Finite element analysis for a strong earthquake*

5.5 - The design of the *Steno dam, which includes provisions for large relative movements of the abutments,* is an example of a study of the best design solutions for an arch dam (Gilg et al 1987).

The *Steno project* has been developed for a site on the Arakhthos river (Greece), which is crossed by a *very large fault* (Fig. 24). The possibility of movement along this fault was envisaged. Besides the usual contraction joints, the dam has therefore been provided with a system of special joints, as follows:

- a *plane horizontal joint*, separating the arch into an upper and a lower part;

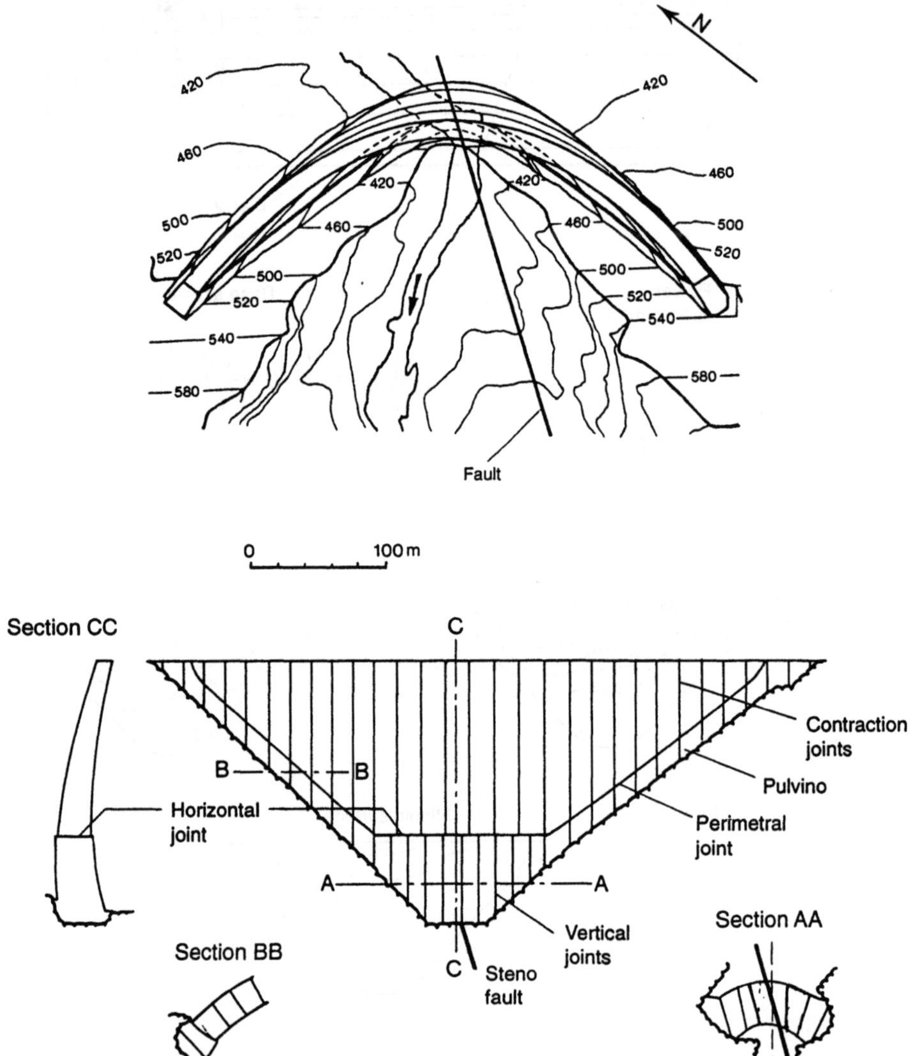

Fig. 24 - *Steno dam. Plan and elevation* (Gilg et al, 1987)

- a *perimetral curved joint*, separating the upper part of the dam from the pulvino;

- several *plane vertical joints*, in the lower part of the dam, one of which corresponds to the fault.

The arch dam for the Steno project was studied by means of several models, namely by *mathematical and scaled models developed at LNEC* (Fig. 25), which allowed the analysis of the dam behaviour for *relative movements of the fault* (Gilg et al 1987, Pedro and Azevedo 1988).

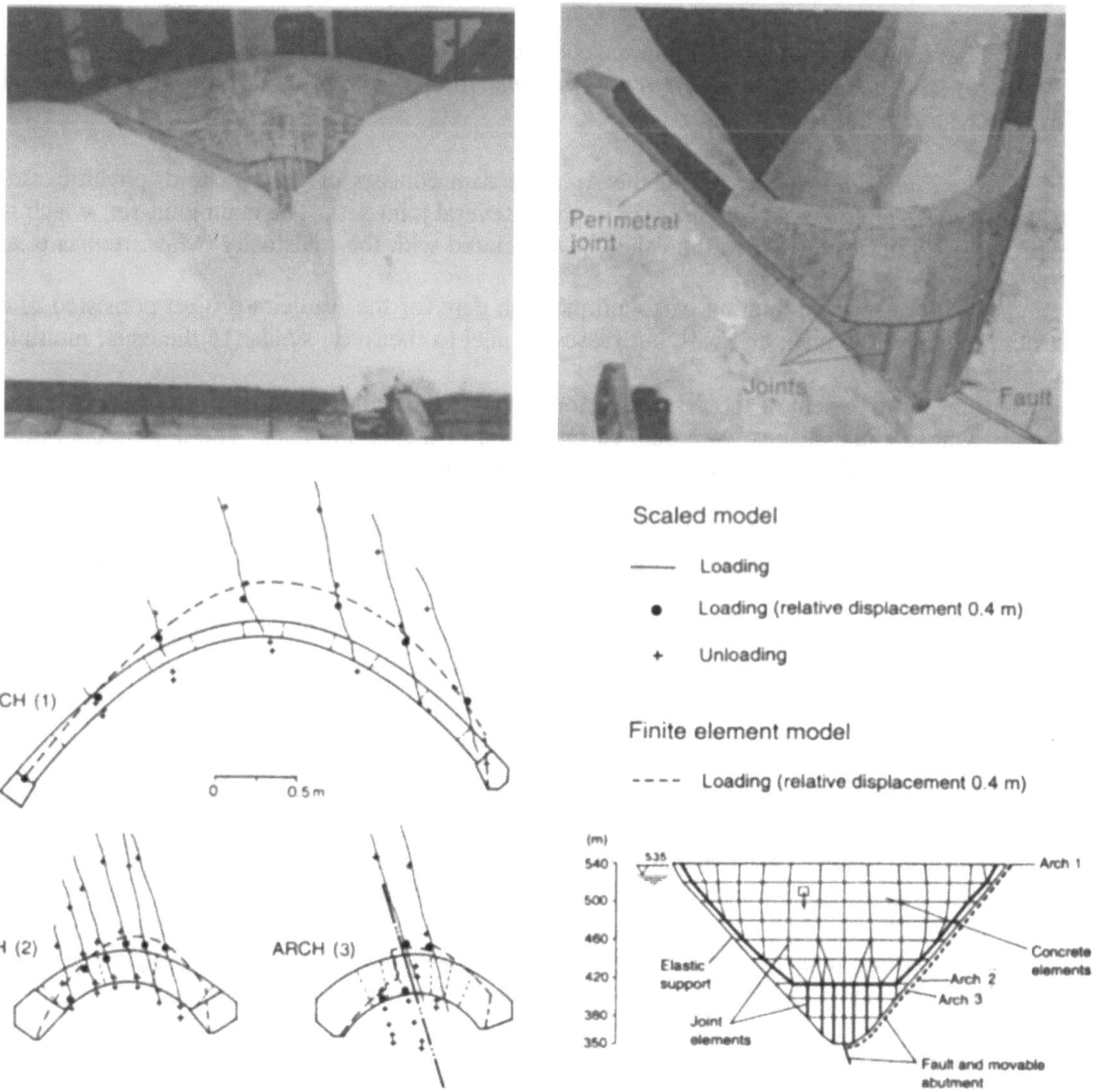

Fig. 25 - *Steno dam. Scaled and finite element model analyses* (Pedro and Azevedo 1988)

The studies of this models showed that an arch dam with the special joints indicated above would be able to support the following relative displacements of the abutments:

- up *to 0.1 m, without significant punctual failures;*

- at least, up to *0.4 m* with important movements at the special joints, and with large compressive stresses and cracking in some zones, but *without a global failure.*

5. 6 - Another example of studies of the best solutions for a dam project is the case of the Aguieira *multiple arch dam* (Heitor and Pedro 1997, Pedro 1997, Oliveira and Pedro 1986).

The Aguieira dam (Fig. 11) was constructed on the Mondego river, near Coimbra (Portugal), and belongs to a *multipurpose scheme* for power production, flood control and irrigation.

The *foundation* rock mass of the Aguieira dam consists of graywackoid, phyllitic, and schistose rocks alternating sequences, crossed by several joint sets. The main joint set, which is approximately perpendicular to the valley, is associated with the schistosity (Mascarenhas et al 1985).

The *first shape* definition of a multiple arch dam for the Aguieira project consisted of a structure, straight in plan, and with buttresses parallel to the river, similar to the usual multiple arch dams.

The improvement of the first definition began with the introduction of *double curvature arches,* whose initial shapes were defined by means of membrane tests. The double curvature improved the behaviour of the arches near the foundation.

The improvement of the stability of the buttresses with regard to their foundation was obtained by means of a *curved lay-out, that had buttresses in a radial position*, against the river banks.

The *buttresses* were also *shorter and thicker* than those of the first shape definition, in order to *place the spillways on them,* without increase of their volume. To achieve this goal, the buttresses were displaced in the upstream direction, and a *gravity crest* was used to close the upper arches.

The *stiffness of the gravity crest was reduced* by means of vertical non-grouted joints to improve the behaviour of the arches under the static loads. The *mass of the gravity crest was also reduced* to improve the dynamic behaviour of the arches. This reduction of the mass had the additional advantage of improving the behaviour under water actions.

The *sealing and drainage systems* of the foundation were also considerably improved by means of a foundation gallery, besides the usual grout curtains.

The development of the above design solutions for the Aguieira multiple arch dam was supported by several *studies of both finite element and scaled models* carried out at LNEC: *structural models* for the analysis of the dam behaviour under static and dynamic actions (Heitor and Pedro 1977, Pedro 1977) (Fig. 26); and *hydraulic models* for the analysis of the

Plaster – diatomite scaled model for static analysis

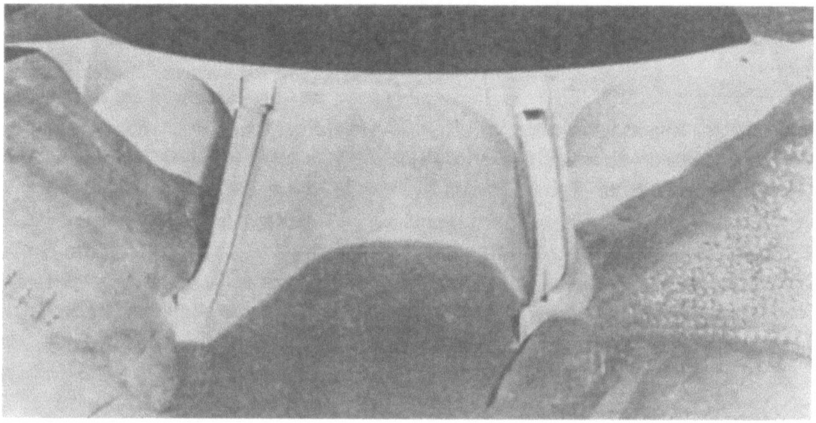

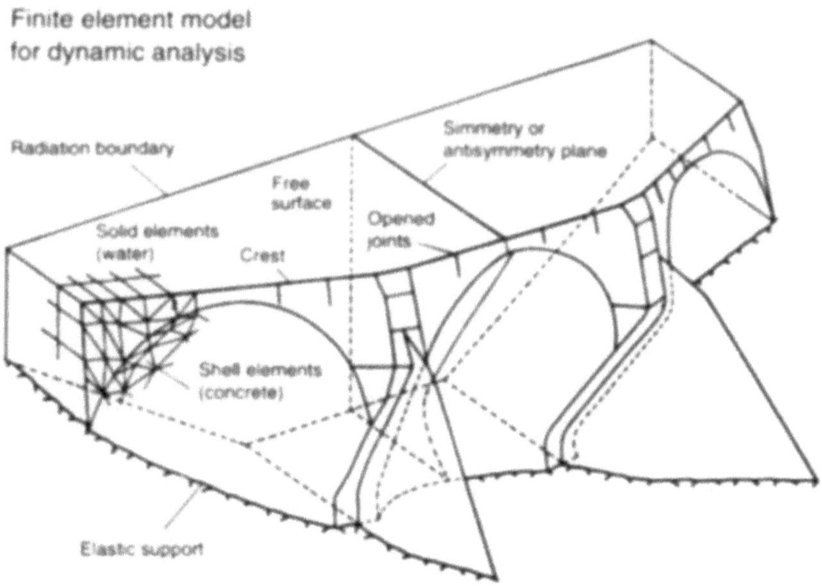

Fig. 26 - *Aguieira dam. Scaled and finite element models* (Heitor and Pedro 1977, Oliveira and Pedro 1986)

behaviour of the dam foundation, the spillways, and the hydraulic circuit (Mascarenhas et al 1985, Ramos 1980).

The Aguieira dam's behaviour has been tested for static and dynamic actions, and a *comprehensive monitoring* has been systematically carried out. The dam is in a *good condition*, and has shown an *appropriate behaviour*.

5
LEGAL ASPECTS CONCERNING DAM SAFETY

1 - The *safety and performance of dams* have always been a concern for those involved in their design, construction and operation, and *became subjects of public interest* in the last decades, particularly due to:

- the *high potential risk* associated with the construction of some dams (big dimension of the reservoir, an upstream location in populated areas, unfavourable foundations, utilisation of new technologies, etc.);

- the *occurrence of a few accidents, and the detection of incidents*, which required expensive repair works;

- a general concern of the public with regard to *environmental aspects*.

This situation has led to an *enforcement of legal requirements* for the safety and performance of dams (Fanelli 1994).

Revisions and developments of the existing regulations, standards and codes of practice have been made in several countries, among others, the USA (USCOLD 1987), China (LDSSC 1988), Portugal (Sanches 1983, CSOPT 1990) and Great Britain (CIRIA 1993).

The standards on safety and performance of dams are *specific for each country*, but, even so, they show *some common tendencies* (which may be associated to ICOLD recommendations), such as:

- the great relevance given to the *organisation of the safety and performance control* in the course of the dam's lifetime, particularly to the definition of the responsibilities of the concerned entities (the primary responsibilities being attributed to the Authority, which represents the public interest, and the Owner);

- the *coverage of the design phase*, as well as of *the entire lifetime*, including the construction, the first filling and operation of the reservoir, and even the abandonment and demolition of the works;

- the recommendations concerning a *frequent inspection* of the works, the installation and operation of an adequate *monitoring system*, as well as the periodic *submission of reports* concerning the evaluation of the condition and behaviour of the works to the Authority;

- the implementation of *warning and alert systems* in the major works, to be operated in extreme cases (exceptional natural events, sabotage, etc.), in order to mitigate the eventual effects of those events (the integrated use of monitoring methods and warning systems is only considered in few countries, for scarcely populated regions).

6.2 - The *organisation of the safety and performance control of dams,* and the role of the different concerned entities are very important. As *example*, the main aspects of this *organisation in Portugal* (CSOPT 1990), are described below (Fig. 27).

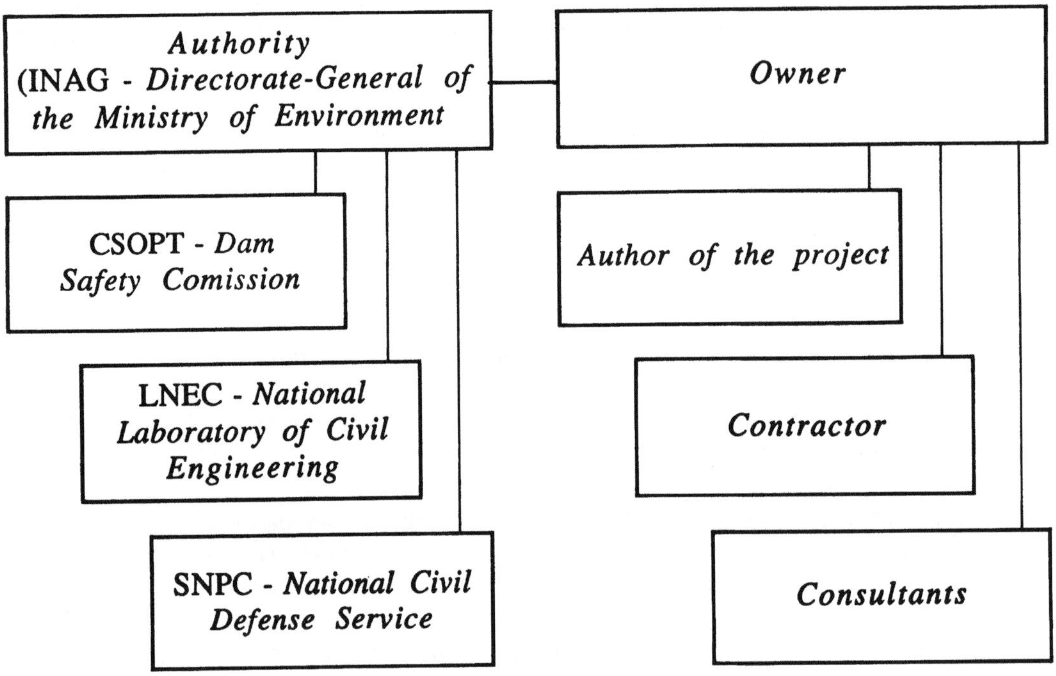

Fig. 27 - *Organisation of the safety and performance control of dams in Portugal* (CSOPT 1990)

The *Authority* is responsible for the approval of the projects, for the commissioning of the works, and for the control of their condition in the course of the entire lifetime.

In the current implementation of its responsibilities, the *Authority receives technical support from the LNEC*, the *SNPC* (National Civil Defence Service, Ministry of Defence), and the *Dam Safety Commission*. The Dam Safety Commission, integrated in the CSOPT (High Council for Public Works and Transports), includes representatives of the Authority, the LNEC, the SNPC, the Universities, the Portuguese National Commission on Large Dams, the Portuguese Engineer Association, the EDP (holding of the Portuguese electricity companies), and the Ministry of Agriculture. Besides giving *advice to the Authority* and *preparing standards and recommendations*, it may propose immediate measures.

The *Owner* is responsible for the preparation of the project and its submission (for approval) to the Authority as well as for the execution of the works (in accordance with the approved projects), and, with the support of LNEC, for the *preparation and implementation of a monitoring plan*. The *monitoring plan* has to cover the construction phase, the first filling, and the operation of the reservoir and include *procedures for the normal conditions, and for exceptional occurrences or an anomalous behaviour* of the works.

In the case of *abandonment or demolition* of the works, the Owner must previously get the approval of the Authority, for the respective project.

REFERENCES

ASCE (American Society of Civil Engineering), 1956 - *Symposium of arch dams*. Knoxville, Tennessee.

ASCE/USCOLD (Committee on Failures and Accidents to Large Dams),1975/88 - *Lessons from dam incidents, USA*. New York.

AUCOLD (Austrian National Committee on Large Dams), 1991 - *Large dams in Austria*. Vienna.

Azevedo MC, Ferreira ME, Costa CP, 1979 - *Influence of the spillway on the safety of arch dams*. LNEC paper 524, Lisbon.

Azevedo MC, Ferreira MJE, Costa CP, 1988 - *The use of geomechanical physical models at LNEC*. LNEC paper 731, Lisbon.

Bister D, Bonnet P, Denis B, Beauchamp R, Dubois P, Goguel B, 1991 - Contribution to monitoring of French concrete dam affected by swelling or shrinkage. Application to adult dams (Chambon, Vouglans) and to fresh concrete (RCC). ICOLD Congress, Vienna.

BNCOLD (British National Committee on Large Dams), 1983 - *Dams in the UK1963 - 1983*. The Institution of Civil Engineers, London.

Borges JF, Castanheta MC, 1983 - *Structural safety*. LNEC, Lisbon.

BRCOLD (Brazilian Committee on Large Dams), 1982 - *Barragens no Brasil*. (Dams in Brazil). Rio de Janeiro.

BRCOLD (Technical Committee on Dam Safety), 1986 - *Segurança de barragens. Recomendações para a formulação e verificação de critérios e procedimentos.* (Dam Safety Recommendations for the formulation and validation of criteria and procedures). Rio de Janeiro.

Bulota G, Im O, Larivière R, 1991 - *Le barrage Daniel Johnson: un vieillissement prematuré* (Daniel Johnson dam: an early ageing). ICOLD Congress, Vienna.

Câmara RC, 1989 - *Modelos para estudo do comportamento dinâmico de barragens de betão pelo método dos elementos finitos* (Finite element models for dynamic analysis of concrete dams). LNEC, Lisbon.

CEB (Euro-international Concrete Committee, Joint Committee on Structural Safety), 1988 - *General principles on reliability for structures.* Bulletin 191 (September). Lausanne.

CIRIA (Construction Industry Research and Information Association), 1968 - *Symposium on arch dams.* The Institution of Civil Engineers, London.

CIRIA, 1993 - *Dam safety in Europe. A report on the OSTEMS mission* to Europe. London.

CNCOLD (Chinese National Committee on Large Dams), 1987 - *Large dams in China. History, achievement, prospect.* Beijing.

Combelles J, 1991 - *Vieillissement des barrages et méthodes de réparation.* (Ageing of dams and repair methods). General Report, ICOLD Congress, Vienna.

CSOPT (High Council for Public Works and Transports), 1990 - *Portuguese regulations for safety of dams; Code of practice for design; Code of practice for construction; Code of practice for operation*; and *Code of practice for observation and inspection.* Lisbon.

Fanelli M, 1990 - *Observed behaviour of arch dams.* In Arch Dams. Balkema, Rotterdam.

Fanelli M, 1994 - *Safety assessment and improvement of existing dams.* General Report, ICOLD Congress, Durban.

Fernandes JS, Molico AA, 1970 - *Cabora Bassa arch dam. Study of different criteria for a design solution.* ICOLD Congress, Montreal.

Ferreira MJE, 1983 - *Modelos geomecânicos* (Geomechanical models). In Desenvolvimentos recentes no domínio da Mecânica das Rochas. LNEC, Lisboa.

FNCOLD (French National Committee on Large Dams), 1985 - *Dams in France.* Paris.

Geringer JJ, 1994 - *The evolution and development of RCC dams in South Africa.* Hydropower & Dams (November). Sutton.

Gicot O, 1981 - *Le récent comportement du barrage de Zeuzier* (the recent behaviour of the Zeuzier dam). Wasser, Energie, Luft (Vol. 72, N° 7/8). Baden.

Gilg B, Indermaur W, Matthey F, Pedro JO, Azevedo MC, Ferreira MJE, PPC Greece, 1987 - *Special design of Steno arch dam in Greece, in relation with possible fault movements*. ICOLD Symposium, Beijing.

Glover R, 1957 - *Arch dams: review of experience*. ASCE Journal of Power Division (April). Ann Arbor, Michigan.

Gomes AS, Matos DS, 1985 - *Quantitative analysis of dam monitoring results. State of the art, application and prospect*. ICOLD Congress, Lausanne.

Guerreiro M, Fernandez CR, Guerreiro MJ, Gomez LG, 1991 - *Causes of failure of the Fonsagrada dam*. ICOLD Congress, Vienna.

Hatano T, Takahashi T, 1957 - *The stability of an arch dam under earthquakes*. Central Research Institute of Electric Power Industry, Tokyo.

Haws ET, 1991 - *Environmental issues in dam projects*. General Report, ICOLD Congress, Vienna.

Heigerth G, Melbinger R, Obernhuber P, Tschernutter P, 1994 - *Assessing and improving the safety of existing dams in Austria*. ICOLD Congress, Durban.

Heitor AM, Pedro JO, 1977 - *Estudo de uma barragem de abóbadas múltiplas pelo método dos elementos finitos* (Study of a multiple arch dam by the finite element method). LNEC paper 484. Lisbon.

Hollingworth F, Hooper DG, Geringer JJ, 1989 - *Roller compacted concrete arched dams*. Water Power (November). Sutton.

ICOLD (Committee on the Foundations of Large Dams and on Underground Works; Committee on Failures and Accidents to Large Dams other than in Connection with their Foundations; Committee on Failures and Accidents to Large Dams), 1974/81 - *Lessons from dam incidents*. Paris.

ICOLD (Committee on Analysis and Design of Dams), 1978/82/87 - *Finite element methods in analysis and design of dams*. Bulletin 30a, Paris.

ICOLD (Committee on the Dictionary, the Glossary, and the World Register of Dams), 1978/95 - *Technical dictionary on dams*. Paris.

ICOLD (Committee on Deterioration of Dams and Reservoirs), 1982 - *Automated observations for the safety control of dams*. Bulletin 41, Paris.

ICOLD (Committee on Deterioration of Dams and Reservoirs), 1984 - *Deterioration of dams and reservoirs*. Paris.

ICOLD (Committee on the Dictionary, the Glossary, and the World Register of Dams), 1984/88 - *World register of dams*. Paris.

ICOLD (Committee on Dam Safety), 1987 - *Dam safety. Guidelines*. Bulletin 59, Paris.

ICOLD (Committee on Analysis and Design of Dams), 1988 - *Dam design criteria. Phylosophy of choice*. Bulletin 61, Paris.

ICOLD (Committee on Dam Safety), 1988a - *Understanding the need for dams and their safety*. Paris.

ICOLD (Committee on Monitoring of Dams and Their Foundations), 1989 - *Monitoring of dams and their foundations. State-of-the-art*. Bulletin 68. Paris.

ICOLD, 1992/93/94 - *Status of dam construction. Yearly inquires*. Internal report. Paris.

ICOLD, (Ad hoc committee on statistical interpretation of failures of dams), 1994 - *Final report*. Paris.

ICOLD (Committee on Dam Ageing), 1994a - *Ageing of dams and appurtenant works*. Bulletin 93, Paris.

IRCOLD (Iranian National Committee on Large Dams), 1985 - *News on Iran's large dams*. Bulletin (June). Tehran.

JPCOLD (Japanese National Committee on Large Dams), 1973 - *Earthquake design features in Japan*. Tokyo.

JPCOLD, 1979/82/84/88/90/91/94 - *Dams in Japan*. Tokyo.

LDSSC (Large Dam Safety Supervision Centre, Ministry of Energy of China), 1988 - *Interim statute for dam safety management of hydropower station*. Hangzhou.

Lemos JV, 1987 - *A distinct element model for dynamic analysis of jointed rock with applications to dam foundations and fault motion*. University of Minnesota, Minneapolis, Minnesota.

LNEC, 1955 - *Symposium on the observation of structures*. RILEM symposium. Lisbon.

LNEC, 1957/57a/68/68a/68b/77 - *Détérmination expérimentale de la forme du barrage de Vajont* (Experimental study on the shape of Vajont dam); *Estudo experimental da barragem do Salto-Funil. Determinação de formas pelo método da membrana* (Experimental study of the Salto-Funil dam. Membrane tests); *Experimental study of Karoon dam. Membrane tests; Estudo experimental da barragem de Odivelas. Ensaios de membrana* (Experimental study of the Odivelas dam. Membrane tests); *Estudo experimental da barragem da Aguieira. Ensaios de* membrana (Experimental study of the Aguieira dam. Membrane tests); *Estudo experimental da barragem de Odelouca. Ensaios de membrana* (Experimental study of the Odelouca dam. Membrane tests). Internal reports. Lisbon.

LNEC, 1962 - *Symposium on the use of computers in civil engineering*. Lisbon.

LNEC, 1963 - *Symposium on concrete dam models*. Lisbon.

LNEC, 1986 - *8th European conference on earthquake engineering*. Lisbon.

LNEC, 1987 - *Conferência Ibero-Americana sobre aproveitamentos hidráulicos* (Ibero-American Conference on Hydraulic Works). Lisbon.

LNEC, 1993 - *A short course on dam foundations in rock masses*. Lisbon.

Mascarenhas AT, Sousa LR, Martins CS, 1985 - *Uplift, seepage control and observed behaviour of Aguieira dam foundation*. ICOLD Congress, Lausanne.

Müller R, Pougatsch H, 1985 - *Maintenance and restauration. Zeuzier dam*. In Barrages Suisses. Swiss National Committee on Large Dams. Lausanne.

Navarro JLG, Aracil JJ, 1945 - *Saltos de água y presas de embalse* (Weirs and dams). Escuela de Ingenieros de Caminos, Canales y Puertos, Madrid.

Naylor DJ, Staag KG, Zienkiewcz OC, 1975 (Ed.) - *Criteria and assumptions for numerical analysis of dams*. University College of Swansea. Swansea.

NBWE (National Board of Waters and Environment of Finland), 1987 - *Dam safety code of practice*. Helsinki.

NWREA (Norwegian Water Resources and Energy Administration), 1986 - *The Norwegian regulation for planning construction and operation of dams*. Norwegian University Press, Oslo.

Okamoto S, Yoshida M, Nakayama K, 1964 - *Observation of dams during earthquakes*. ICOLD Congress, Edinburgh.

Okamoto S, 1973 - *Introduction to earthquake engineering*. University of Tokyo Press, Tokyo.

Oliveira ER, Pedro JO, 1986 - *The rise and decline of structural analysis as a research topic in structural engineering*. Conference on Recent Advances in Simulation of Complex Systems, Tokyo.

Oliveira ER, Pedro JO, Pina CA, Lemos JV, 1992 - *Non-linear and time-dependent analysis in three-dimensional structures*. International Conference on Education, Practice and Promotion of Computational Methods, Using Small Computers, Dalian (China).

Oliveira ER, Pedro JO, 1993 - *LNEC research plan 1992-95*. LNEC paper 794. Lisbon.

Pedro JO, 1973 - *Finite element stress analysis of plates, shells and massive structures*. International Course on Structural Concrete, LNEC, Lisbon.

Pedro JO, 1977 - *Dimensionamento das barragens abóbada pelo método dos elementos finitos*. (Finite element stress analysis of arch dams) LNEC paper 479. Lisbon.

Pedro JO, Pereira JJ, 1979 - *Models for seismic analysis of dams*. ICOLD Congress, New Delhi.

Pedro JO, Azevedo MC, 1982 - *Criteria and models for safety evaluation of concrete dams*. ICOLD Congress, Rio de Janeiro.

Pedro JO, Gomes AF, Câmara RC, 1984 - *Stress evaluation in concrete dams. The example of Varosa arch dam*. International Conference on Safety of Dams, Coimbra.

Pedro JO, Pina CA, 1985 - *Finite element analysis of arch dams up to failure*. International Conference on Finite Elements in Computational Mechanics, Bombay.

Pedro JO, Câmara RC, 1986 - *Coupled models for dynamic analysis of arch dams*. European Conference on Earthquake Engineering, Lisbon.

Pedro JO, Silva AF, Pereira J, Florentino CA, Pinho JS, Câmara R, Silva H, 1986 - *Seismic monitoring of concrete dams and reservoirs. Portuguese experience*. European Conference on Earthquake Engineering, Lisbon.

Pedro JO, 1987 - *A engenharia das barragens de betão em Portugal. Estruturas mais importantes e alguns estudos*. (Concrete dam engineering in Portugal. Major structures and some studies). Ibero-American Conference on Hydraulic Works. Lisbon.

Pedro JO, Pina CA, 1987a - *Finite element analysis of plates, shells and massive structures*. International Conference on Education, Practice and Promotion of Computational Methods Using Small Computers, Guangzhou.

Pedro JO, Azevedo MC, 1988 - *Arch dam on active fault*. ICOLD Congress, San Francisco, (LNEC paper 722).

Pedro JO, Pina CA, Baptista AL, 1989 - *Finite element analysis of an arch dam taking into account time and rupture effects*. International Conference on Computational Plasticity. Models, Software and Applications, Barcelona. (LNEC paper 756).

Pedro JO, Pina CA, Oliveira SB, 1989a - *Structural safety of a cracked arch dam*. ICOLD Symposium, Copenhagen.

Pedro JO, 1990 - *Ageing of large structures. Evaluation by computational methods*. Lecture, International Conference on Education, Practice and Promotion of Computational Methods, Using Small Computers, Macao. (LNEC paper 761).

Pedro JO, Pina CA, Ramos JM, 1991 - *Ageing evaluation for concrete and masonry dams. Portuguese experience*. ICOLD Congress, Vienna.

Pedro JO, Neves EM, Ramos CM, 1992 - *Activity of the Laboratório Nacional de Engenharia Civil in the field of large dams*. LNEC paper 788. Lisbon.

Pedro JO, Lemos JV, 1993 - *Modelling in safety evaluation*. General Report, EUROCK'93 Symposium, Lisbon.

Pedro JO, Mascarenhas AT, Silva HS, 1993a - *Ageing of concrete dam foundations.* EUROCK '93 Symposium, Lisbon. (LNEC paper 797).

Pedro JO, Mascarenhas AT, Sousa LR, Rodrigues LF, Silva HS, Castro AT, 1993b - *Rock mass foundation deterioration of Portuguese concrete dams. The Varosa and Venda Nova case histories.* EUROCK'93 Symposium, Lisbon.

Pedro JO, 1993c - *Ageing of dams and appurtenant works*-Indian Journal of Power and River Valley Development (August-September). Calcutta.

Pina CA, Câmara R, 1987 - *Non-linear analysis of arch dams.* 4th Congress of Theoretical and Applied Mechanics, Swansea. (LNEC paper 717).

Pina CA, 1988 - *Modelos de elementos finitos para análise de tensões em barragem de betão* (Finite element models for the stress analysis of concrete dams). LNEC, Lisboa.

Pina CA, Costa CP, Câmara R, Lemos TV, Baptista AL, Oliveira S, Gomes JP, 1990 - *Análise não-linear de estruturas laminares e maciças* (Non-linear analysis of plates, shells and massive structures). Portuguese Conference on Structural Engineering, Lisbon. (LNEC paper 772).

Pina CA, Costa CP, 1993 - *Failure analysis of rock foundations by experimental and numerical methods.* A Short Course on Dam Foundations in Rock Masses, Lisbon (LNEC paper 799).

Pina CA, Costa CA, Oliveira SB, Gomes JP, 1994 - *Safety evaluation of a cracked arch dam by experimental and numerical methods.* LNEC paper 800. Lisbon.

PNCOLD (Portuguese National Committee on Large Dams), 1985 - *Seepage and uplift measurements in the safety control of arch dams. The examples of Varosa and Bouçã dams.* ICOLD Congress, Lausanne.

PNCOLD, 1985a - *Cracking and repair works in Cabril dam.* ICOLD Congress, Lausanne.

PNCOLD, 1992 - *Large dams in Portugal.* Lisbon.

Quintela AC, Cardoso JL, Mascarenhas JM, 1987 - *Aproveitamentos hidráulicos romanos a sul do Tejo* (Roman hydraulic works in southern Portugal). Directorate-General of Hydraulic Resources, Ministry of the Plan and Administration of the Territory, Lisbon.

Ramos CM, 1980 - *Estabilidade das grades metálicas de protecção dos circuitos hidráulicos* (stability of the protection trashracks in hydraulic circuits). Simposium on the Use of Water, Lisbon.

Ramos JM, Pinho JS, 1986 - *A new method of quantitative analysis of dam displacements.* International Conference on Mathematical Methods and Experimental Measurements. Porto Caras.

Ramos JM, Castro AT, Baptista AL, 1992 - *Analysis of swelling processes in concrete dams*. LNEC paper 778. Lisbon.

Ramos JM, 1994 - *Fiabilidade e monitoramento de barragens de betão* (Reliability and monitoring of concrete dams). LNEC, Lisboa.

Rocha M, Serafim JL, Silveira AF, 1958 - *A method of quantitative interpretation of the results of the observation of concrete dams*. ICOLD Congress, New York.

Rocha M, Serafim JL, Ferreira ME, 1960 - *The determination of the safety factor of arch dams by means of models*. International Colloquium on Models of Structures, Madrid. (LNEC paper 163).

Rocha M, 1964 - *Mechanical behaviour of rock foundations of concrete dams*. ICOLD Congress, Edinburgh.

Rocha M, Silveira A, 1964a - *The use of models to determine temperature stresses in arch dams*. LNEC paper 230. Lisbon.

Rocha M, Azevedo MC, Pereira MC, 1964b - *Note on the study of large span multiple arch dams*. LNEC paper 237. Lisbon.

Rocha M, 1965 - *Statement of the physical problem of the arch dams*. LNEC paper 263. Lisbon.

Rocha M, 1965a - *Structural model techniques. Some recent developments*. In Stress Analysis. John Wiley, Chichester.

Rocha M, Silveira AF, Azevedo MC, Ferreira ME, 1969 - *The use of model tests in the determination of the safety factor in arch dams*. Water Power (December). Sutton. (LNEC paper 433).

Rocha M, 1973 - *Mecânica das Rochas* (Rock Mechanics). LNEC, Lisbon.

Rocha M, 1974 - *Present possibilities of studying foundations of concrete dams*. ISRM Congress, Denver.

Rocha M, 1978 - *Analysis and design of the foundations of concrete dams*. General Report, International Symposium on Rock Mechanics Applied to Dam Foundations, Rio de Janeiro.

Rydzewski JR, 1965 (Ed.) - *Theory of arch dams*. (University of Southampton), Pergamon Press, New York.

Schnitter N, 1994 - *A history of dams. The useful pyramids*. Balkema, Rotterdam.

Serafim JL, Pedro JO, 1965 - *Seismic study of arch dams*. ASCE, Structural Division (April). Ann Arbor, Michigan.

Serafim JL, 1973 - *A engenharia de barragens em Portugal* (Dam engineering in Portugal). Electricidade, N° 90, Lisbon.

Serafim JL, Oliveira CS, 1987 - *Effects caused by earthquakes on dams*. ICOLD Symposium, Beijing.

Serafim JL, Clough RW, 1990 (Ed.) - *Arch dams*. Balkema, Rotterdam.

Severn RT, Taylor PR, 1965 - *Earthquake effects on an arch dam by response spectra methods*. Symposium on Vibrations in Civil Engineering, Institution of Civil Engineers, London.

Severn RT, Jeary AP, Ellis BR, 1980 - *Forced vibration tests and theoretical studies on dams*. Proceedings of the Institution of Civil Engineers (September, June), London.

Silveira AF, Pedro JO, 1964 - *Quantitative interpretation of the results obtained in the observation of concrete dams*. ICOLD Congress, Edinburgh. (LNEC paper 253).

Silveira AF, Azevedo MC, Esteves Ferreira MJ, Costa CP, 1979 - *High density and low strength materials for geomechanical models*. International Colloquium on Physical Geomechanical Models, Bergamo.

Silveira AF, 1990 - *Some considerations on the durability of dams*. Water Power (February). Sutton.

Silveira AF, Pina CA, Costa PC, Direito FT, 1991 - *Influence of foundation heterogeneity on the safety of arch dams*. ICOLD Congress, Vienna.

Smith NAF, 1992 - *The heritage of Spanish dams*. Colegio de Ingenieros de Caminos, Canales y Puertos, Madrid.

SPCOLD (Spanish National Committee on Large Dams), 1976 - *Grandes presas. Experiencias españolas en su projecto y construction*. (Large dams. Spanish experience in design and construction). Madrid.

SPCOLD, 1993 - *Selección de presas españolas 1973-1993 (Selection of Spanish dams 1973-1993)*. Madrid.

Stagg KG, Zienkiewicz OC, 1968 (Ed.) - *Rock Mechanics in Engineering Practice*. John Wiley, Chichester.

Swanson AA, Sharma RP, 1979 - *Effects of the 1971 San Fernando earthquake on Pacoima arch dam*. ICOLD Congress, New Delhi.

SWCOLD (Swiss National Committee on Large Dams), 1985 - *Swiss dams. Monitoring and maintenance*. Lausanne.

Takahashi T, Tsutsumi H, Mashuko Y, 1959 - *Behaviour on the vibration of arch dams. Kamishiiba arch dam. Central Research Institute of Electric Power Industry, Tokyo.*

TECHNOPROMEXPORT, 1988 - *High dams of the Soviet Union*. Moscow.

Tischler W, Schlosser J, 1985 - *Foundation sealing of Schlegeis arch dam*. ICOLD Congress, Lausanne.

Urbistondo R, Yges L, 1985 - *Presa de El Atazar. Tratamiento de fissuras del paramento de aguas arriba* (El Atazar dam. Treatment of the upstream face cracks) Revista de Obras Publicas (May-Jun). Madrid.

USBR (US Bureau of Reclamation), 1938 - *Trial load method of analysing arch dams*. Denver, Colorado.

USBR, 1956 - *Treatise on Dams*. Denver, Colorado.

USCOLD (US Committee on Large Dams), 1982 - *Dam safety practices and concerns in the United States*. Denver, Colorado.

USCOLD, 1988 - *Development of dam engineering in the United States*. Pergamon Press, New York.

US Committee on Safety Criteria for Dams, 1985 - *Safety of dams. Flood and earthquake criteria*. National Academy Press, Washington.

US Engineering Foundation, 1928/31/33 - *Arch dam investigation*. New York.

Wittke W, 1984 - *Rock mechanics*. Springer, Vienna.

World Bank, 1993 - *Supervision of dams by governmental authorities world-wide*. Washington, Colombia.

Zienkiewicz OC, Holister GS, 1965 (Ed.) - *Stress analysis*. (University College of Swansea), John Wiley, Chichester.

PART 2

MODELS AND METHODS OF ANALYSIS

by

M. Fanelli

National Power Board, ENEL

Italy

1. GENERALITIES AND CRITICAL CONSIDERATIONS

1.1 Arch Dams: peculiar features/advantages, structural subtypes

Within the wider class of concrete dams, arch dams, although a minority in terms of relative numbers (fig. 1), form an important category. In fact, whenever local conditions (valley shape, mechanical properties of foundations) are favorable, they constitute a potentially advantageous option over other structural types. This is due to the fact that arch dams can be designed to withstand the design loads with an overall concrete volume that is, generally speaking, substantially less than for a gravity dam spanning the same valley. In other words, the material resistance is more effectively exploited.[1].

In addition to that, it can be said that arch dams are generally better suited than gravity (or buttress) dams to resist adverse exceptional events, like seismic loads and overtopping.

Every advantage, however, has to be paid for. In the case of arch dams, the price to pay is a greater attention and care in the characterization and treatment of the foundations; stricter demands on the quality of concrete and rocky abutments; sometimes (double-curvature dams) erection constraints; well-planned layout, execution and treatment (grouting) of construction joints; more accurate formwork and geometrical tracing during monolith erection; last but not least, more elaborate design and analysis procedures, involving the use of mathematical models. (It is worth mentioning in passing that such mathematical models are used to good avail also in back-analysis and interpretations of in-service behavior, or in the assessment of the effects of exceptional events).

Now within the subclass of 'arch dams', at a closer look, one can easily make finer distinctions. In narrow canyons or gorges one can find (or design, in the case of a new dam) so-called 'pure arch' solutions (fig. 2), in which the load-carrying capacity is for the greater part assigned to horizontal arches working essentially in compression.

In wider valleys the so-called 'cantilever' effect, i.e. the load share taken by the bending of vertical elements, takes on a greater importance, and correspondingly it is advantageous to look for a geometrical shape that allows the 'dead weight' effect to counteract the tensile stresses that, in some critical areas, are inevitably associated with

---

[1] This economical advantage subsists, albeit to a lesser degree, also in comparison with buttress dams (which on the other hand are known to be somewhat more prone to cracking and require more elaborate formwork).

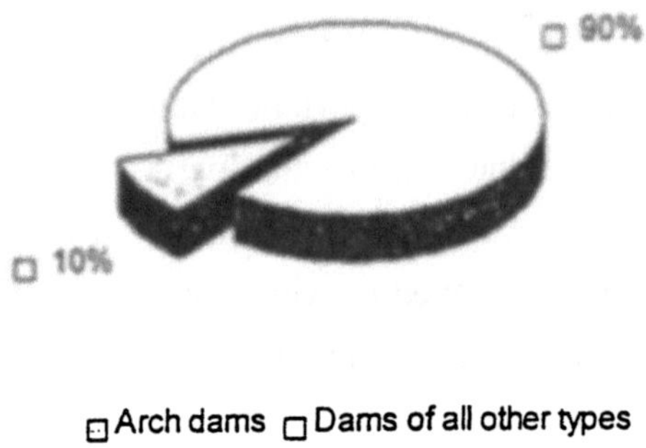

Fig. 1
Percentage of arch dams on dams of all types (H>30m or V>10^6 m^3)

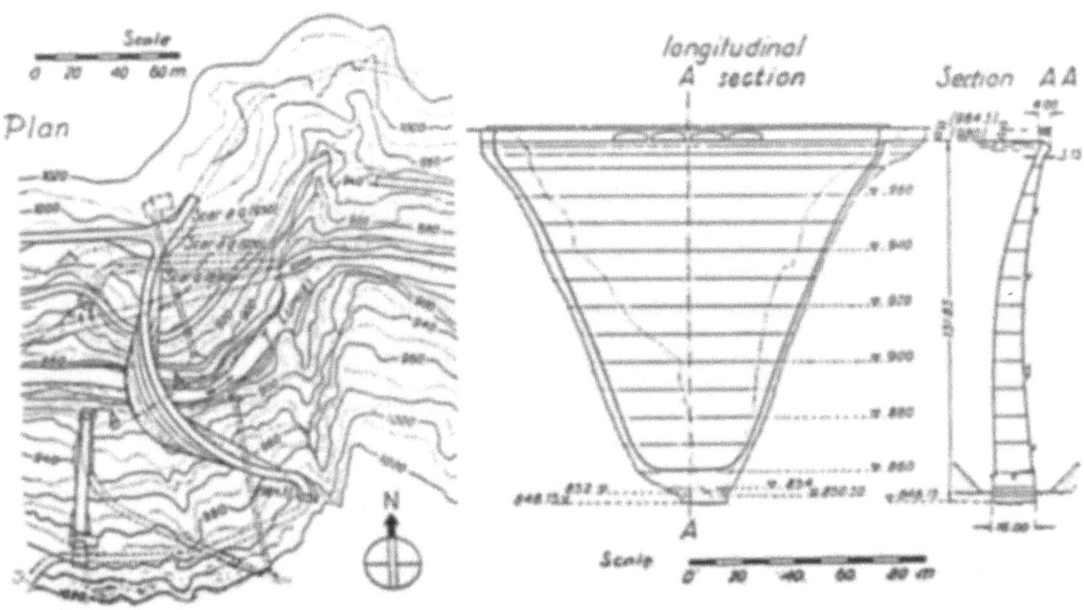

Fig. 2
"Pure arch" dam (Lumiei dam-Italy)

any appreciable degree of cantilever bending. In these cases it is more correct to speak of 'arch-gravity' dams (fig. 3), and the thicknesses are greater than in a pure arch solution, especially near the basis. The relative 'cantilever' and 'arch' stiffness (to which it is to be added the 'torsional' stiffness) affect the distribution of loads between horizontal and vertical (and inclined) elements, so that the optimal shape of the dam can depart - in a more or less pronounced way - from the 'circular arch' shape that is prevalent in the case of narrow gorges.

In particular, if one compares U-shaped with V-shaped wide valleys, it can be shown that the global cantilever stiffness is about four times greater (for a same crest length and with the same thickness distribution) in the case of V-shape valleys. So, in this case the cantilevers (and especially the shorter ones in the valley sides) will take on a proportionally greater share of the loads, so that the load distribution along the arches will present a peak at the crown and a gradual decrease toward the abutments. The rational design guideline which consists in assigning the arches a midline shape as close as possible to a funicular curve of the arch-carried loads will thus, in this case, lead to a distribution of curvature radii which will present a minimum at crown and a gradual increase toward the abutments. In fact, in many V-shaped valley we find arch shapes that are parabolic (fig. 4), elliptic (with the greater axis along the valley axis) or logarithmic-spiral shaped, i.e. in all cases with radii gradually increasing from crown to the abutments. In the case of intermediate width of the valley, and when one pursues the greatest possible volume economy (as well as the fuller use of the material resistance capabilities), one can find a third type of arch dams: the so-called 'double curvature' or 'dome' type (fig. 5), in which the vertical sections (cantilevers) are assigned a pronounced curvature (much more than what is needed, in an arch-gravity solution, to exploit the dead-weight effect as explained above).

This shape is intended to bring into play not only the bending stiffness of the cantilevers, but also their compressional stiffness. The cantilevers are then called, in a way, to act as vertical arches [2]. However, since at crest - as a consequence of the free horizontal boundary - the compressional state of the cantilevers must of necessity tend to zero, this secondary arch effect is not usually very pronounced. This type, in effect, after enjoining a brief period of favor in the years from the late '30s to the early '60s, is now less and less used; all the more so, that in order to realize high curvatures in the vertical sections one is obliged to have severe building constraints (the monoliths are

[2] The aim being to achieve very thin shells, resisting mainly by 'membrane' 2-D compressional action, and competitive in volume with pure arch solutions.

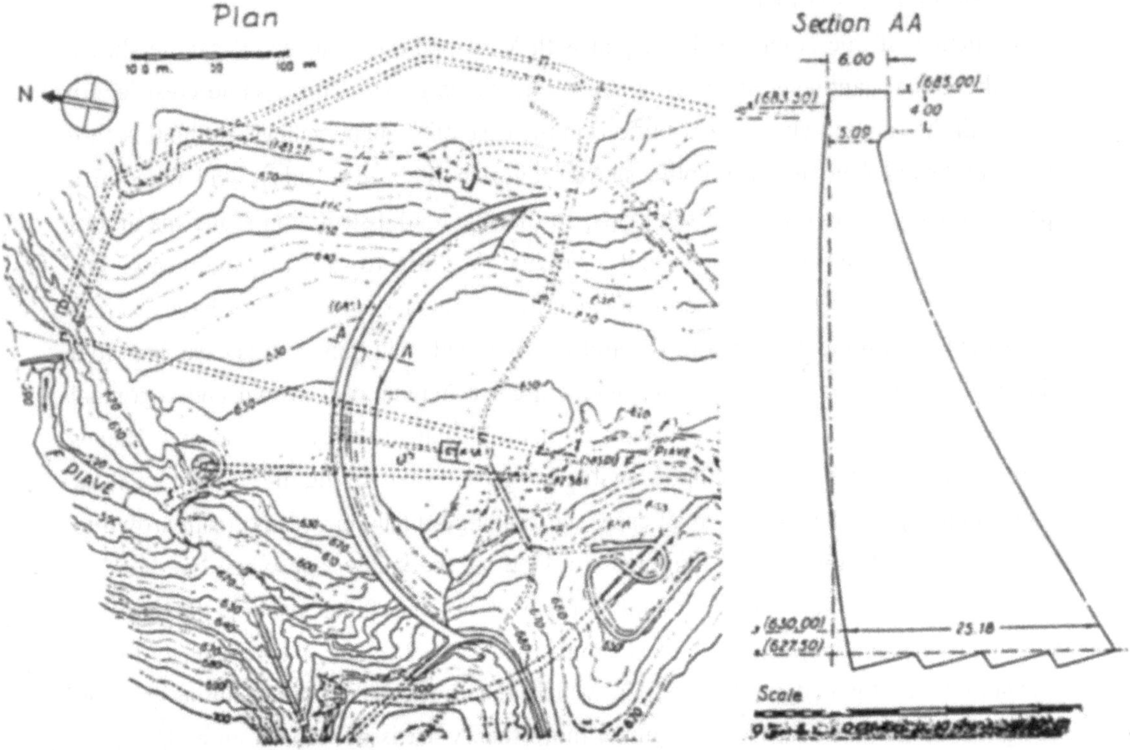

Fig. 3
Arch-gravity dam (Pieve di Cadore dam - Italy)

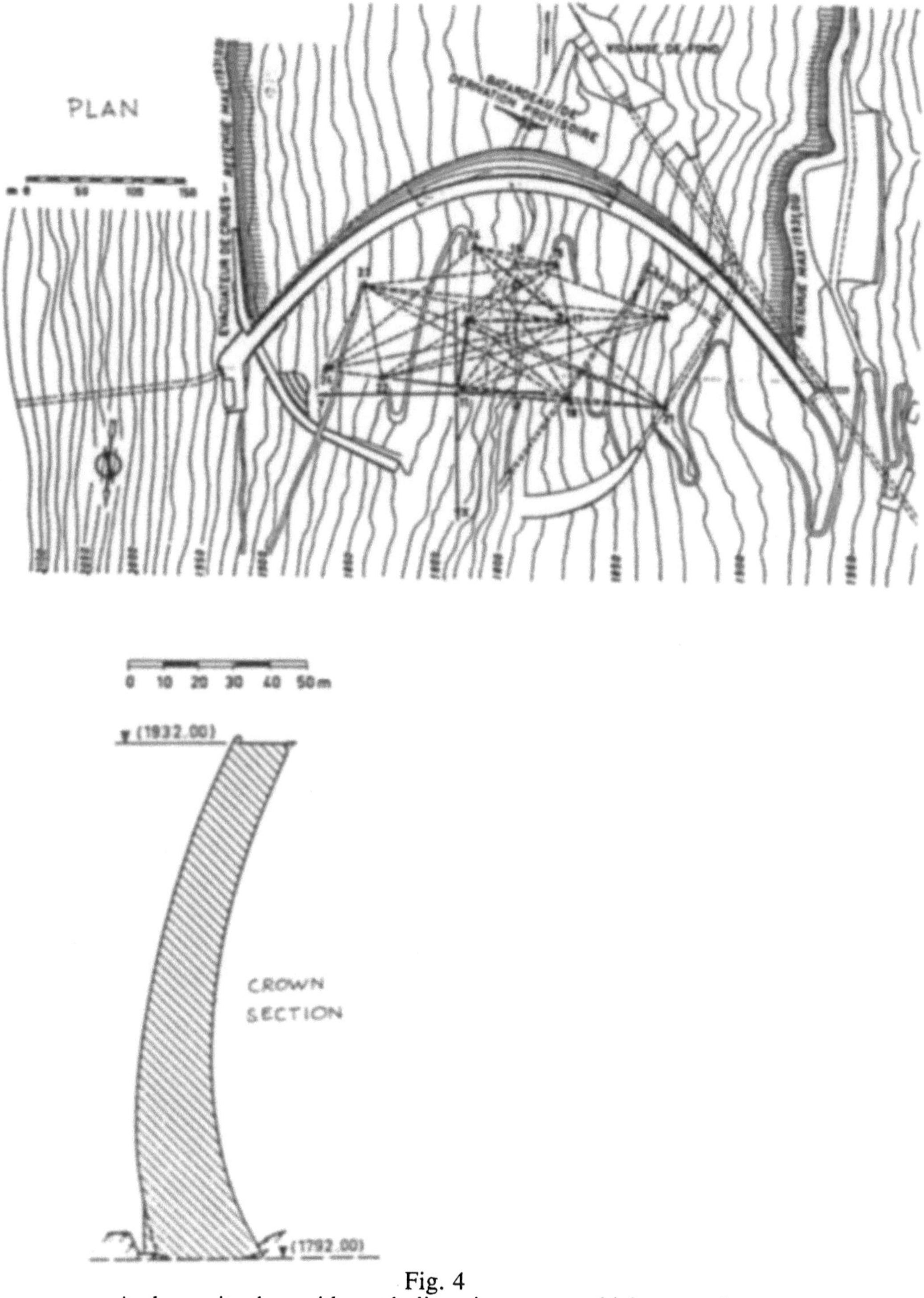

Fig. 4
Arch-gravity dam with parabolic-axis, constant-thickness arches
(Valle di Lei-Switz.)

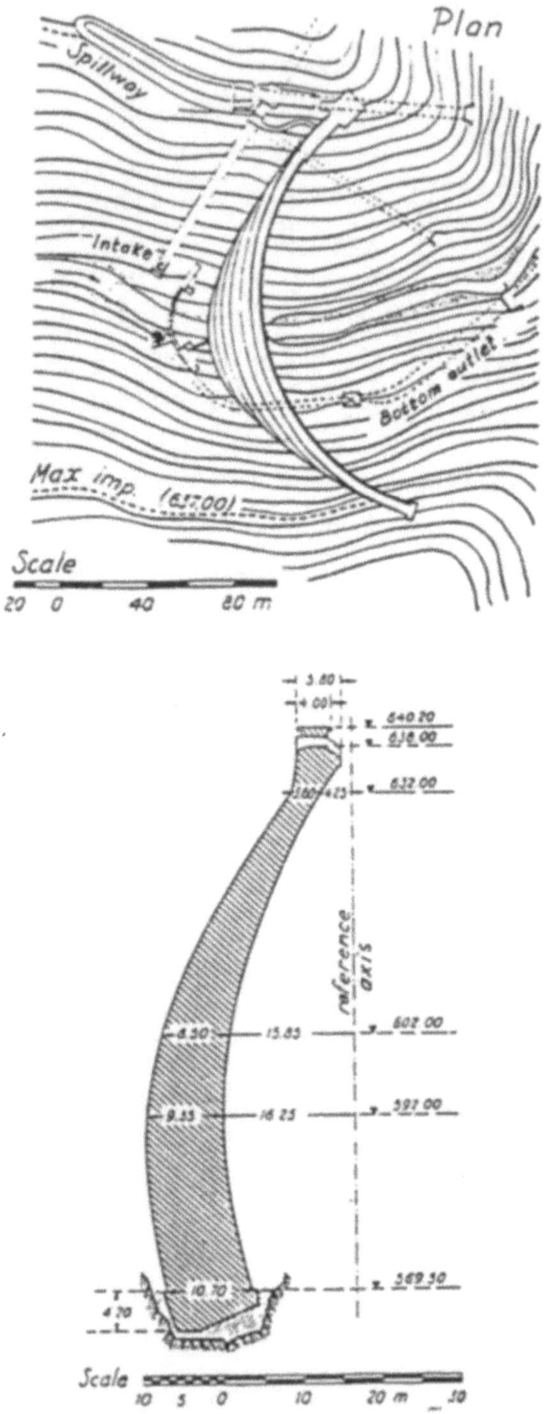

Fig. 5
Double curvature arch dam (Osiglietta dam-Italy)

not free-standing without high vertical tensile stresses, so that grouting of adjacent monolith joints has to be carried out at short intervals during monolith erection, and the latter has to be carried out according to strict planning, namely ensuring that at any given epoch the difference in elevation among the individual monolith tops is negligible; a considerable part of the weight of overhanging blocks it thus carried by arch action).

The other important sub-class of arch dams is to be found in the so-called 'multiple arch' structural type, which is used in very wide, relatively low valleys (very high witdth-to-height ratios, say $C/H > 5 \div 6$). The idea is to subdivide this large width into shorter spans, each of which can then accommodate a slender 'pure arch' or 'double curvature' solution. Of course, artificial structures (buttresses) are needed to take the abutment reaction of these multiple arches at the intermediate division points (other than the valley sides, where the reactions are taken directly by the abutment rock masses), see fig 6. This type is now somewhat out of fashion; in several cases it has shown sensitivity to cracking (either in the buttresses or in the arch-shaped 'barrels').

The idea to make the dam shape not entirely dependent on the valley profile can be discerned, in a much more moderate degree, in the device known as 'peripheral joint' or 'pulvino', (fig. 7) quite frequently used in arch gravity (or sometimes also in double-curvature) arch dams. This type of design is claimed to allow several aims to be pursued:

i) to design symmetrical, or near-symmetrical, dams in moderately unsymmetrical valleys, by regularizing with a concrete 'artificial rock' mass the irregular abutment profile;
ii) to provide a pre-arranged discontinuity along the zones of probable vertical, or sub-vertical, tensile stresses (especially on the upstream side), so that instead of natural cracking, and uplift generation by pressurized water infiltration, at unpredictable locations, there will be controlled joint opening (the waterstops that are a necessary ingredient of the peripheral joint preventing water from entering the joint and generating uplift forces);
iii) to provide a better distribution (and lower peak values) of abutment stresses on the natural rock foundation[3].

---

[3] This conception has been intensely discussed, because it is not easy to show in a conclusive way that all of the intended benefits do indeed take place, or that they are worth the additional costs incurred.

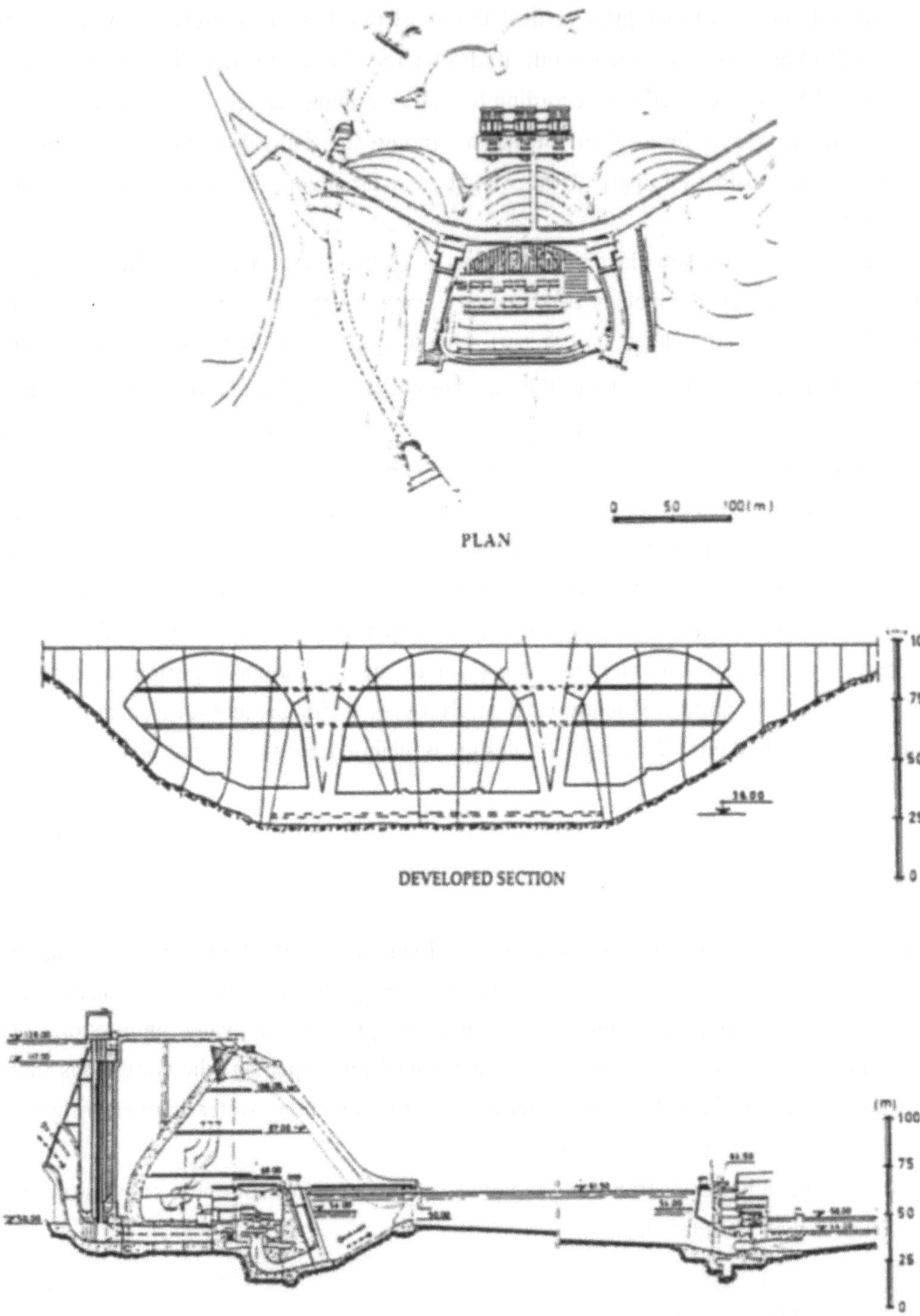

Fig. 6
Multiple-arch dam (Aguieira-Portugal)

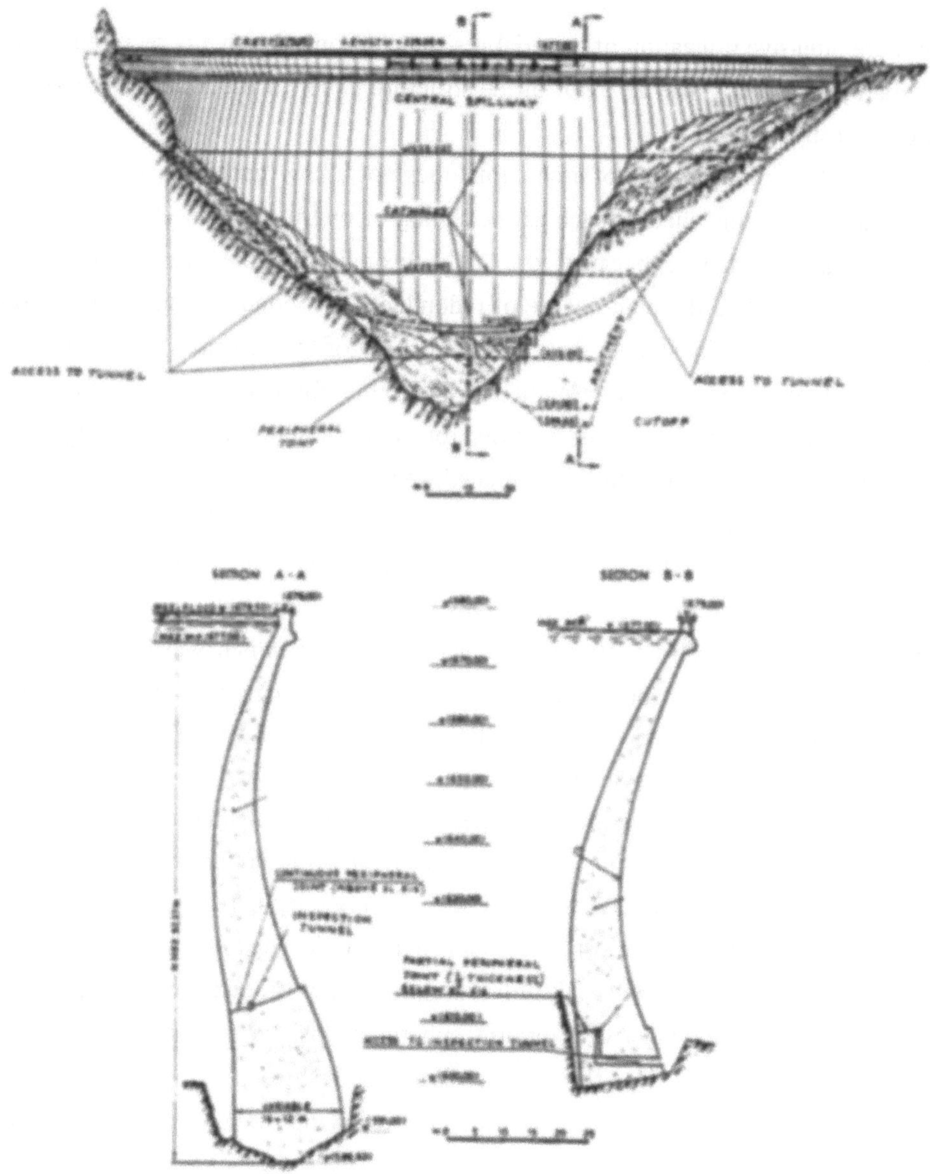

Fig. 7
Peripheral joint (Val Gallina dam, Italy)

1.2 Models for arch dams behavior: an overall view

The more intensive utilization of material resistance properties that is required from an arch dam demands that the behavior of the structure under foreseeable service loads be closely investigated.

Also, the relative slenderness of arch dams does not allow, generally speaking, to set up 'global equilibrium' checks (such as the sliding or overturning checks so commonly used for gravity dams), because the high flexibility of the slender arches and cantilevers, as well as the high degree of redundancy of boundary constraints, make a 'rigid-body' approach invalid, so that a detailed study of the deformation field becomes necessary. (Exceptions to this remark can be found for ancient gravity dams with curved plan and insufficient cross-section, for which a suitably modified 'rigid-body' equilibrium analysis can sometimes be applied, see fig.8).

So design, or back-analysis of behavior, must necessarily be based, for arch dams, on a continuum mechanics type of approach, including the constitutive laws both for concrete and foundation rock. Within this general framework, a wide variety of alternative mathematical models can be set up, a few of which will be illustrated in some depth in the present short course (and each one pertaining to a more or less well-defined field of application). However, before reaching this stage of building up a mathematical model that is suitable for the case in hand (and in order to effect a rational choice among the various possible options), it is essential to have in mind a conceptual-intuitive model of the structural behavior of an arch-like structure subject to the type of boundary conditions (B.C.) and of service loads that are to be found for dams, i.e. structures that must withstand hydrostatic, dead-weight, thermal and possibly seismic loads (not to speak of less frequent types of load such as, e.g., surface ice thrust, wave action, underwater or atmospheric explosions, etc.).

Indeed, the main logical steps that can be discerned in the setting up and engineering use of a workable, reliable model to be used in practice in order to analyze an arch dam are the following:

- Conceptual-intuitive model of behavior (based mainly on first principles and engineering experience);

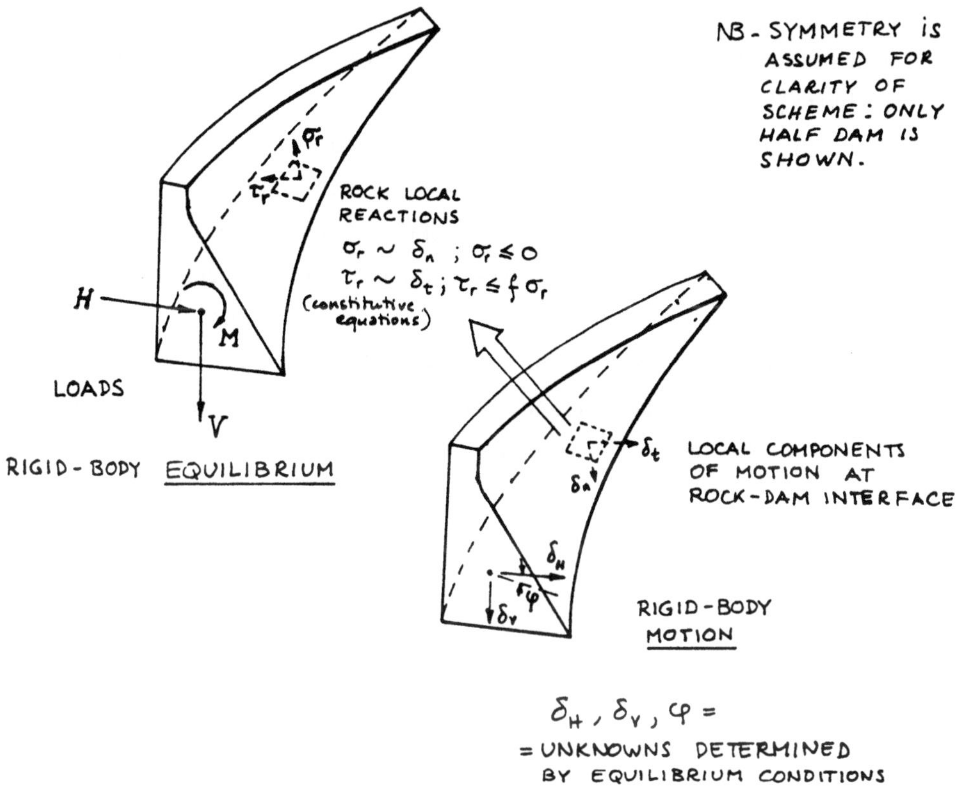

Fig. 8
Simplified scheme of rigid-body analysis for curved-axis gravity dam

- Mathematical model, i.e. the translation of the preceding model into symbolic equations involving the B.C.'s, the external loads, the internal stress-strain state of the material[4];
- Numerical model, translating the preceding one into a numerical procedure from which a suitable software (SW) can be written for implementation on the available hardware (HD)[4];
- Verification (checking, debugging, validation if necessary ...) of the SW[4];
- Use of the SW for the case in hand, so as to obtain quantitative (numerical) result;
- Critical analysis of results, in relation to standards or pre-set requirements;
- Engineering decisions; restart of the process if the design is changed.

It is easy to observe that the most critical steps, as well as the ones in which experience and intuition play a key role, are the first and the last ones. Even the best SW cannot substitute for sound engineering judgment; also let us not forget that a thorough knowledge of the techniques of modeling and mathematical-numerical analysis are only a small - albeit important - part of the engineering process, and a perhaps necessary, but certainly not sufficient, condition for successful design or reliable behavior assessment.

All the more so that, as we shall see in the following section, the 'real' state of stress of a dam in a given situation is, at the present state of the art, not accessible (being influenced, among other things, by a very complex, substantially unknowable initial state deriving from a myriad details of the erection history).

So, every mathematical model has to be intended only as a reference scheme to be supplemented by qualitative elements of knowledge and by analogy with past experiences on like situations.

Besides, the designed artificial structure has to be matched to the natural rock formations constituting its foundations. In this connection, it is worth recalling that in a majority of cases problems, if any, are bound to arise rather from unforeseen (or insufficiently investigated) foundation circumstances than from faulty dam design (fig.9). Since the foundations are supposed to be represented in the mathematical models simply as '*appropriate*' (and in any case, certainly grossly approximated) B.C.'s, this is one more, cogent reason to take the numerical results with all due degree and amount of cautionary, critical detachment. In plain words: the numerical analysis

[4] These three steps can be skipped if the particular type of dam has already been modeled with a proven software.

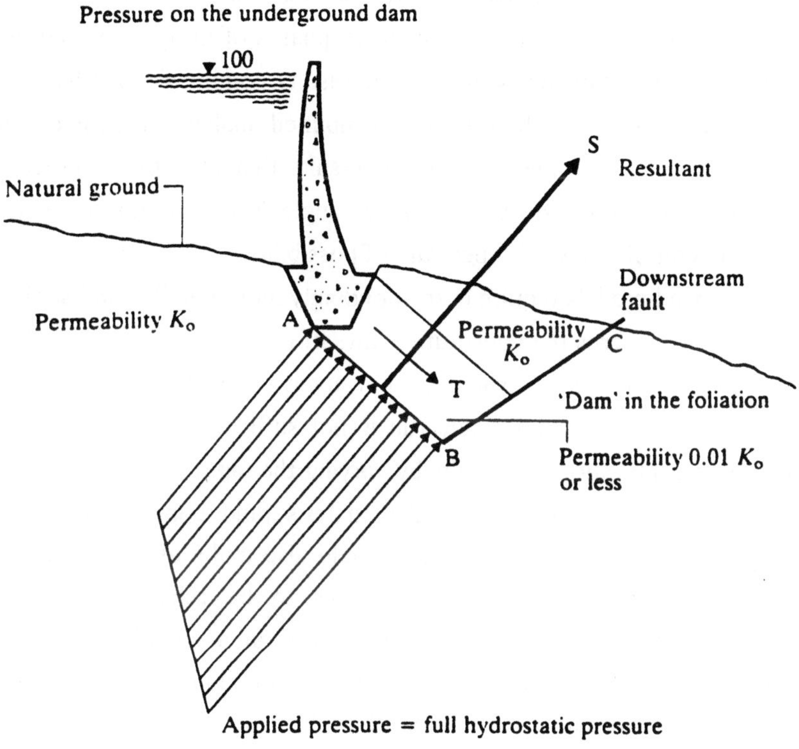

Fig. 9
Mechanism of collapse for Malpasset Dam.
(Source: Water Power and Dam Construction, November 1985)

specialist should never overcome, or overshadow, the experienced dam engineer (assisted by competent geologist and rock-mechanics expert).

In the same spirit, a balanced view should be entertained about the application of 'advanced' methods of analysis, especially non-linear behavior models. In fact, sometimes it is inevitable to use such advanced methods in order to explore possible modes of response to exceptional situations. But one should avoid, on one hand, indiscriminate recourse to complicated models in stages of the engineering process that do not warrant their use (e.g. in the preliminary phases of design); on the other hand, one should also be wary of believing that the results coming from 'advanced' models are necessarily 'more realistic' than those of simplified analyses. Non-linear models, in particular, can give non unique solutions or results that are quite sensitive to initial conditions and load path history, so that a single 'run' will not give a sufficient complete impression of the whole spectrum of possibilities [5].

Moreover, advanced models require correspondingly more detailed and sophisticated input data (e.g. as concerns the constitutive laws of concrete and rock, or the response spectrum of the seismic input), which are not always available at the moment of the analysis. Again, in such cases a single run is not sufficient and one should perform heavy parametric explorations of results by varying the input data, with corresponding increases in the cost and duration of the analysis (and complications in synthetic appraisal of results). The mention of costs should also make us aware that the choice of the analysis model/method to be used has to be related also to its 'cost-effectiveness': to use a costly, sophisticated model in a situation where a simple, economical one would be sufficient is certainly not a sound engineering practice.

Lastly, the critical appraisal of the results of sophisticated methods of analysis often requires to use acceptability criteria which are not yet sufficiently firmly established or supported by experience, in contrast to the case of simpler methods whose results, admittedly more schematic, are easier to appraise because there is a vast lore of past experience about their practical meaning and about the range of ' acceptable' values (all taken in the proper context and within the knowledge that the whole process does not necessarily portray physical reality in all of its details), see fig.10.

To conclude this qualitative discussion let us emphasize that computations are never an end in themselves. They are but tools to achieve certain aims, e.g. to demonstrate compliance with official standards, or to give the designer a reasonable assurance that

[5] Not to mention the fact that, e.g., in F.E. analyses important results are (or can be) mesh-dependent near singular points. Great skill is required to evaluate such results.

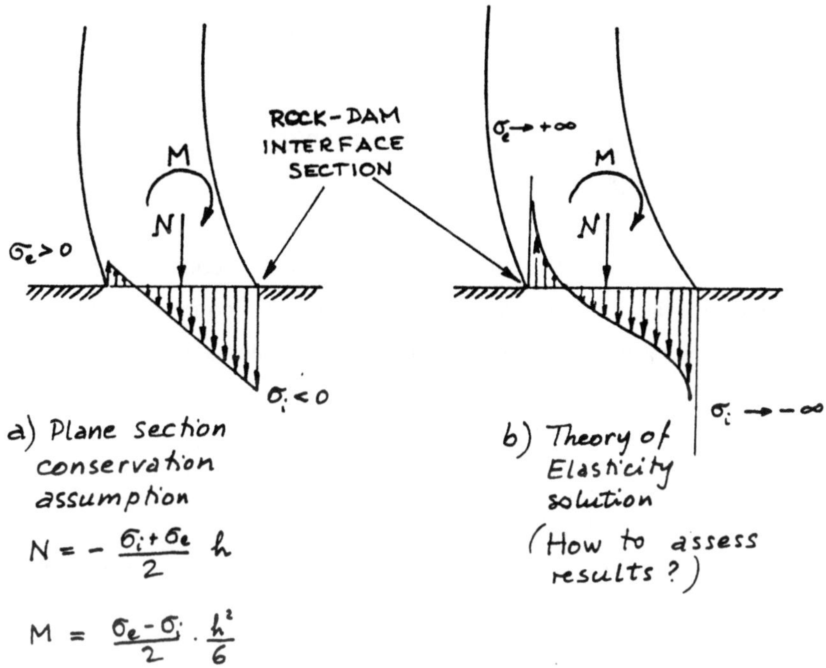

Fig. 10
Physical meaning of "exact" solutions may be not clear

the dam will perform satisfactory (and without excessive risks to third parties) during all of its intended service life [6], or to interpret the behavior of an existing dam. The aim of the exercise should of course direct the choice of the right modeling option, rather that letting the availability of a particular SW tool influence the type and scope of numerical analysis. Sensitivity and experience of the dam engineer provide, once again, the key guidelines in this respect, in the absence of formal rules linking the object of analysis to the selection of the 'right' mathematical treatment to be adopted.

1.3 The basic stress/strain (or internal action/displacements) fields, and their approximate representation

All of the different models that can be adopted relate to a common basic scheme. Each of them can, in fact, be derived from this ideally 'complete' schematization by neglecting certain variables, which can be conditionally assumed to have negligible influence on the results, or by making ad-hoc simplifying assumptions.

A general space-time reference frame (fig. 11) allows to describe the geometry of the structure (and foundations!) and the distribution of loads. The spatial reference frame can be external to the structure (a Cartesian trihedron, e.g.), or -more rarely- an 'intrinsic' reference frame (fig. 12) laid on the structure itself, using the tools of the 'absolute differential geometry'. The dam can be described (thinking of it, for the moment, as a monolithic solid) either by assigning the two surfaces (upstream and downstream facings) or by defining the 'middle sheet' and the local distribution of thicknesses [7] (fig. 13).

On this geometrical background, equilibrium equations can be written involving the stress variables (unknown) and the external loads (known data). They take on the form of partial differential equations. Compatibility conditions concerning the strain (or displacement) variables are usually implicitly satisfied and need not be written for the structure proper, unless one wants to describe special local conditions (formations of discontinuities, like cracks, etc.); they must be explicitly laid down, however, for the boundaries where the structure is joined to the foundations. This leads us to consideration of the boundary conditions in general (among which, as distinct from the foundation/structure interface constraints, one can include both 'surface traction' and

[6] It is now commonly assumed that this should cover a period of at least 100 yrs.

[7] The foundations are usually described by the contour lines of sound rock (after abutment excavations).

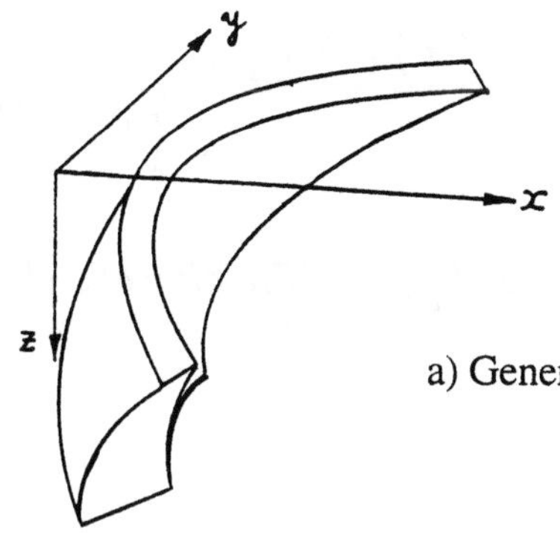

Fig. 11
a) General Cartesian 3-D reference frame

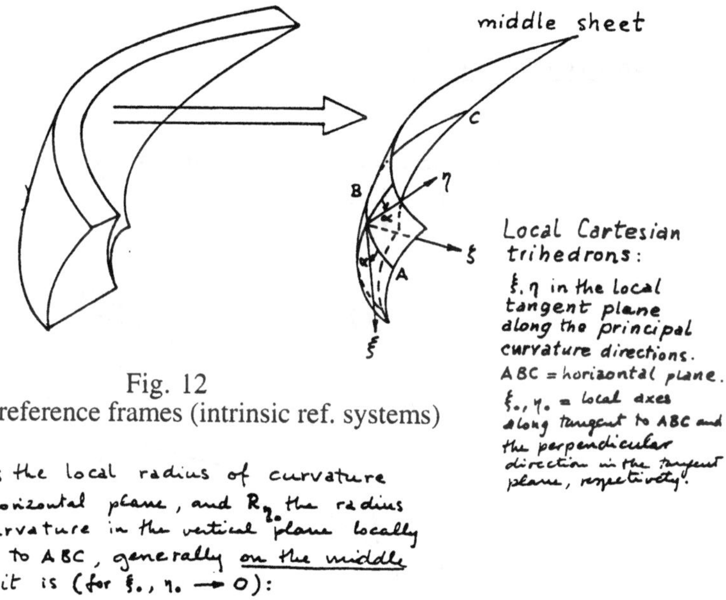

Fig. 12
b) Local reference frames (intrinsic ref. systems)

If R_{ξ_0} is the local radius of curvature in the horizontal plane, and R_{η_0} the radius of curvature in the vertical plane locally normal to ABC, generally on the middle sheet it is (for $\xi_0, \eta_0 \to 0$):

$$\zeta \cong \frac{\xi_0^2}{2R_{\xi_0}} + \frac{\eta_0^2}{2R_{\eta_0}} + \frac{\xi_0\eta_0}{R_{\xi_0\eta_0}} + \cdots$$

The local principal curvature directions (geodetics) ξ, η are defined by the angle α (see figure) such that:

$$\mathrm{tg}\, 2\alpha = \frac{\frac{1}{R_{\xi_0\eta_0}}}{\frac{1}{2R_{\xi_0}} - \frac{1}{2R_{\eta_0}}}$$

; the principal radii R_ξ, R_η then are given by:

$$\frac{1}{R_\xi} = \frac{1}{2R_{\xi_0}} + \frac{1}{2R_{\eta_0}} + \frac{1}{2}\sqrt{\left(\frac{1}{R_{\xi_0}} - \frac{1}{R_{\eta_0}}\right)^2 + \frac{4}{R_{\xi_0\eta_0}^2}} \; ,$$

$$\frac{1}{R_\eta} = \frac{1}{2R_{\xi_0}} + \frac{1}{2R_{\eta_0}} - \frac{1}{2}\sqrt{\left(\frac{1}{R_{\xi_0}} - \frac{1}{R_{\eta_0}}\right)^2 + \frac{4}{R_{\xi_0\eta_0}^2}} \; .$$

$$x_m = x_m(y, z) = x_0(z) + f_z(y)$$
$$y_m = y \quad , \quad z_m = z$$

Middle sheet equation

$$h = h(y,z) = h_0(z) + g_z(y)$$

Thickness distribution law.

Upstream surface:

$$x_u = x_m - \frac{h}{2} \frac{1}{\sqrt{1+\left(\frac{\partial f_z}{\partial y}\right)^2}} \quad , \quad y_u = y_m + \frac{h}{2} \frac{\partial f_z/\partial y}{\sqrt{1+\left(\frac{\partial f_z}{\partial y}\right)^2}}$$

Downstream surface:

$$x_d = x_m + \frac{h}{2} \frac{1}{\sqrt{1+\left(\frac{\partial f_z}{\partial y}\right)^2}} \quad , \quad y_d = y_m - \frac{h}{2} \frac{\partial f_z/\partial y}{\sqrt{1+\left(\frac{\partial f_z}{\partial y}\right)^2}}$$

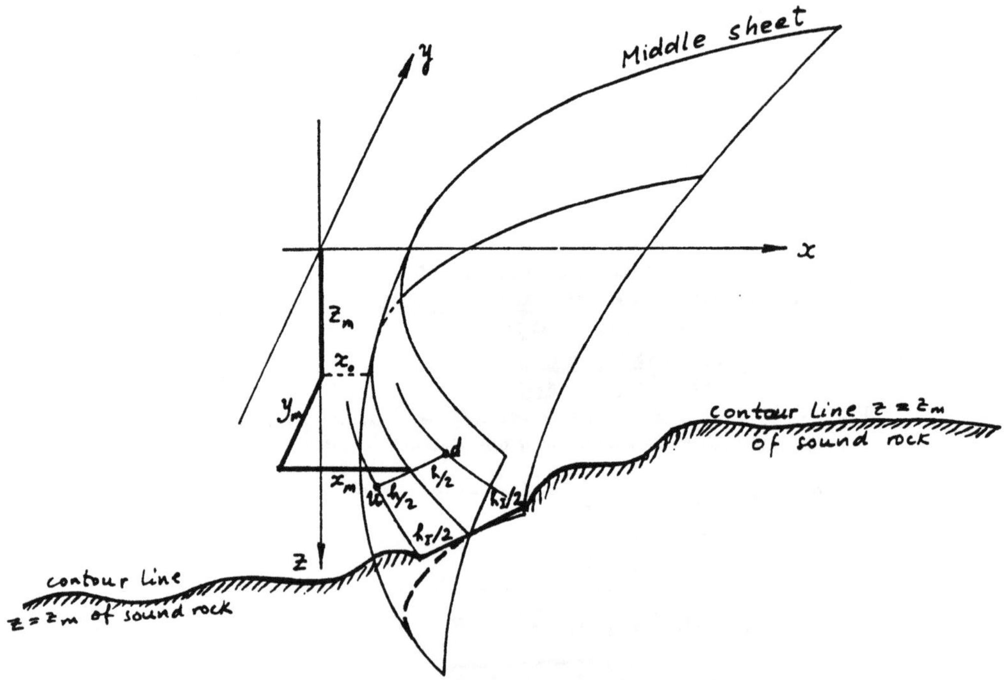

Fig. 13
Geometrical definition of an arch dam and its foundations

thermal conditions on the external surfaces) (fig. 14). In the case of dynamic problems, of course also initial conditions need be specified.

"Constitutive equations", i.e. the assumed relationship between stresses and strains (e.g. Hooke's law) allow formal closure of the equilibrium problem.

The indefinite equilibrium equations can be written either in a strictly local form, e.g. the partial differential equations of the Theory of Elasticity [8], or in an integral form, by integrating the stress components, and their successive moments, across the thickness. While the first form is the basis of the Finite Element method, this second form is particularly suitable for thin shells, and is more in line with traditional engineering approaches. In fact, it uses as unknown variables the global components of internal action (fig. 15) (normal and tangential forces for unit length, bending and torsional moments for unit length) which, although possessed of a univocal physical meaning only under restrictive assumptions (linear distribution of normal stresses across the thickness, deriving from the plane section conservation hypothesis), yet generally allow traditional resistance criteria to be applied, thus maintaining a welcome continuity with the past engineering practice embodied in a vas number of successful, experience-proven structures conceived and built before the advent of the more modern methods of analysis[9] .

The problem is in this way simplified and substantially scaled down by one dimension (from a 3-D problem one is left with a substantially 2-D one). Also the relevant components of displacement variables that need be considered can be reduced to the displacements and rotations of the middle sheet (fig. 16). Some of the classical approximate methods of analysis, e.g. the Ritter method or the Trial Load Method (T.L.M.), try to go one step further by treating the dam not as a collection of 2-D shell elements, but as a grid of 1-D 'beam' elements: the horizontal 'arches' and the vertical, or sub-vertical, 'cantilevers' (fig. 17). This global, or macroscopic, approach has the advantage of making immediately available all of the methodology and results of classical beam theory; the drawback, on the other hand, is that in the thicker portions

[8] If referred to a general "external" reference frame, e.g. a Cartesian trihedron, then the conditions on "surface tractions" are not the simplest possible. Conversely, if an "intrinsic" reference grid laid on the dam external surfaces (or on the middle surface) is used, the "surface traction " conditions take a simple form, but the indefinite equilibrium equations for stress components are more complex, involving the principal curvatures and torsion of said surfaces.

[9] Most of the existing standards of acceptability criteria are, implicitly or explicitly, based on this type of schematization. By contrast, if the "microscopic", or strictly local, point of view is adopted (3-D F.E.M.) then acceptability criteria have to be set for the local stresses, and one can say that there is not yet sufficient practical experience to do that with certainty of success, avoiding both excessive conservatism and the risk of reduction of safety margins.

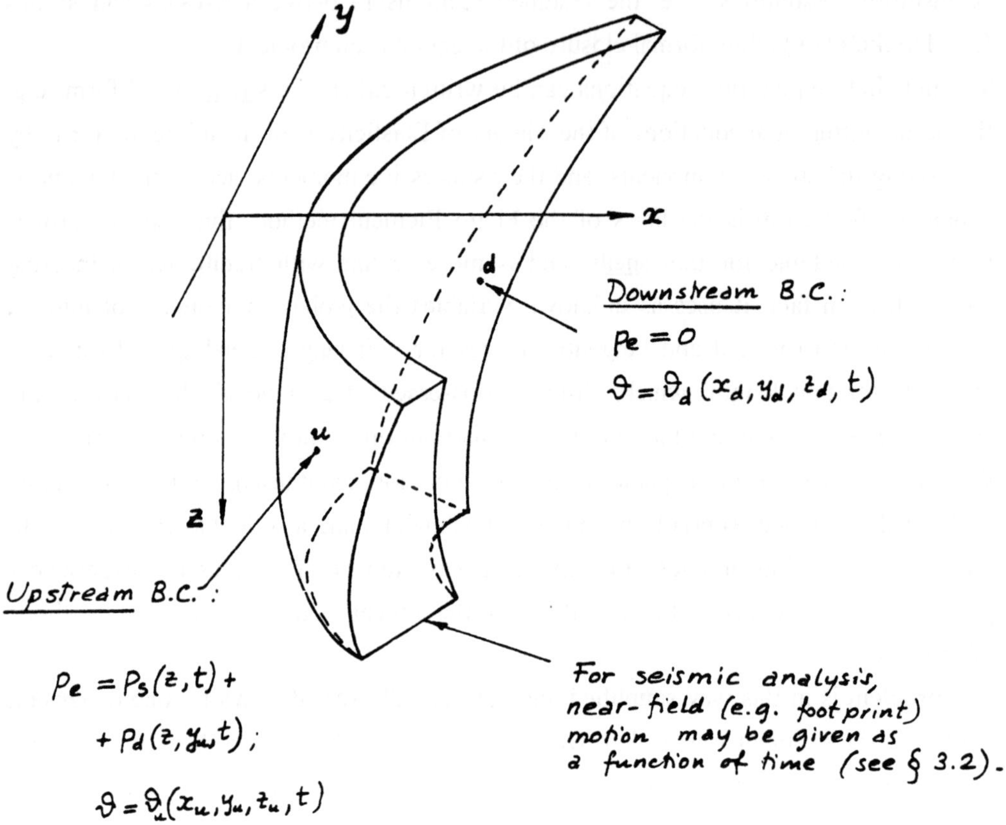

Fig. 14
B.C. for arch dam analysis

Fig. 15
Shell model: global and local internal actions

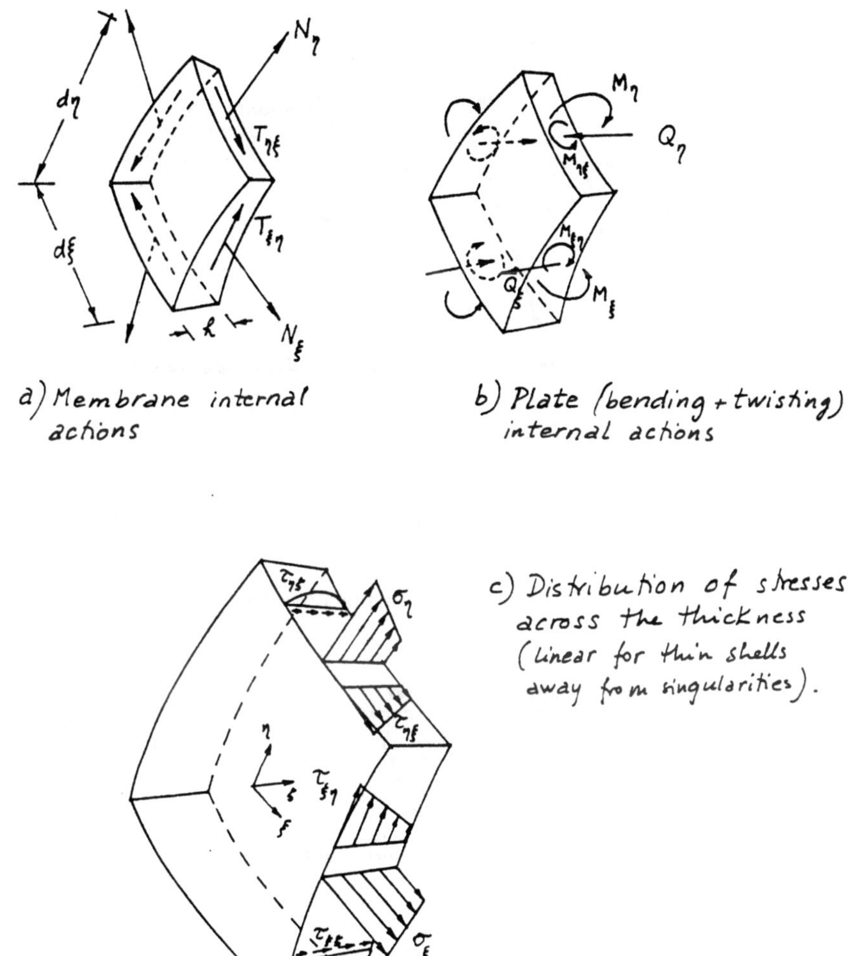

$$N_\xi = \int_{-\frac{h}{2}}^{h/2} \sigma_\xi \, d\zeta \quad ; \quad T_{\xi\eta} = \int_{-\frac{h}{2}}^{h/2} \tau_{\xi\eta} \, d\zeta \quad ; \quad M_\xi = \int_{-\frac{h}{2}}^{h/2} \sigma_\xi \, \zeta \, d\zeta \quad ; \quad M_{\xi\eta} = \int_{-\frac{h}{2}}^{h/2} \tau_{\xi\eta} \, \zeta \, d\zeta$$

$$N_\eta = \int_{-\frac{h}{2}}^{h/2} \sigma_\eta \, d\zeta \quad ; \quad T_{\eta\xi} = \int_{-\frac{h}{2}}^{h/2} \tau_{\eta\xi} \, d\zeta \quad ; \quad M_\eta = \int_{-\frac{h}{2}}^{h/2} \sigma_\eta \, \zeta \, d\zeta \quad ; \quad M_{\eta\xi} = \int_{-\frac{h}{2}}^{h/2} \tau_{\eta\xi} \, \zeta \, d\zeta$$

$$-Q_\xi = \int_{-\frac{h}{2}}^{\frac{h}{2}} \tau_{\xi\zeta} \, d\zeta \quad ; \quad -Q_\eta = \int_{-\frac{h}{2}}^{h/2} \tau_{\eta\zeta} \, d\zeta \, .$$

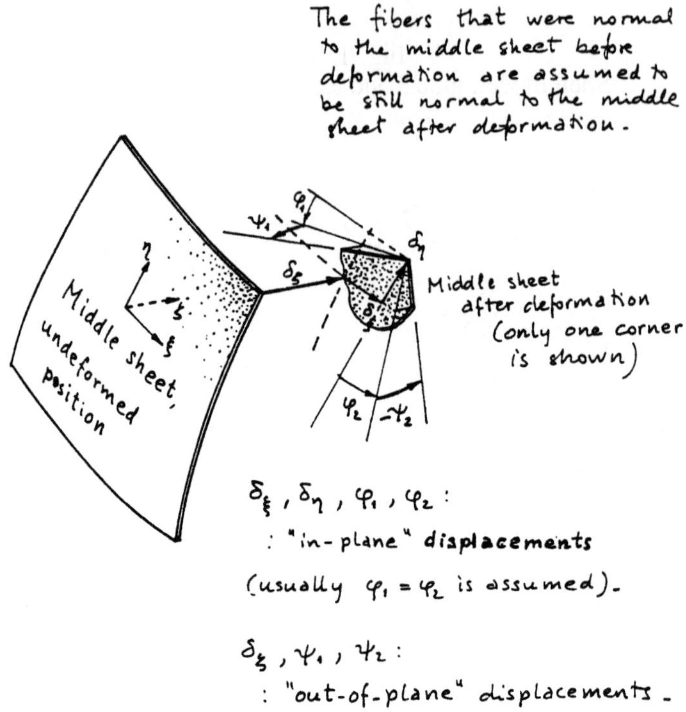

Fig. 16
Shell model: displacement variables

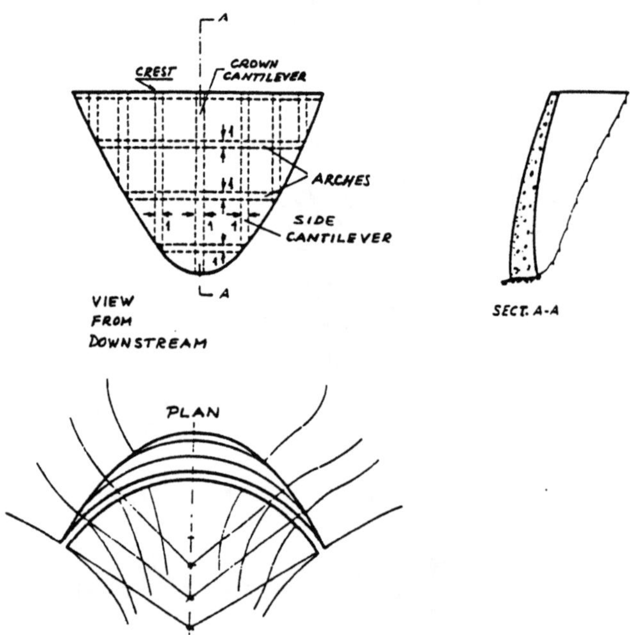

Fig. 17
Arches/ cantilevers grid

of the dam, or near singular regions (such as the interface with foundations) these results are, strictly speaking, no longer faithfully representative of the real situation. In particular, the Navier assumptions (plane section conservation, linear distribution of normal stresses across the thickness, etc.) can no longer be taken as a good approximation. However, since many of the old arch dams were designed using this kind of modeling, it is reasonable to admit that this approach leads to acceptable results, even if this goal is reached working with 'conventional' variables which are somewhat detached from a direct physical counterpart [10] .

Another advantage of the intermediate ('shell') approximation, which can easily be carried over to the 'arch-and-cantilever-grid' one, is the possibility of clearly putting in evidence, and differentiating among, the several kinds of local resisting (load-carrying) capacities of the arch dam (fig. 18):

- membrane action (arches and, if there is a second curvature in the vertical plane, also cantilevers);
- plate action (bending of arch and cantilever elements; torsion of the shell elements);
- tangential shear resistance (activated by the local geometrical torsion of the shell).

Each of these different load-carrying capacities has a clear intuitive meaning, which helps the designer in forming a well-defined mental picture of how the arch dam reacts to the imposed loads and transmits them to the rocky foundation masses. This kind of structural insight is an invaluable component of the dam engineer's abilities, and its development should be encouraged by every available means. In this context, the 'more advanced' methods of analysis cannot be said to have a comparable effectiveness: there the engineer is really in control only in the phase of data gathering and introduction; the intermediate mathematical machinery takes then up, with hundreds of non-engineering variables, essentially as a 'black box', and the engineer is given back control once again in the later, final phase of result assessment. Of course experience and sensitivity can be gained also in this way, but the process is a somewhat slower and less direct one.

---

[10] It is worth recalling, on the other hand, that more elaborate methods can give questionable results for the singular regions here considered. Thus, linear F.E.M.s give mesh-dependent results, and would even give indefinitely larger stresses with finer and finer meshes near re-entrant corners (foundation interface). If one considers, furthermore, the uncertainties about the initial stress state and the influence of construction and load path history, the "greater accuracy" of very elaborate methods of analysis is not to be taken in the sense of an assured greater faithfulness to reality, but in a somewhat different meaning.

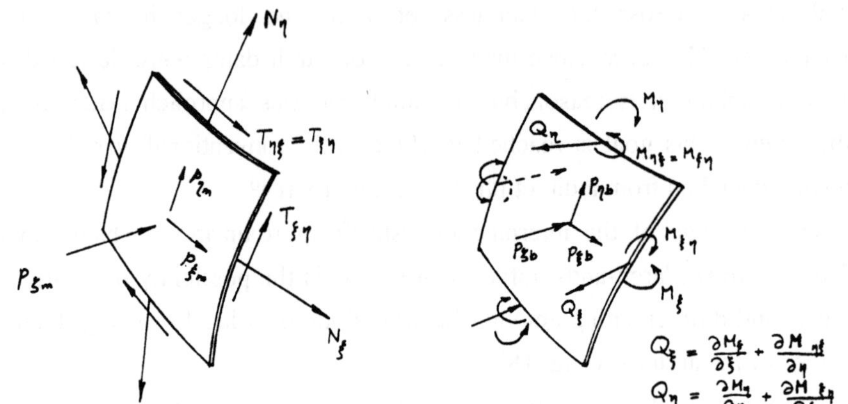

$$Q_\xi = \frac{\partial M_\xi}{\partial \xi} + \frac{\partial M_{\eta\xi}}{\partial \eta}$$

$$Q_\eta = \frac{\partial M_\eta}{\partial \eta} + \frac{\partial M_{\xi\eta}}{\partial \xi}$$

$$-p_{\zeta m} = \frac{N_\xi}{R_\xi} + \frac{N_\eta}{R_\eta} + 2\frac{T_{\xi\eta}}{R_{\xi\eta}} \qquad p_{\zeta b} = \frac{\partial^2 M_\xi}{\partial \xi^2} + \frac{\partial^2 M_\eta}{\partial \eta^2} + 2\frac{\partial^2 M_{\xi\eta}}{\partial \xi \partial \eta}$$

$$p_{\xi m} = -\frac{\partial N_\xi}{\partial \xi} - \frac{\partial T_{\eta\xi}}{\partial \eta} \qquad p_{\xi b} = -\frac{Q_\xi}{R_\xi}$$

$$p_{\eta m} = -\frac{\partial N_\eta}{\partial \eta} - \frac{\partial T_{\xi\eta}}{\partial \xi} \qquad p_{\eta b} = -\frac{Q_\eta}{R_\eta}$$

a) Membranal carrying capacity. b) Plate carrying capacity.

a+b) $p_\zeta = -\left(\frac{N_\xi}{R_\xi} + \frac{N_\eta}{R_\eta} + 2\frac{T_{\xi\eta}}{R_{\xi\eta}}\right) + \frac{\partial^2 M_\xi}{\partial \xi^2} + \frac{\partial^2 M_\eta}{\partial \eta^2} + 2\frac{\partial^2 M_{\xi\eta}}{\partial \xi \partial \eta}$; $p_\xi = -\frac{\partial N_\xi}{\partial \xi} - \frac{\partial T_{\eta\xi}}{\partial \eta} - \frac{Q_\xi}{R_\xi}$; $p_\eta = -\frac{\partial N_\eta}{\partial \eta} - \frac{\partial T_{\xi\eta}}{\partial \xi} - \frac{Q_\eta}{R_\eta}$

Fig. 18
Equilibrium of global shell internal actions and external loads

2. ANALYSIS & BACK-ANALYSIS: DIFFERENT MODELS & METHODS

2.1 Simplified finite difference (F.D.) methods

These are now somewhat out of fashion, as a consequence of the success and diffusion of the general-purpose F.E. methods. However, as it has been pointed out earlier, they have still considerable engineering interest and can profit from extensive experience. Their didactic value suggests to illustrate in some detail at least the more important of them.

Traditionally, these methods use a mix of beam theory and F.D. (finite difference) formulation in order to express local equilibrium between internal and external forces. Keeping the same simplifying assumptions, nothing stands in the way of using a different formulation (a 'weak' one) by means of 'variational-like' global quantities (e.g. the total potential energy), expressed as a function of the chosen unknowns (displacement, strain or stress variables). In this latter case, local equilibrium conditions are no longer satisfied (as opposed to global equilibrium which is always preserved), but a judicious choice of the unknowns and of their probable spatial distribution (shape functions) allows to greatly reduce the number of degrees of freedom (d.o.f.). Thus, even in an up-to-date context, such methods can still find a place in the panoply of tools used by present-day dam engineers. For instance, they may be of great assistance in the early stages of design, where a clear overall picture of structure behavior is needed, and local details can be overlooked, or better set apart for later stages of the design process.

Among these simplified methods, we have chosen to illustrate the Ritter central-cantilever model and the very well known, and still used, Trial Load Method (T.L.M.) developed by the USA Bureau of Reclamation (B.o.R.) in the early 1930's.

The drawback of these methods resides, above all, in their inability to take into account local effects, non-linear distribution of stresses across the thickness (departure from Navier's assumption), grossly approximate representation of foundations compliance via the so-called Vogt coefficients based on Boussinesq's elastic half-space theory (see further on).

These simplified methods were originally developed for elastic linear analysis; although one can conceive adaptations or extensions for non-linear analysis, in the following we will illustrate only the original, linear-elastic formulations (unless otherwise specified). For completeness, it will be necessary to begin with a recall of the theory of the isolated arch and of the Tölke method (axially symmetrical reservoirs).

2.1.1 Recall of the 'isolated arch' theory

In the historical context of arch dam analysis, the first model used (as soon as the methods of solid mechanics allowed quantitative computations to be carried out, i.e. towards the middle of the XIXth century) was that of the 'isolated arches'. According to this model, indeed, every 'arch' which can be ideally cut out from the dam by two horizontal planes separated by a unit vertical interval is thought to behave independently from all the others and, thus, to carry the whole of the hydrostatic pressure acting at its level (besides all the other actions, be they thermal, seismic or other, that take place at the same level) (fig.19). Guidi, Bresse, Ippolito and others contributed to the perfection of the isolated arch model which was widely used until the years 1920's. This very simple model is considerably removed from reality, because physical interconnections between adjacent arches causes a considerable redistribution of load, as deeper analysis and experimental evidence have amply demonstrated. On the one hand, however, the static analysis of an isolated arch is a necessary component of the rational formulation of any more complete theory. On the other hand, the 'isolated arch' model can be thought of as a sort of 'limit analysis' approach in which all the loads are ideally increased to the point of producing a diffused state of horizontal cracks (due to insufficient resistance to bending of the vertical elements). In this limit state (which is, one should note, only one among many other possibilities) the régime of the arch dam would, indeed, be quite well represented by the 'isolated arch' model .
In the following paragraphs we will present the theory of the isolated arch on compliant foundations, which will also prepare the ground for the more advanced 'classical' arch-cantilever models. For simplicity we will consider circular-axis arches under constant external pressure, but similar developments can be carried out for other arch shapes.

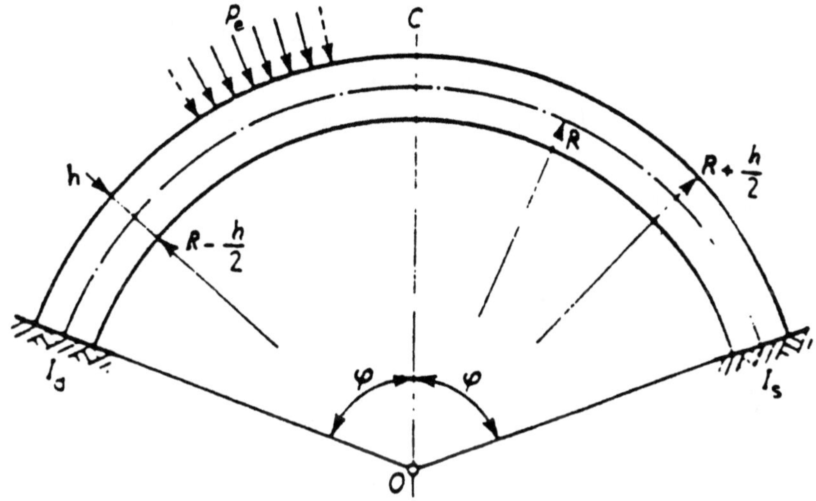

Fig. 19
Isolated arch (constant thickness and radius)

The circular arch of constant thickness under symmetric loads and with compliant abutments

Let us consider a circular arch of radius R (referred to the middle fiber), radial thickness h, angular aperture 2φ (fig. 20). The thickness in the direction perpendicular to the plane of the arch is taken to be 1m; the elastic (Young) modulus of the arch material will be denoted by E_c, that of the rocky abutments by E_r. The Poisson effect will be disregarded for simplicity sake, so $\nu_c = \nu_r = 0$

The external load will be taken to be a uniform pressure on the extrados, whose amount will be denoted by p_e.

In the case of a dam, the usual component of p_e will be:

$$p_e = \gamma_a z, \tag{2.1.1.1}$$

γ_a being the specific weight of water, z the depth of the arch middle plane under the water free surface. In the case of seismic action (acting on a horizontal direction lying along the symmetry plane of the arch), p_e will include a hydrodynamic component $p_d(z,t)$:

$$p_e = \gamma_a z + p_d(z,t), \tag{2.1.1.2}$$

whose precise analytical form need not concern us here. (Also, the inertial forces associated with the movement of the arch own masses during the seismic event will have be considered; but let us differ this question to a later stage of our study, and consider $p_e(z)$ simply as a pseudo-static radial force per unit length of the upstream boundary of our arch of unit thickness).

If our circular arch were complete ($2\varphi = 2\pi$) and unconstrained, and if $h/R \ll 1$, the régime of the arch (fig. 21) would be one of simple, nearly uniform compression :

$$\overline{\sigma}_a = -p_e \cdot \frac{R+\frac{h}{2}}{h} \tag{2.1.1.3}$$

(the negative sign denoting compression),

and the radial displacements would likewise be uniform (apart from possible rigid-body components) and given by the following formula:

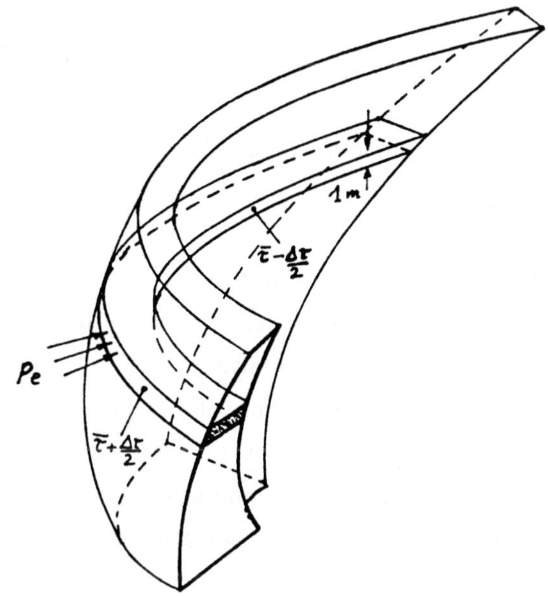

Fig. 20
The "isolated arch" model

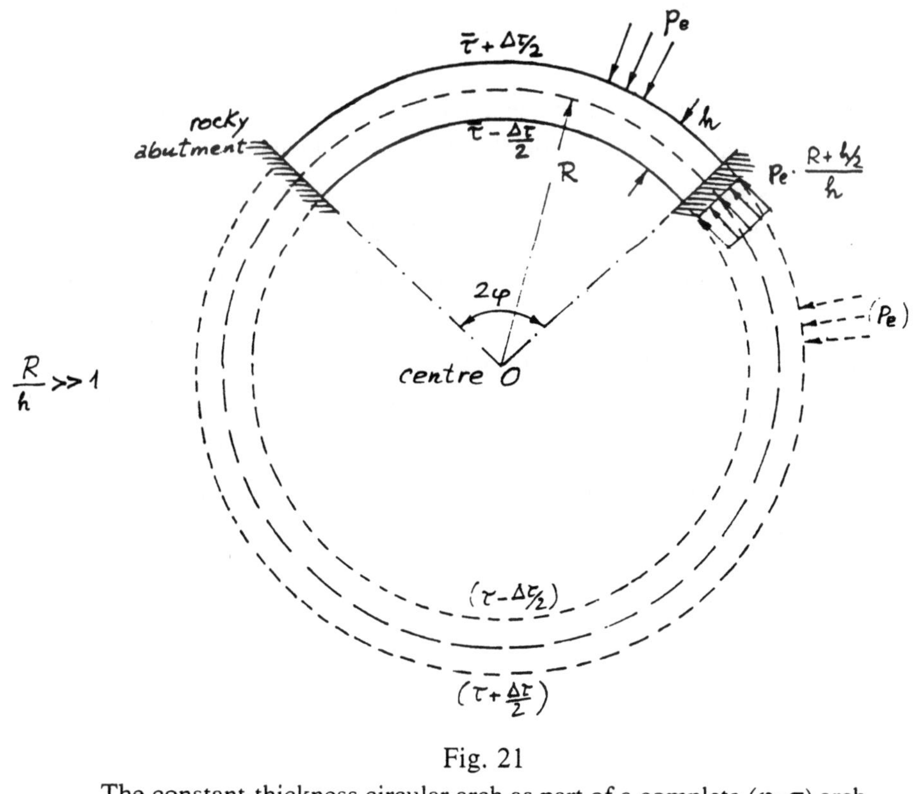

Fig. 21
The constant-thickness circular arch as part of a complete ($\varphi=\pi$) arch

$$\delta = -\bar{\sigma}_a \frac{R}{E_c} \qquad (2.1.1.4)$$

(the negative sign being valid if the positive direction of δ be taken toward the arch interior).

These were, in fact, the 'tube' or 'Mariotte' formulae that were used in the very first quantitative analysis of arch dams. They disregarded completely the influence of the abutment constraints, which we are now going to analyze.

This process will of necessity entail the calculation of redundant reactions, which considerably alter the simple static régime given by (2.1.1.3), as well as the radial displacement pattern given by (2.1.1.4) (not mentioning the fact that tangential displacements, completely absent in the 'tube' régime, will also arise).

Besides the live actions represented by external pressures, we suppose that the arch is subjected to thermal actions given by a uniform temperature increase $\bar{\tau}$ and a thermal linear 'jump' across the thickness h from $\frac{\Delta\tau}{2}$ on the extrados to $-\frac{\Delta\tau}{2}$ on the intrados (fig.22).

The thermal dilatation coefficient will be denoted by α $[(°C)^{-1}]$. Beyond the 'free-body equilibrium' state of uniform compression (2.1.1.3), the redundants occasioned by the abutment constraints can be reduced to a reaction H passing through the 'elastic barycenter' of the arch and a bending moment M (fig. 23).

The compatibility conditions that allow to determine the two unknowns H, M are obtained by expressing the equality of elastic rotation, and of chord-wise elastic displacement, for the arch and for the rocky abutment.

For the latter, in order to be in line with the formulation of the more advanced models, we will adopt the following relations, adapted from the Vogt theory (see further on, 2.1.3), between local abutment reactions and local elastic displacements:

tangential displacement (toward the interior of abutment):

$$\delta_N = C_{NN} \frac{N_I}{E_r},$$

with $N_I = -(\bar{\sigma}_a h + H \cos\varphi);$ (2.1.1.5)

radial displacement:

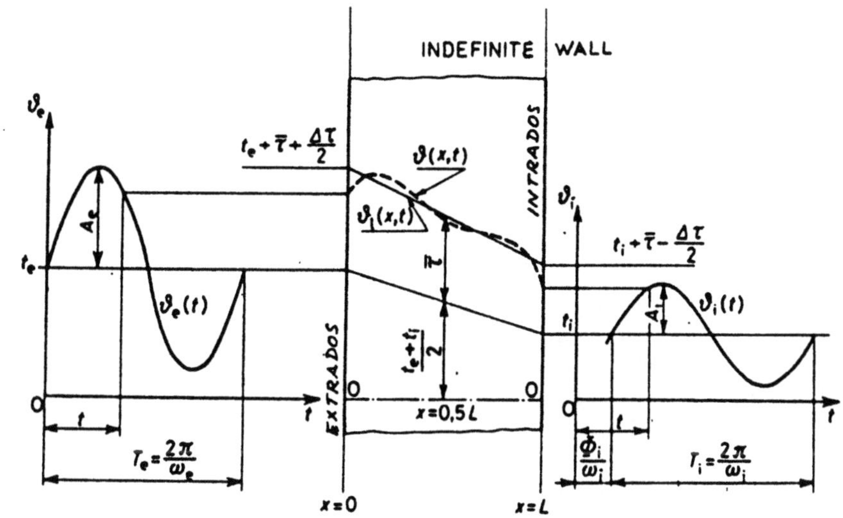

$$\bar{\tau} = \frac{t_e + t_i}{2} + A_e \cdot a_1 \cdot \sin(\omega t - \Phi_1) + A_i \cdot a_1 \cdot \sin(\omega t - \Phi_i - \Phi_1)$$
$$\Delta\tau = t_e - t_i + A_e \cdot a_2 \cdot \sin(\omega t - \Phi_2) - A_i \cdot a_2 \cdot \sin(\omega t - \Phi_i - \Phi_2)$$
$$(\omega_e = \omega_i = \omega)$$

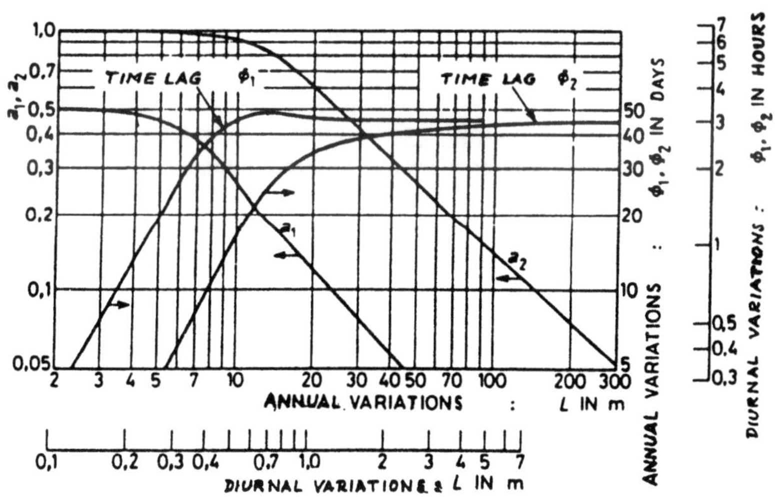

Fig. 22
Linearized thermal variations for sine variations on the facings

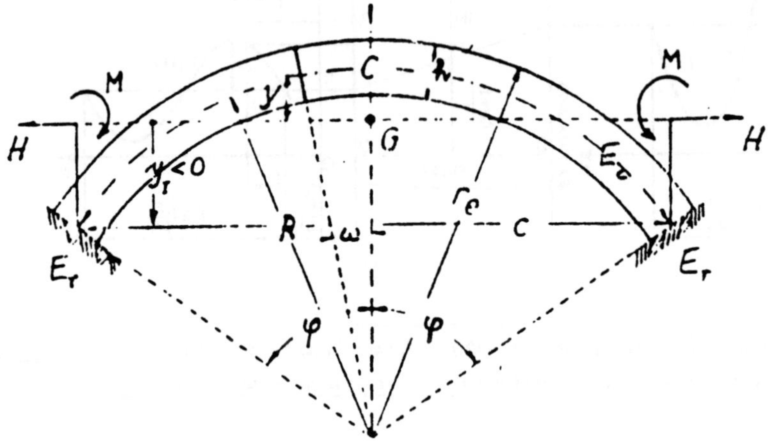

Fig. 23
Isolated arch: reference system and redundants

$$\delta_T = C_{xx}\frac{T_l}{E_r} + C_{\alpha x}\frac{M_l}{E_r h},$$

with $\quad {}_l = H\sin\varphi,\ M_l = -(M + Hy_l);$ (2.1.1.6)

rotation (clockwise for right-bank abutment seen from above):

$$\vartheta = C_{\alpha x}\frac{T_l}{E_r h} + C_{\alpha\alpha}\frac{M_l}{E_r h^2},$$

with $\quad C_{NN}, C_{xx}, C_{\alpha x}, C_{\alpha\alpha}$ dimensionless coefficients (2.1.1.7)

The order of magnitudes for usual aspect ratios of the foundation 'footprint' are (fig.24):

$$C_{NN} \cong 2.5;\quad C_{xx} \cong 2.8;\quad C_{\alpha x} \cong 0.7;\quad C_{\alpha\alpha} \cong 5.5$$

Let us define the following geometric dimensionless integrals:

$$I_1 = R\int_{-\varphi}^{\varphi}(y - y_l)\frac{ds}{J},$$

where $\quad J = \frac{h^3}{12},\quad ds = Rd\omega,\quad y_l = \dfrac{R\cos\varphi - \int_{-\varphi}^{\varphi}\frac{\cos\omega}{J}ds}{\int_{-\varphi}^{\varphi}\frac{ds}{J}};$ (2.1.1.8)

$$I_2 = \int_{-\varphi}^{\varphi} y^2\frac{ds}{J}; \tag{2.1.1.9}$$

$$I_3 = \int_{-\varphi}^{\varphi}\frac{ds}{h}; \tag{2.1.1.10}$$

$$I_4 = \int_{-\varphi}^{\varphi}\sin^2\omega\frac{ds}{h}; \tag{2.1.1.11}$$

$$I_5 = \frac{1}{h_l}\int_{-\varphi}^{\varphi}(y - y_l)\frac{ds}{h}; \tag{2.1.1.12}$$

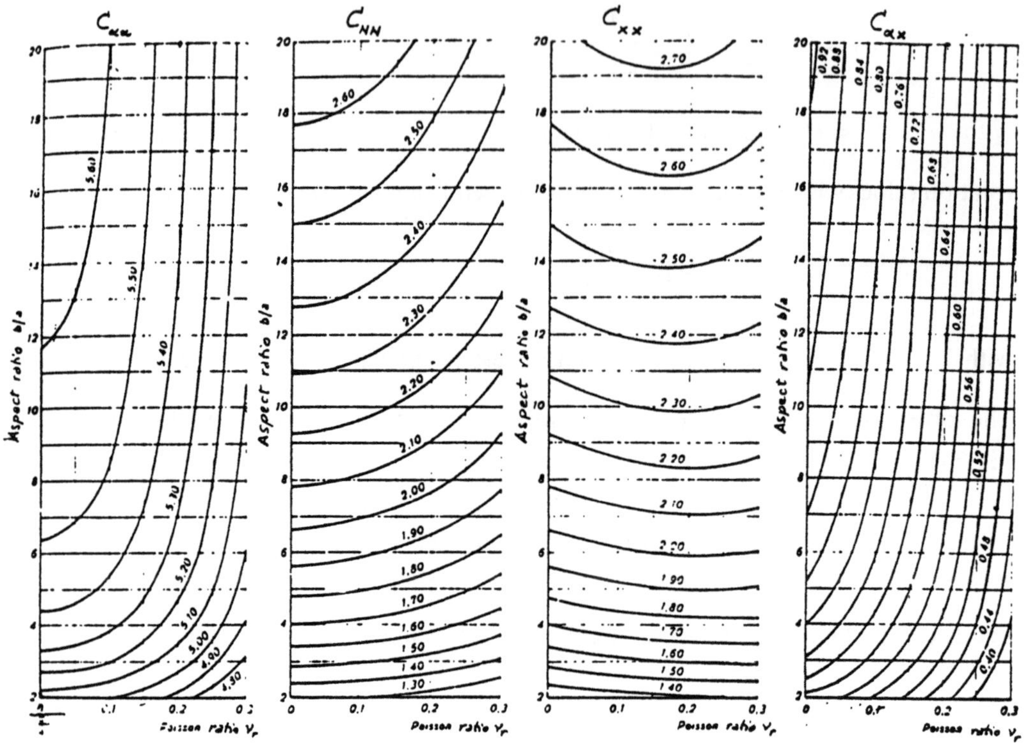

Vogt coefficients as functions of b/a, ν_r.

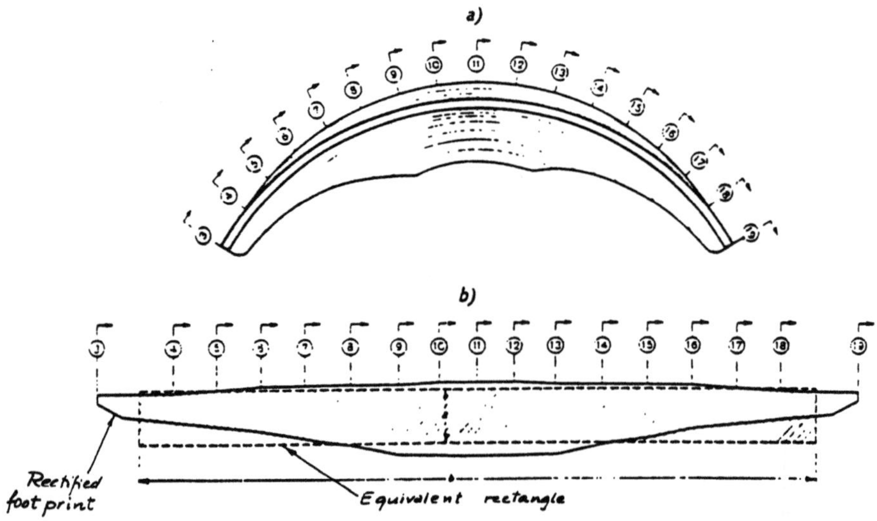

Fig. 24
Abutment compliance: Vogt coefficients

$$I_6 = R^2 \int_{-\varphi}^{\varphi} \frac{ds}{h}; \qquad (2.1.1.13)^{11}$$

The compatibility equations above referred to can be written as follows, by assuming: $\chi \cong 1.25$ and $G_c \cong \dfrac{E_c}{2}$:

i) compatibility of chord-wise displacements:

$$\frac{1}{E_c}\left[\overline{\sigma}_a \cdot 2R\sin\varphi + \frac{M}{R} I_1 + H\left(I_2 + I_3 + 1.5 I_4\right) + \alpha E_c (\overline{\tau}\cdot 2R\sin\varphi - \Delta\tau \cdot h_l I_5)\right] =$$

$$= -\frac{2}{E_c}\left\{ C_{NN}\frac{\overline{\sigma}_a h_l + H\cos\varphi}{\dfrac{E_r}{E_c}}\cos\varphi + \left[C_{xx}\frac{H\sin\varphi}{\dfrac{E_r}{E_c}} - C_{\alpha x}\frac{\dfrac{M}{R}+\dfrac{Hy_l}{R}}{\dfrac{E_r}{E_c}\dfrac{h_l}{R}}\right]\sin\varphi \right\};$$

$$(2.1.1.14)$$

ii) compatibility of abutments rotations:

$$\frac{1}{2E_c}\left(\frac{M}{R} I_6 - E_c\,\alpha\Delta\tau R I_3\right) = \frac{1}{E_c}\left[-C_{\alpha\alpha}\frac{\dfrac{M}{R}+\dfrac{Hy_l}{R}}{\dfrac{E_r}{E_c}\left(\dfrac{h_l}{R}\right)^2} + C_{\alpha x}\frac{H\sin\varphi}{\dfrac{E_r}{E_c}\dfrac{h_l}{R}}\right] \qquad (2.1.1.15)$$

Then the following relations can be obtained:

$$\frac{M}{R} = Q\frac{h_l}{R}H + W\frac{h_l}{R}E_c\alpha\Delta\tau h_l, \qquad (2.1.1.16)$$

---

[11] In the case of costant thickness it is:

$$I_1 = 24R^3\frac{\sin\varphi - \varphi\cos\varphi}{h^3};\ I_2 = 12R^3\frac{\varphi + \sin\varphi\cos\varphi - 2\dfrac{\sin^2\varphi}{\varphi}}{h^3};\ I_3 = 2\frac{R\varphi}{h};\ I_4 = R\frac{\varphi - \dfrac{\sin 2\varphi}{2}}{h};$$

$$I_5 = 2R^2\frac{\sin\varphi - \varphi\cos\varphi}{h^2};\quad I_6 = 24R^3\frac{\varphi}{h^3};\quad y_l \approx R\left(\cos\varphi - \frac{\sin\varphi}{\varphi}\right) < 0$$

with: $$Q = 2\frac{C_{\alpha x}\sin\varphi - C_{\alpha\alpha}\dfrac{\cos\varphi - \sin\varphi/\varphi}{h_l}R}{\dfrac{E_r}{E_c}\left(\dfrac{h_l}{R}\right)^2 I_6 + 2C_{\alpha\alpha}},\qquad(2.1.1.17)$$

$$= \frac{\dfrac{E_r}{E_c}\cdot I_3}{\dfrac{E_r}{E_c}\left(\dfrac{h_l}{R}\right)^2 I_6 + 2C_{\alpha\alpha}};\qquad(2.1.1.18)$$

$$\overline{\sigma}_a R\cdot A + E_c\,\alpha\,\overline{\tau}\cdot R\cdot B + E_c\,\alpha\,\Delta\tau\cdot h_l\cdot C = H\cdot D + \frac{M}{R}\cdot F,\qquad(2.1.1.19)$$

with:

$$A = \frac{E_r}{E_c}\cdot 2\sin\varphi + 2C_{NN}\cos\varphi,\quad B = \frac{E_r}{E_c}\cdot 2\sin\varphi,\quad C = -\frac{E_r}{E_c}\cdot I_5,$$

$$D = -2C_{NN}\cos^2\varphi - 2C_{xx}\sin^2\varphi + 2C_{\alpha x}R\frac{\cos\varphi - \dfrac{\sin\varphi}{\varphi}}{h_l}\sin\varphi - \frac{E_r}{E_c}\left(I_2 + I_3 + 1.5I_4\right),$$

$$F = 2C_{\alpha x}R\frac{\sin\varphi}{h_l} - \frac{E_r}{E_c}I\qquad(2.1.1.20)$$

So that from (2.1.16) and (2.1.19) it obtains:

$$H = \overline{\sigma}_a R\cdot F_1 + E_c\,\alpha\,\overline{\tau}\,R\cdot F_2 + E_c\,\alpha\,\Delta\tau\,h_l\cdot F_3,\qquad(2.1.1.21)$$

$$F_1 = \frac{A}{L},\quad F_2 = \frac{B}{L},\quad F_3 = \frac{P}{L},$$

with: (2.1.1.22)

$$L = D + QF\frac{h_l}{R},\quad P = C - WF\frac{h_l}{R},$$

non-dimensional functions of φ, $\frac{R}{h}$, $\frac{E_r}{E_c}$ as Q and W; M is then given by (2.1.1.16).

In the case of rigid abutments $\left(E_c/E_r \to 0\right)$ and no thermal loads ($\overline{\tau} = \Delta\tau = 0$) one gets:

$$H = -\overline{\sigma}_a R\cdot\frac{2\sin\varphi}{I_2 + I_3 + 1.5I_4}\ ;\quad \frac{M}{R} = 0$$

For shallow ($\varphi \ll 1$), very slender ($h/R \to 0$), arches:

$$H \to \frac{15}{4}\overline{\sigma}_a\, h \cdot \frac{h^2}{R^2 \varphi^4} = -\frac{15}{4}\frac{\overline{\sigma}_a R}{\varphi} \cdot \left(\frac{h}{R\varphi}\right)^3$$

Then the following stress régime is obtained in any section of the arch, identified by the ordinate $y = R\left(\cos\omega - \frac{\sin\varphi}{\varphi}\right)$ above the elastic barycenter:

$$\sigma_e = \overline{\sigma}_a + \frac{H\cos\omega}{h} - \left(M + Hy\right)\frac{6}{h^2} \qquad \text{at the extrados,} \quad (2.1.1.23)$$

$$\sigma_i = \overline{\sigma}_a + \frac{H\cos\omega}{h} + \left(M + Hy\right)\frac{6}{h^2} \qquad \text{at the intrados (fig. 25).}$$

(2.1.1.24)

At mid section one would also have:

$$\tau_{max} = \frac{3}{2} H \frac{\sin\omega}{h} \qquad \text{(in a radial direction).} \quad (2.1.1.25)$$

It remains to be seen how the redundants H, M and the thermal loads $\overline{\tau}$, $\Delta\tau$ affect the displacements régime (fig. 26), in particular the crown radial displacements δ_C [12].
It is seen that δ_c can be expressed as follows (with $x_l = R\sin\varphi$):

$$\delta_C = -\frac{\overline{\sigma}_a \cdot R}{E_c}\left(1 - \cos\varphi\right) + \delta_N \sin\varphi + \delta_T \cos\varphi + \vartheta \cdot x_l - \frac{1}{E_c}\int_0^{\varphi} \frac{M + Hy}{J} \cdot x ds +$$
$$+\frac{\chi}{G_c}\int_0^{\varphi} \frac{T}{h}\cos\omega \cdot ds - \frac{1}{E_c}\int_0^{\varphi} \frac{N'}{h}\sin\omega ds + \alpha\left[\overline{\tau}R\left(\cos\varphi - 1\right) + \Delta\tau\int_0^{\varphi}\frac{x}{h}ds\right],$$

(2.1.1.26)

with $T = H\sin\omega$, N' = axial load from redundants = $H\cos\omega$, $x = R\sin\omega$

(2.1.1.27)

Taking as before $\chi \cong 1.25$ and $G_c \cong \frac{E_c}{2}$, it comes successively:

[12] δ_c is taken positive in a downstream direction, i.e. toward the arch center. In the following developments (as in the formulae for H and M/R) the abutment thickness h_l has been evidenced, as distinct from the arch thickness h, to allow for local thickness increases at the abutments. Also, in this way the formulae can be used for variable-thickness, circular-axis arches, by properly evaluating the integrals $I_1 \ldots I_{11}$, see (2.1.1.8) to (2.1.1.13) and (2.1.1.30), and y_l.

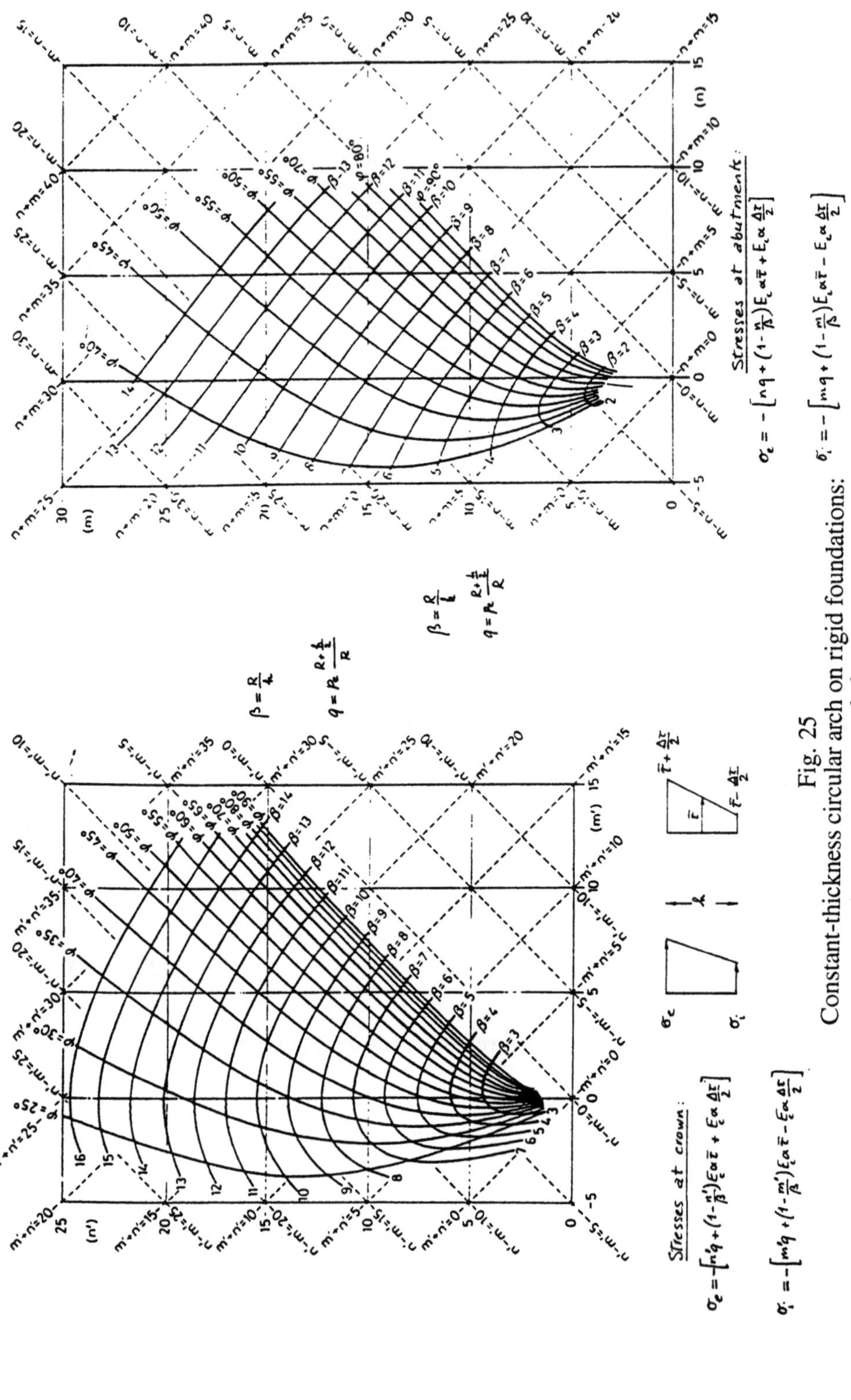

Fig. 25
Constant-thickness circular arch on rigid foundations:
stresses at crown and abutments

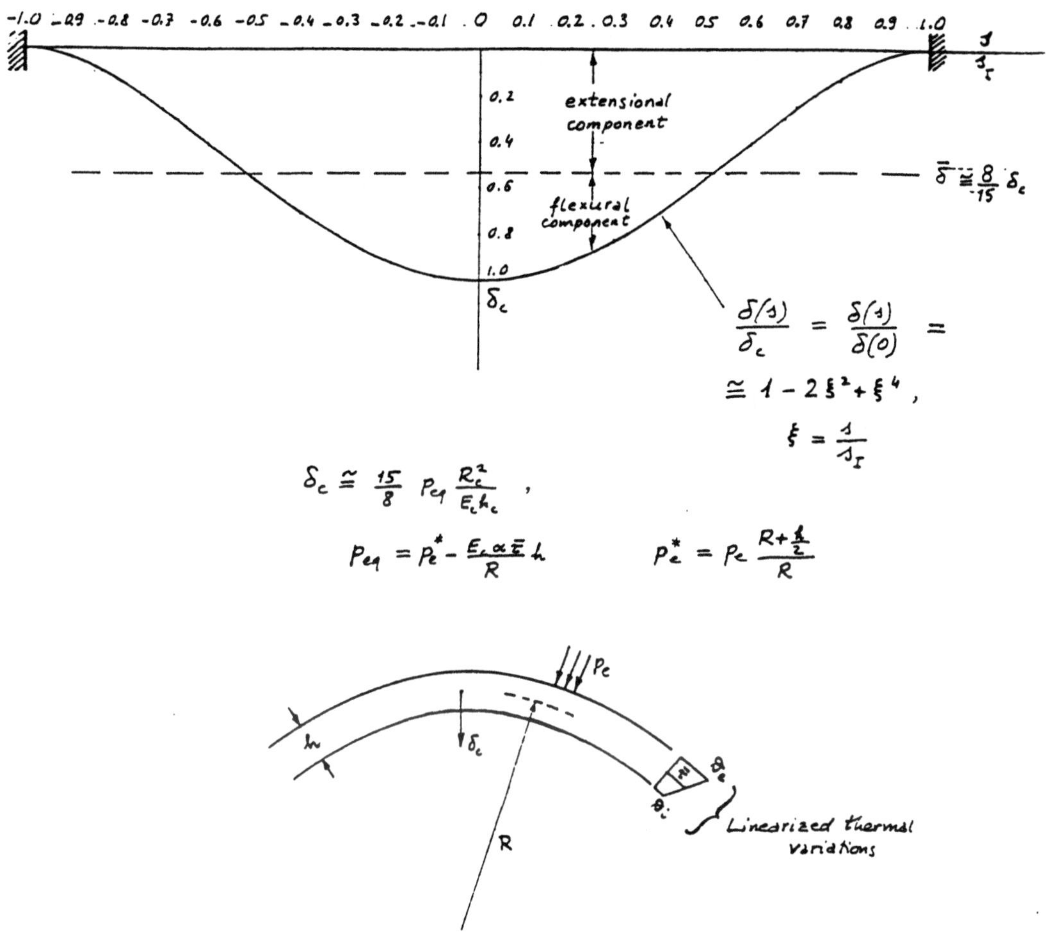

Fig. 26
Elastica of a shakow slender arch with constant thickness and built-in ends, under constant radial pressure

$$\delta_c \cong -\frac{\overline{\sigma}_a \cdot R}{E_c}\left(1-\cos\varphi\right) - C_{NN}\frac{\overline{\sigma}_a h_l + H\cos\varphi}{E_r}\sin\varphi +$$

$$+\left[C_{xx}\frac{H\sin\varphi}{E_r} - C_{\alpha x}\frac{M+Hy_l}{E_r h_l}\right]\cos\varphi + x_l\left[C_{\alpha x}\frac{H\sin\varphi}{E_r h_l} - C_{\alpha\alpha}\frac{M+Hy_l}{E_r h_l^2}\right] +$$

$$-\frac{1}{E_c}\int_0^{\varphi}\frac{M+HR\left(\cos\omega - \frac{\sin\varphi}{\varphi}\right)}{J}R\sin\omega\cdot R d\omega + 1.5\frac{H}{E_c}\int_0^{\varphi}\frac{R\sin 2\omega}{2h}d\omega + \qquad (2.1.1.28)$$

$$+\alpha\left[\overline{\tau}R\left(\cos\varphi - 1\right) + \Delta\tau\int_0^{\varphi}\frac{R\sin\omega}{h}R d\omega\right];$$

$$\delta_c \cong -\overline{\sigma}_a \cdot R\left[C_{NN}\frac{h_l\sin\varphi}{E_r R} + \frac{1-\cos\varphi}{E_c}\right] +$$

$$+H\left[\frac{C_{xx}-C_{NN}}{2E_r}\sin 2\varphi - \frac{C_{\alpha x}}{E_r h_l}\left(y_l\cos\varphi + x_l\sin\varphi\right) - C_{\alpha\alpha}\frac{x_l y_l}{E_r h_l^2} +\right.$$

$$\left.-\frac{1}{E_c}\int_0^{\varphi}\frac{R^3}{J}\left(\cos\omega - \frac{\sin\varphi}{\varphi}\right)\cdot\sin\omega d\omega + 1.5\int_0^{\varphi}\frac{R\sin 2\omega}{2h}\cdot d\omega\right] +$$

$$+\frac{M}{R}\left[-C_{\alpha x}\frac{R\cos\varphi}{E_r h_l} - C_{\alpha\alpha}\frac{x_l R}{E_r h_l^2} - \frac{1}{E_c}\int_0^{\varphi}\frac{R^3\sin\omega}{J}d\omega\right] +$$

$$+\frac{R}{E_c}\left[E_c\,\alpha\,\overline{\tau}\left(\cos\varphi - 1\right) + E_c\,\alpha\,\Delta\tau\int_0^{\varphi}\frac{R\sin\omega}{h}\cdot d\omega\right]; \qquad (2.1.1.29)$$

where H, M/R are given as functions of $\overline{\sigma}_a R$, $E_c \alpha \overline{\tau} R$, $E_c \alpha\Delta\tau h_l$ by equations (2.1.1.21), (2.1.1.16) respectively.

By posing:

$$I_7 = 12R^2 \int_{-\varphi}^{\varphi} \frac{\cos\omega}{h^3} ds; \quad \text{for constant thickness circular arches:} \quad I_7 = \frac{12R^3}{h^3}\sin\varphi$$

$$I_8 = 6R^2 \int_{-\varphi}^{\varphi} \frac{\sin 2\omega}{h^3} ds; \quad \text{"} \quad I_8 = \frac{3R^3}{h^3}(1-\cos\varphi)$$

$$I_9 = \int_{-\varphi}^{\varphi} \frac{\sin 2\omega}{2h} ds; \quad \text{"} \quad I_9 = \frac{R}{h}\cdot\frac{1-\cos\varphi}{4}$$

$$I_{10} = 12R^2 \int_{-\varphi}^{\varphi} \frac{\sin\omega}{h^3} ds; \quad \text{"} \quad I_{10} = \frac{12R^3}{h^3}(1-\cos\varphi)$$

$$I_{11} = \int_{-\varphi}^{\varphi} \frac{\sin\omega}{h} ds; \quad \text{"} \quad I_{11} = \frac{R}{h}(1-\cos\varphi)$$

(2.1.1.30)

$$a = 1-\cos\varphi + C_{NN}\frac{E_c}{E_r}\frac{h_l}{R}\sin\varphi,$$

$$b = \frac{C_{XX}-C_{NN}}{2\,{}^{E_r}\!/_{E_C}}\sin 2\varphi - \frac{C_{\alpha X}}{h_l\,{}^{E_r}\!/_{E_C}}(y_l\cos\varphi - x_l\sin\varphi) - C_{\alpha\alpha}\frac{x_l y_l}{h_l^{2}\,{}^{E_r}\!/_{E_C}} - I_8 +$$

$$+1.5 I_9 + \frac{\sin\varphi}{\varphi} I_{10},$$

$$c = -\frac{C_{\alpha X}}{h_l\,{}^{E_r}\!/_{E_C}} R\cos\varphi - C_{\alpha\alpha}\frac{x_l R}{h_l^{2}\,{}^{E_r}\!/_{E_C}} - I_{10},$$

$$d = \cos\varphi - 1,$$

$$e = \frac{R}{h_l} I_{11}, \tag{2.1.1.31}$$

it comes finally

$$E_c\,\delta_c = -\overline{\sigma}_a \cdot R \cdot a + H \cdot b + \frac{M}{R}\cdot c + E_c\,\alpha\,\overline{\tau}\cdot R\cdot d + E_c\,\alpha\,\Delta\tau\cdot h_l\cdot e, \tag{2.1.1.32}$$

or, by taking into account expressions (2.1.1.21), (2.1.1.16) for H and M/R respectively,

$$E_c\,\delta_c = -\overline{\sigma}_a \cdot R \cdot f_1 + E_c\,\alpha\,\overline{\tau}\cdot R\cdot f_2 + E_c\,\alpha\,\Delta\tau\cdot h_l\cdot f_3, \tag{2.1.1.33}$$

with:

$$f_1 = a - \left(b + c\frac{h_l}{R}Q \right) F_1 \ ,$$

$$f_2 = d + \left(b + c\frac{h_l}{R}Q \right) F_2 \ , \qquad (2.1.1.34)$$

$$f_3 = e + c\frac{h_l}{R}W + \left(b + c\frac{h_l}{R}Q \right) F_3$$

for Q, W see (2.1.1.17), (2.1.1.18); for F_1 , F_2 , F_3 , see (2.1.1.22). Note that f_1 , f_2 , f_3, are dimensionless functions of φ, R/h , E_c/E_r .

So $E_c\delta_c$ is a function of these quantities (fig. 27); through Vogt coefficients: C_{NN} , $C_{\alpha\alpha}$, $C_{\alpha x}$, C_{xx}, it is also a function of the 'aspect ratio' of the foundation 'footprint'.

For rigid abutment ($E_c/E_r \rightarrow 0$), no thermal loads ($\tau = \Delta\tau = 0$) and shallow (φ=0) slender ($h/R \rightarrow 0$) arches, one gets $E_c\delta_c \rightarrow -\frac{15}{8}\bar{\sigma}_a R$.

(Note that for $\varphi = \pi$, i.e. for a complete circle constrained at the downstream apex, (fig. 28),

$E_c\delta_c = -2\ \bar{\sigma}_a R\ ;\ H = M/R = 0$

From application of the above formulae it is easy to conclude that for an isolated arch under uniform external pressure (and with negative temperature variations $\bar{\tau} < 0$, $\Delta\tau < 0$) the maximum bending moment occurs usually near abutments[13] , and causes the maximum horizontal tensile stresses (if any) at the upstream face. This circumstance justifies the frequent practice of providing thicker sections near the abutments, with a downstream kind of flaring, or fillet.

Quite often the geometric shapes that realize such conditions are the so-called 'two centers', or the 'three centers', arches (fig. 28). The treatment is not dissimilar from that of the constant-thickness circular arch, apart from the evaluation of more cumbersome geometric integrals.

---

[13] Unless the ratio $E_r\ /\ E_c$ be excessively low, in which case the worst situation occurs at crown (with $\tau < 0$, $\Delta\tau < 0$)

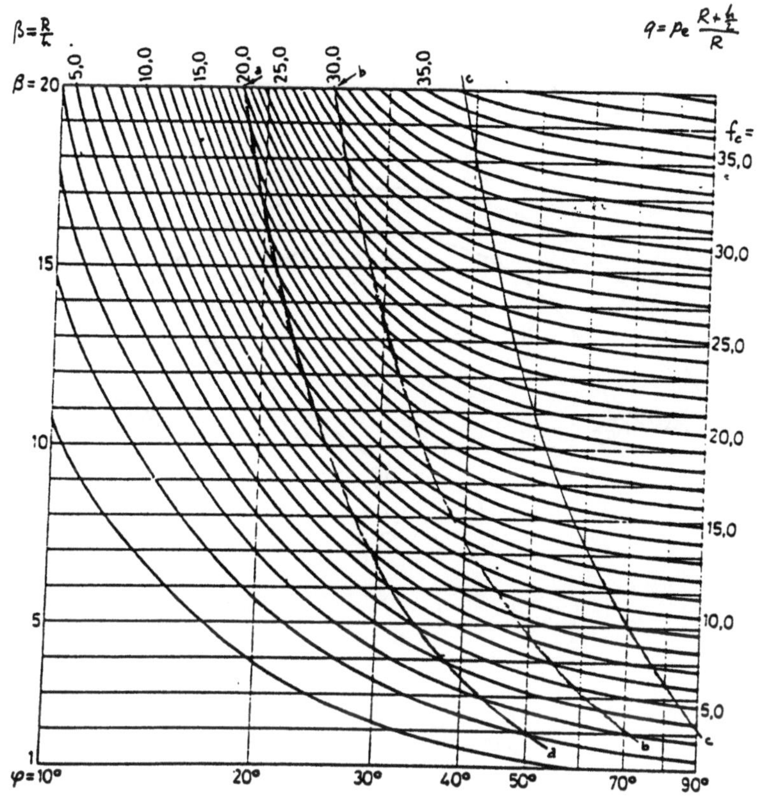

Fig. 27

Values of f_c as a function of φ, R/h; $\delta_c = \frac{qR}{E_c}(1 - E_c\alpha\frac{F}{q\beta})fc$

Line a : $\delta_c = 1.\frac{\bar{\sigma}_a}{E_c}.R$

Line b : $\delta_c = 1.5 . \frac{\bar{\sigma}_a}{E_c}.R$

Line c : $\delta_c = 1.8 . \frac{\bar{\sigma}_a}{E_c}.R$ $\quad (\equiv \frac{15}{8}\frac{\bar{\sigma}_a}{E_c}R)$

} All curves for rigid abutments $(\frac{E_c}{E_r} \to 0)$

$q = p_e.\frac{R + \frac{h}{2}}{R}$

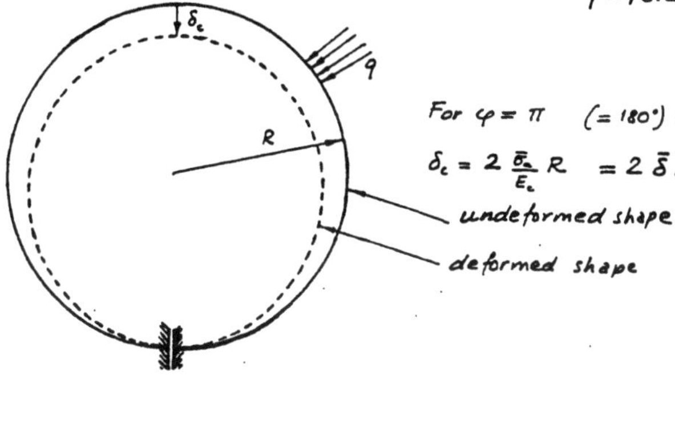

For $\varphi = \pi$ $(= 180°)$:

$\delta_c = 2\,\frac{\bar{\sigma}_a}{E_c}\,R = 2\,\bar{\delta}.$

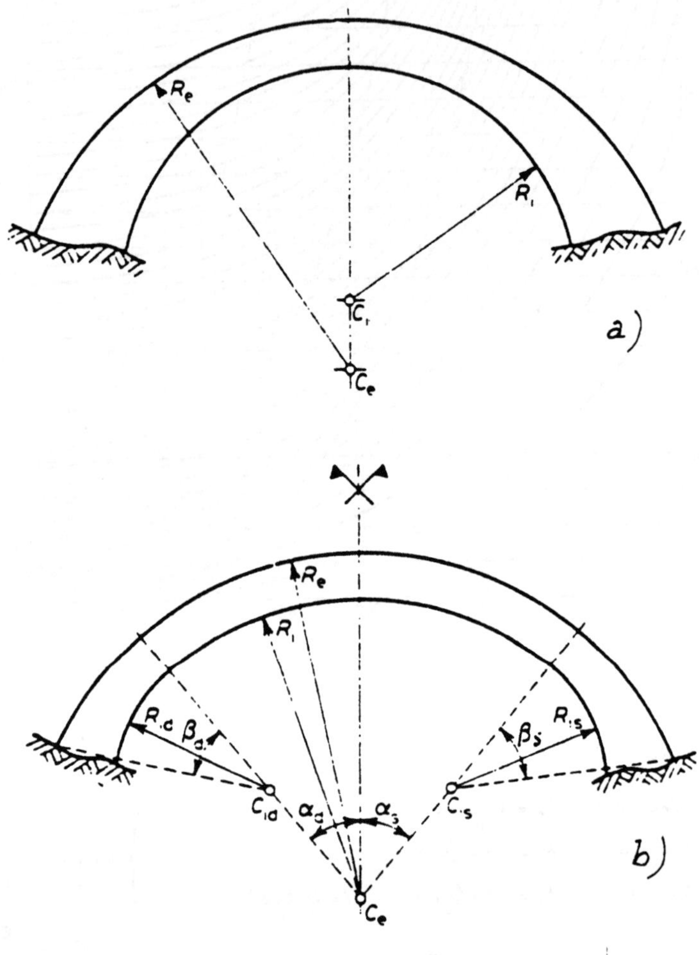

Fig. 28
Variable-thickness circular arches:
a) two-centers arch
b) three-centers arch

2.1.2 The rotationally symmetric reservoir: the Tölke model

In this model one considers that the dam is, ideally, a portion of a rotationally symmetric reservoir (with water on the exterior) (fig. 29). This of course poses some limitations on the dam shape: the arches must have circular axis and constant thickness (variable with elevation z), and the line of centers should be vertical.

The live loads (external hydrostatic pressure, azimuthally uniform temperature variations) being also rotationally symmetric, the régime of displacements will be purely radial. It is therefore easy to recognize that in this ideal reference structure no relative rotations occur between the sections of two overlying 'arches' belonging to a common vertical plane; as a consequence, no torsional actions will come into play.

Likewise, the régime of the 'arches' will be one of pure compression (no bending moment, no shear) and, if one denotes by p_a the uniform radial pressure borne by the generic arch of radius R and thickness h, one will have(fig. 30):

$$\bar{\sigma} = -p_a \frac{R + h/2}{h}, \quad \sigma_e = \sigma_i = \bar{\sigma}_a \quad ^{14} \tag{2.1.2.1}$$

$$\delta_c = -\frac{\bar{\sigma}_a}{E_c} R = \frac{p_a}{E_c} \frac{R + h/2}{h} R \tag{2.1.2.2}$$

Since in general p_a , R and h will be variable with elevation z, the vertical sections of the rotationally symmetric reservoir ('cantilevers') will in general be subject to bending. Assuming as positive the bending moments $M(z)$ that cause a vertical tensile stress at the extrados, it will be:

$$M(z) = E_c J \frac{\partial^2 \delta_c}{\partial z^2} \qquad \left(J = \frac{h^3}{12} \right) \tag{2.1.2.3}$$

As a consequence, the cantilever will carry a radial load (fig. 31):

$$p_m = \frac{\partial^2 M}{\partial z^2} \tag{2.1.2.4}$$

---

14 Under the assumption of slender arches, i.e. $h / R \ll 1$.

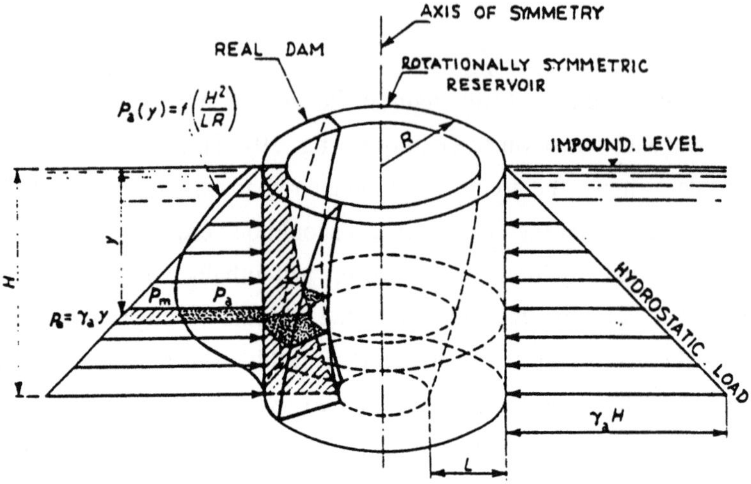

Fig. 29
Arch dam as a part of a rotationally symmetric reservoir: the Tölke model

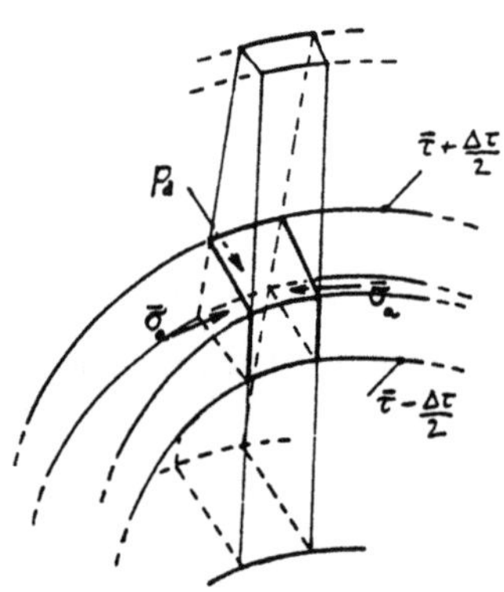

Fig. 30
The Tölke model: actions on a generic arch element

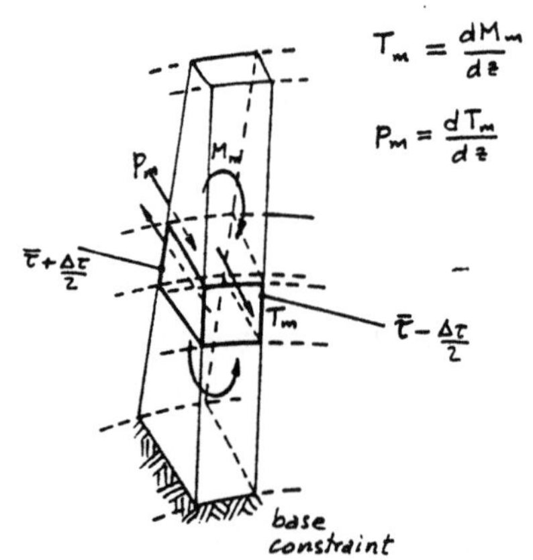

Fig. 31
The Tölke model: actions on a generic cantilever element

It is easy to see that no other internal actions exist which can contribute to the equilibrium of an element ($dz \times Rd\omega \times h$) of the rotationally symmetric structure. Therefore the indefinite differential equilibrium equation will be for the live external loads (hydrostatic pressure plus dynamic pressures due to earthquake) (fig. 32):

$$p_e = \gamma_a z + p_d(z,t) = p_a + p_m$$

or, by (2.1.2.2), (2.1.2.4),

$$p_e(z) = \gamma_a z + p_d = E_c \left[\frac{h\delta_c}{R\left(R + \frac{h}{2}\right)} + \frac{\partial^2}{\partial z^2}\left(J \frac{\partial^2 \delta_c}{\partial z^2} \right) \right] \qquad (2.1.2.5)^{15}$$

which is the central part of the Tölke model. Of course, to solve for δ_c (which will then allow determination of horizontal and vertical stresses, see further on), this must be complemented by B.C.'s.

For the reference rotationally symmetric structure, these concern only the cantilever (fig. 33). At the top one has to express the conditions of zero bending moment and zero shear (free edge):

$$M(z) = 0 \quad , \quad \frac{\partial M}{\partial z} = 0 \qquad \text{for } z{=}z \text{ (crest)}, \qquad (2.1.2.6)$$

which after (2.1.2.3) become:

$$\frac{\partial^2 \delta_c}{\partial z^2} = 0 \quad , \quad \frac{\partial}{\partial z}\left(J \frac{\partial^2 \delta_c}{\partial z^2} \right) = 0 \qquad \text{for } z{=}z \text{ (crest)}. \qquad (2.1.2.7)$$

At bottom, the compatibility conditions between radial displacement, and rotation in a vertical plane, of the cantilever and of the rocky foundation, will have to be written down. By representing the foundation compliance through the Vogt coefficients $C_{\alpha\alpha}$,

[15] The z axis is assumed to be directed downwards and to have its origin at crest; the reservoir is supposed to be completely full up to the crest

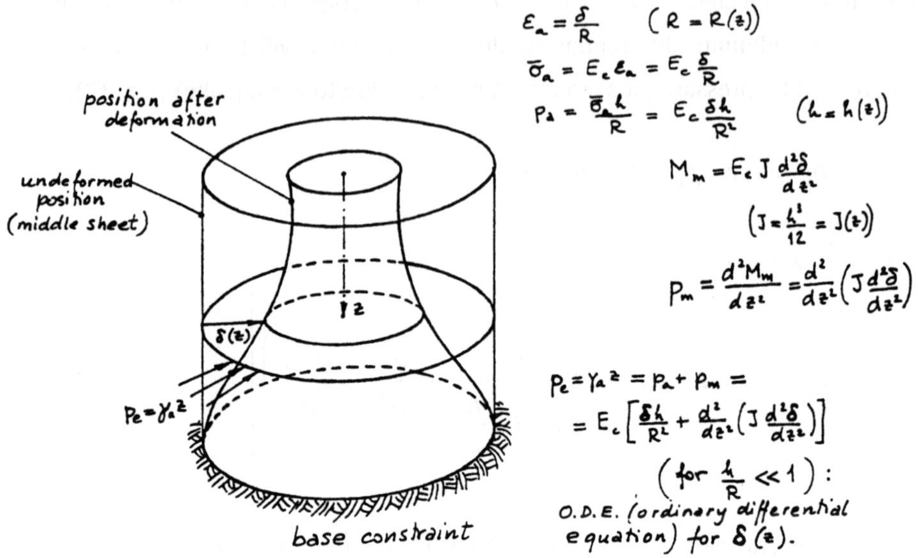

Fig. 32
The Tölke model: equilibrium and compatibility equation

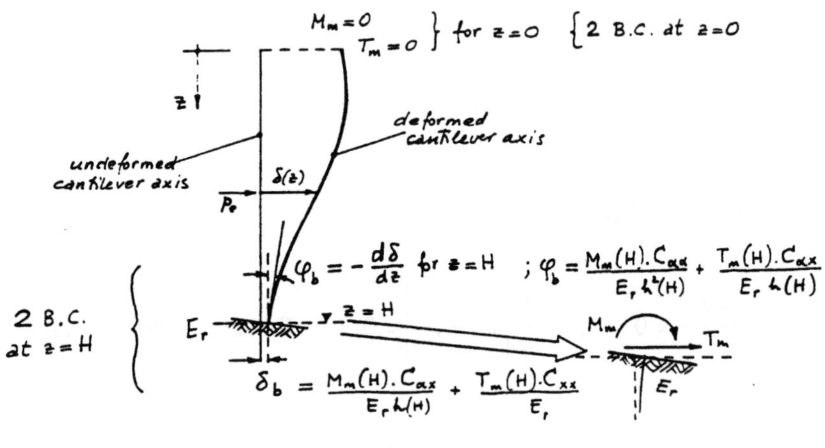

Fig. 33
The Tölke model: B.C. for δ(z)

$C_{\alpha x}$, C_{xx} as before (in this case C_{NN}, does not come into play) it is easy to express these two conditions:

$$\delta_c = M\frac{C_{\alpha x}}{E_r h} + T\frac{C_{xx}}{E_r} \qquad \text{for } z = z\,(\text{bottom}), \qquad (2.1.2.8)$$

where $= \dfrac{\partial M}{\partial z}$;

$$-\frac{\partial \delta_c}{\partial z} = M\frac{C_{\alpha\alpha}}{E_r h^2} + T\frac{C_{\alpha x}}{E_r h} \qquad \text{for } z = z\,(\text{bottom}).^{16} \qquad (2.1.2.9)$$

By taking into account (2.1.2.3) one gets the explicit bottom B.C.'s:

$$\delta_c = \frac{E_c}{E_r}\left[J\frac{C_{\alpha x}}{h}\frac{\partial^2 \delta_c}{\partial z^2} + C_{xx}\frac{\partial}{\partial z}\left(J\frac{\partial^2 \delta_c}{\partial z^2}\right)\right], \qquad (2.1.2.10)$$

$$-\frac{\partial \delta_c}{\partial z} = \frac{E_c}{E_r}\left[J\frac{C_{\alpha\alpha}}{h^2}\frac{\partial^2 \delta_c}{\partial z^2} + \frac{C_{\alpha x}}{h}\frac{\partial}{\partial z}\left(J\frac{\partial^2 \delta_c}{\partial z^2}\right)\right], \qquad (2.1.2.11)$$

for $z = z$ (bottom).

The fourth-order differential equation (2.1.2.5), together with the four B.C.'s (2.1.2.6), (2.1.2.7), (2.1.2.10), (2.1.2.11), allow in principle to solve numerically for $\delta_c(z)$. If a closed-form analytical solution is not feasible in the particular case at hand, it is always possible to discretize the problem by expressing all the derivatives as finite differences (F.D.) on a suitably chosen linear (one-dimensional) grid of points lying along the **z**-axis (fig. 34); in this way an algebraic linear system of equations is obtained, with the values of δ_c in the discretization points as unknowns. To be noted that in this case two additional 'fictitious' grid points have to be added above the crest, and two more below the foundation, in order to express the F.D.'s of the order required by the B.C.'s at crest and at bottom. The values of δ_c in these fictitious grid points are determined by the B.C.'s.

Once solved for δ_c, one gets easily for the stresses:

---

[16] It has been assumed in (2.1.2.8 et 9) that the *z* axis is direct downwards; if the contrary convention is adopted, then in (2.1.2.8 et 9) the + sign in front of *T* should be changed to the - sign, and viceversa.

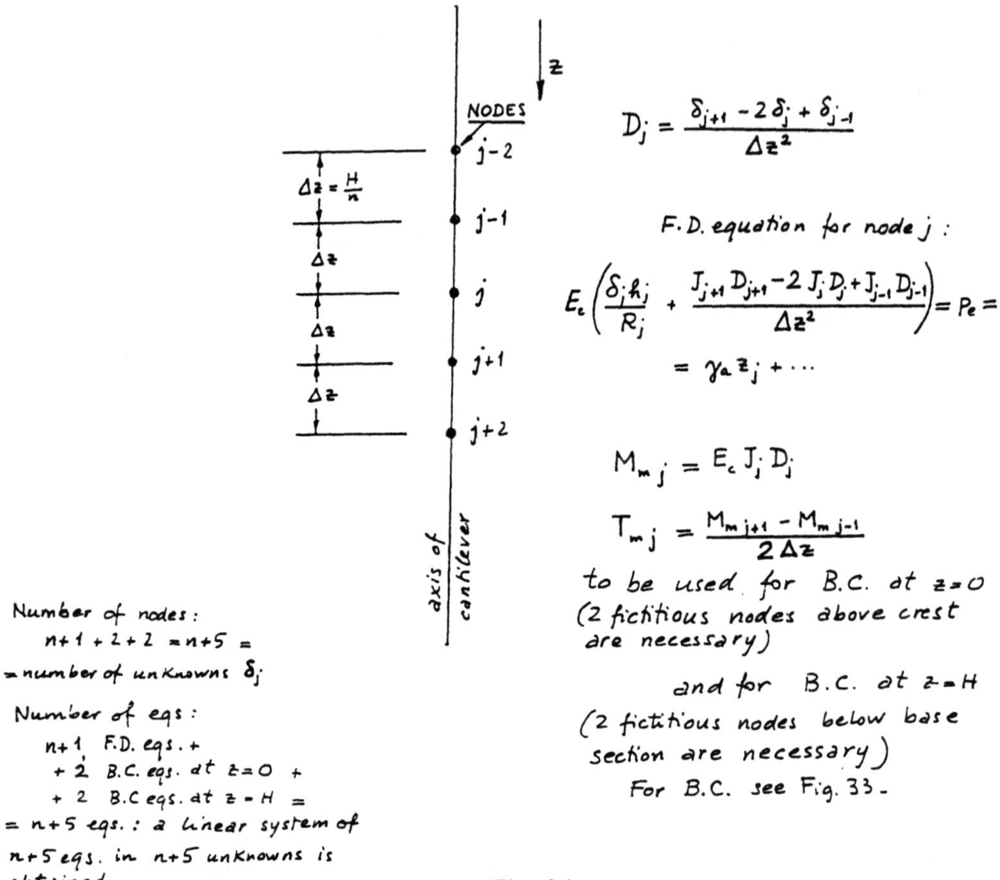

Fig. 34
The Tölke model: the F.D. scheme

i. for the arches (horizontal stresses) $$\overline{\sigma}_a = \sigma_e = \sigma_i = -\frac{E_c}{R}\delta_c\,, \qquad (2.1.2.12)$$

ii. for the cantilevers (vertical stresses) $$\sigma_m = \sigma_{me} = -\sigma_{mi} = \frac{E_c h}{2}\frac{\partial^2 \delta_c}{\partial z^2}\,, \qquad (2.1.2.13)$$

which completes the analysis of the reference rotationally symmetric structure.
The dead weight is usually assumed to be carried by the cantilevers; the relevant stresses should then be evaluated for the isolated cantilever and added to (2.1.2.13).
As for the thermal loads (fig. 35), the following considerations apply (for simplicity hereafter these loads are considered as acting separately from live loads; the relative effects should then be superposed):
a) for a rotationally symmetric distribution of a linear temperature 'jump' across the thickness h from $+\frac{\Delta\tau}{2}$ on the extrados (upstream, exterior of axially symmetric reservoir) to $-\frac{\Delta\tau}{2}$ on the intrados (downstream, interior of axially symmetric reservoir)[17] no additional deformations or displacements would be produced in the isolated arches, the only effect being the additional horizontal stresses:

$$\Delta\sigma_i = -\Delta\sigma_e = E_c\frac{\alpha\Delta\tau}{2}; \qquad (2.1.2.14)$$

however, the connection between overlying arches provided by 'cantilevers' introduces a régime of displacements which affects also the arches, see b).

b) for a rotationally symmetric distribution of a radially uniform temperature variation $\overline{\tau}$ (which, however, can be variable with elevation z), it must be considered that in place of (2.1.2.2) one should write:

$$\delta_c = -(\overline{\sigma}_a\frac{R}{E_c} + \alpha\overline{\tau}R), \qquad (2.1.2.15)$$

and by (2.1.2.1):

[17] $\Delta\tau$ can be variable with elevation: $\Delta\tau = \Delta\tau(z)$

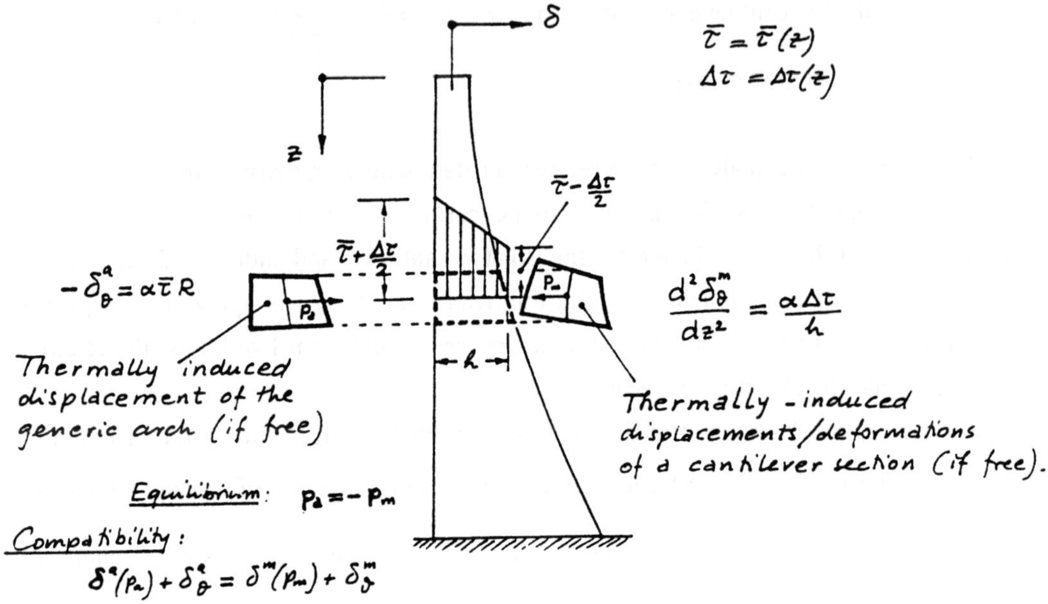

Fig. 35
The Tölke model: the thermal actions

$$p_a = -\frac{h\bar{\sigma}_a}{R+\frac{h}{2}} = \frac{hE_c}{R\left(R+\frac{h}{2}\right)}\left(\delta_c + \alpha\bar{\tau}R\right). \tag{2.1.2.16}$$

In place of (2.1.2.3) one should write, for the cantilever[18],

$$M(z) = E_c J\left(\frac{\partial^2 \delta_c}{\partial z^2} - \frac{\alpha\Delta\tau}{h}\right) \tag{2.1.2.17}$$

Besides, $p_a + p_m = 0$ gives by (2.1.2.16), (2.1.2.4), (2.1.2.17):

$$E_c\left[\frac{h\left(\delta_c + \alpha\bar{\tau}R\right)}{R\left(R+\frac{h}{2}\right)} + \frac{\partial^2}{\partial z^2}\left(J\left(\frac{\partial^2 \delta_c}{\partial z^2} - \frac{\alpha\Delta\tau}{h}\right)\right)\right] = 0 \tag{2.1.2.18}$$

or

$$\frac{h\delta_c}{R\left(R+\frac{h}{2}\right)} + \frac{\partial^2}{\partial z^2}\left(J\frac{\partial^2 \delta_c}{\partial z^2}\right) = \alpha\left[\frac{-h\bar{\tau}}{R+\frac{h}{2}} + \frac{\partial^2}{\partial z^2}\left(J\frac{\Delta\tau}{h}\right)\right], \tag{2.1.2.19}$$

in place of (2.1.2.5) (indefinite differential equation).
The B.C.'s are modified as follows:

$$\frac{\partial^2 \delta_c}{\partial z^2} = \frac{\alpha\Delta\tau}{h}, \frac{\partial}{\partial z}\left[J\left(\frac{\partial^2 \delta_c}{\partial z^2} - \frac{\alpha\Delta\tau}{h}\right)\right] = 0 \quad \text{for } z = z \text{ (crest)}; \tag{2.1.2.20}$$

$$\delta_c = \frac{E_c}{E_r}\left[J\frac{C_{\alpha x}}{h}\left(\frac{\partial^2 \delta_c}{\partial z^2} - \frac{\alpha\Delta\tau}{h}\right) + C_{xx}\frac{\partial}{\partial z}\left(J\left(\frac{\partial^2 \delta_c}{\partial z^2} - \frac{\alpha\Delta\tau}{h}\right)\right)\right]$$

---

[18] It is here supposed for simplicity that the cantilever axis is near-vertical, so that any free-body horizontal displacement caused in the cantilever by uniform temperature variations $\bar{\tau}$ is negligible.

$$-\frac{\partial\delta_c}{\partial z}=\frac{E_c}{E_r}\left[J\frac{C_{\alpha\alpha}}{h^2}\left(\frac{\partial^2\delta_c}{\partial z^2}-\frac{\alpha\Delta\tau}{h}\right)+\frac{C_{\alpha x}}{h}\frac{\partial}{\partial z}\left(J\left(\frac{\partial^2\delta_c}{\partial z^2}-\frac{\alpha\Delta\tau}{h}\right)\right)\right]$$

for $z = z$ (bottom)[19] (2.1.2.21)

After solution of (2.1.2.19) with B.C.'s (2.1.2.20), (2.1.2.21) one has for the stress régime:

- for horizontal stresses:

$$\sigma_e=-\frac{E_c}{R}\left(\delta_c+\alpha\bar{\tau}R\right)-E_c\frac{\alpha\Delta\tau}{2},$$
$$\sigma_i=-\frac{E_c}{R}\left(\delta_c+\alpha\bar{\tau}R\right)+E_c\frac{\alpha\Delta\tau}{2};\qquad(2.1.2.22)$$

- for the vertical stresses:

$$\sigma_m=\sigma_{me}=-\sigma_{mi}=E_c\frac{h}{2}\left(\frac{\partial^2\delta_c}{\partial z^2}-\frac{\alpha\Delta\tau}{h}\right)\qquad(2.1.2.23)$$

It is to be noted that in the Tölke reference model the arches are not externally constrained; this leads to an estimate of the statical régime of the arches that excessively underestimates the possibilities for tensile horizontal stresses. So, a modification commonly adopted consists in computing by the Tölke model the partial pressure $p_a(z)$, see (2.1.2.16), that is carried by the generic arch at elevation z, and then solving for the stress régime of the arch as if it were an isolated arch with its real abutment constraints under external pressure $p_a(z)$ (and thermal variations $\bar{\tau}(z)$, $\Delta\tau(z)$, if any).

This empirical adjustment, although widely used in the 30's, entails some logical inconsistency. In fact, the subdivision of total external pressure $p_e(z)$ among the arch-carried component [$p_a(z)$] and the cantilever-carried component [$p_m(z)$] depends on the assumed dependence of δ_c on $\overline{\sigma}_a$ (and $\bar{\tau}$), see (2.1.2.15), which is based on the assumption of unconstrained arch.

[19] It is assumed that the foundations would tend, if free, to undergo negligible thermal displacements/rotations.

One would be tempted to improve the Tölke model by considering that in the constrained arch it is approximately (see par. 2.1.1):

$$\delta_c \cong -\frac{15}{8}\overline{\sigma}_a \frac{R}{E_c} + K_1 \alpha \overline{\tau} R + K_2 \alpha \Delta\tau R, \qquad (2.1.2.24),$$

with numerical coefficients 15/8 instead of 1; K_1 instead of -1; K_2 instead of 0. But in this way one would again introduce an inconsistency, because the rotations (in a horizontal plane) of two overlying sections of arch would then be different, thus bringing into play a torsional stiffness which is inactive in the Tölke model. If one then tries to go one step further, introducing into the equilibrium equation (2.1.2.4) a term which corresponds to this torsional stiffness, one is led essentially to the Ritter model, which will be treated in the next paragraph[20] (fig. 36).

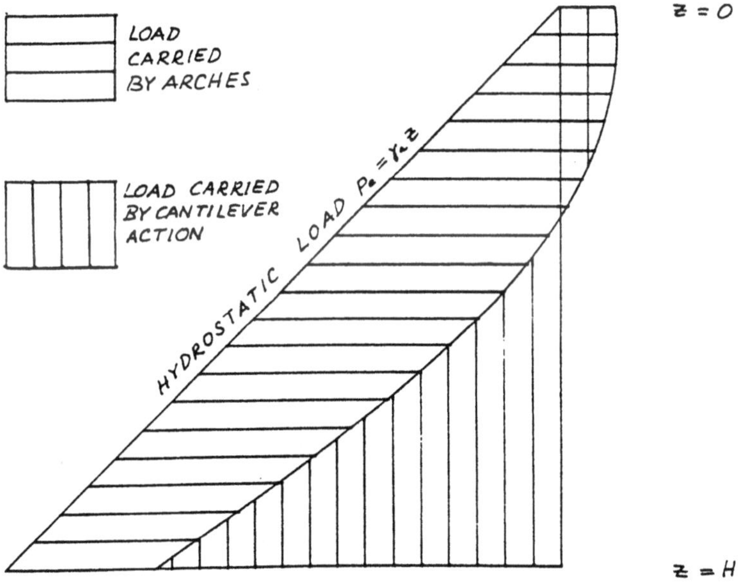

Fig. 36
The Tölke model:
hydrostatic pressure subdivision between arch and cantilever action

---

[20] It could be argued with some reason that in the Tölke method one overestimates the arch stiffness (by a factor~ 15/8), which approximately makes up for neglecting the torsional stiffness. Of course this argument is only qualitative and so the Tölke reference model has not a very clear relationship to the "real" situation, of the arches in particular.

2.1.3 The Ritter (single central cantilever) model

This model takes as a starting point the consideration of the radial equilibrium of a central cantilever, and is best suited to the case of symmetric, slender arch dams.
It could seem surprising, at first sight, that consideration of a single cantilever should lead to a good representation of the static régime, even in the proximity of the central cantilever itself. Indeed, the presence of lateral cantilevers, as already pointed out, can provide for strong differences in the distribution of the arch-borne loads according to the shape of the valley.
However, closer reflection shows that there are good reasons to explain the success of so simple a representation.
In fact, the equilibrium of the central cantilever includes the "elastic-bed-like" reactions of the arches, which are expressed as a suitable linear function of the crown arch displacements. Now, it can be shown that this apparent stiffness of an arch at crown is but little sensitive to the actual load distribution along the arch itself.
In fact, the elastica of a slender arch under normal pressure (fig.37) is of necessity - taking into account the end constraints - composed of an average radial displacement $\bar{\delta}$ attributable to the average compressional state, and of a flexural part, δ_f, which is subjected to rather rigid conditions:

$\delta_f \cong -\bar{\delta}$ at the abutments, if those are sufficiently stiff;

$\varphi = \frac{\partial \delta_f}{\partial s} = 0$ at crown (for symmetric situations) and

$\varphi = \frac{\partial \delta_f}{\partial s} \cong 0$ at the abutments, again if those are sufficiently stiff.

This calls for a rather limited possibility of variation of δ_f (s), so that at a first approximation at crown $\delta_c = \bar{\delta} + \delta_f$ (crown) is quite univocally related to the local values of pressure and curvature: in fact it can be shown that:

$$\delta_c = \delta_{(crown)} \cong \frac{15}{8} p_a \frac{R^2_c}{E_c h_c} = \frac{15}{8}\left(-\frac{\bar{\sigma}_a}{E_c}\right) R_c, \qquad (2.1.3.1)^{21}$$

[21] p_a being the part of the external pressure borne by the arch, and $\bar{\sigma}_a$ = average normal stress across thickness = - $p_a R_c / h_c$. Of course the above is all the more "true" as the objective circumstances are

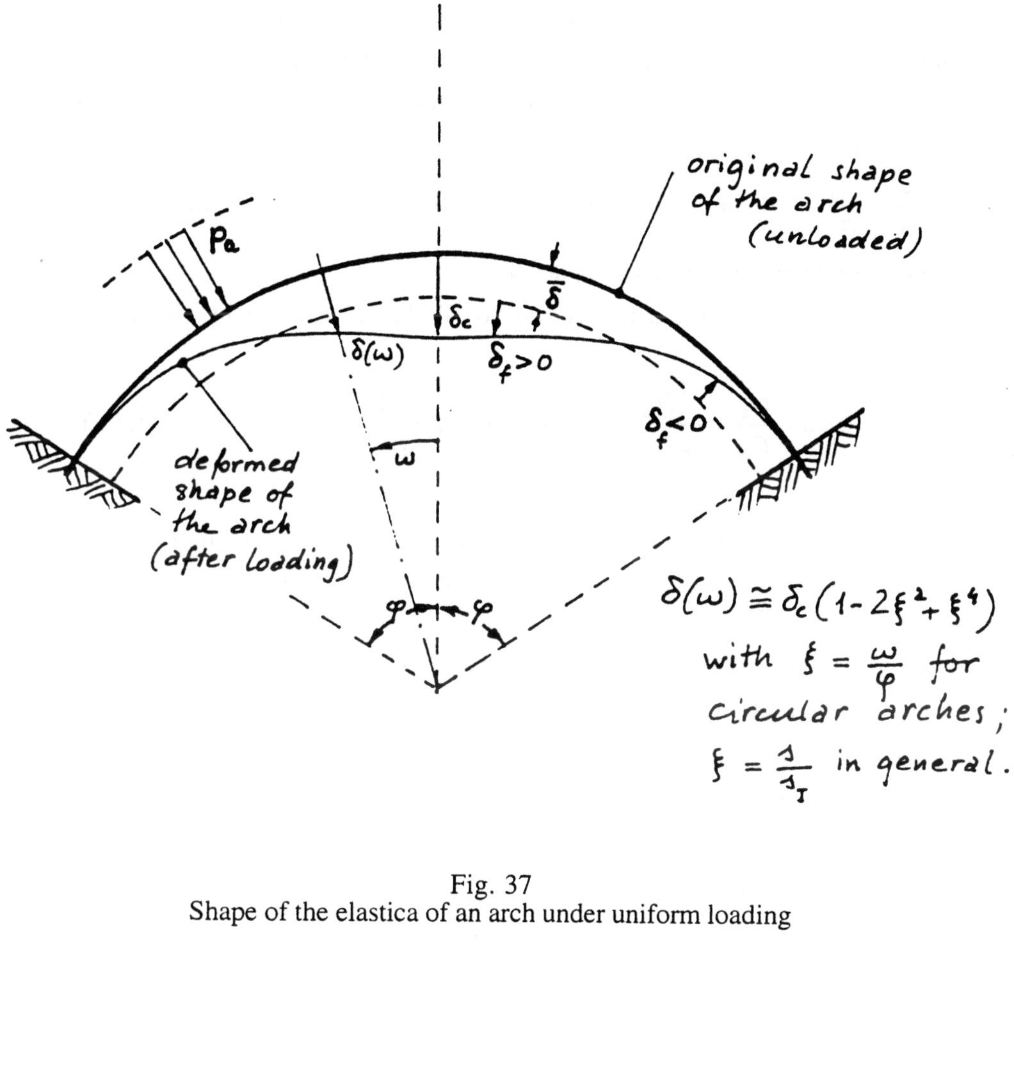

Fig. 37
Shape of the elastica of an arch under uniform loading

E_c being the Young modulus of concrete.

(This can also be seen by considering the deformation energy of the arch and the external work done by pressures during deformation (fig.38): since the radial displacements go down to zero quite rapidly from crown towards the abutments, the external work is not very sensitive to the shape of the elastica, or to the pressure distribution, away from a restricted zone near the crown. The deformation energy, in its turn, is mostly sensitive to the thickness-averaged value of dilatation $\bar{\varepsilon} = -\left(\frac{\delta}{R} - \frac{\partial \delta_t}{\partial s}\right)$, which can be shown to be nearly constant along the arch, given the constraints on tangential displacements δ_t).

If now one reflects that the the elastic "recall force" exerted by the arch on the cantilever is:

$$p_a = -\frac{N_a}{R_c} + \frac{\partial^2 M_a}{\partial s^2} \cong -\frac{\bar{\sigma}_a h_c}{R_c} \cong \frac{8}{15} h_c \frac{\delta_c E_c}{R^2{}_c} \tag{2.1.3.2}$$

proportional to δ_c through an almost fixed constant, it is seen that the fact of considering only the central cantilever cannot cause a very large departure from "reality".

Moreover, a definitely valley-shape-dependent contribution to the equilibrium of the central cantilever comes from the <u>torsional stiffness</u>, which is influenced by the variation of the half-lengths of the arches with elevation, $s_l(z)$. In fact, the contribution at crown of torsional stiffness to the said cantilever equilibrium is:

$$p_t = 2\frac{\partial^2 M_{sz}}{\partial s \partial z} = 2\frac{\partial}{\partial z}\left(\frac{\partial M_{sz}}{\partial s}\right) = 2\frac{\partial}{\partial z}\left[\frac{\partial}{\partial s}\left(\frac{h^3}{12}E_c\frac{\partial \varphi}{\partial z}\right)\right] = 2\frac{\partial}{\partial z}\left[E_c\frac{h^3}{12}\frac{\partial}{\partial z}\left(\frac{\partial \varphi}{\partial s}\right)\right]. \tag{2.1.3.3}$$

Now along the arch one can assume as a first approximation (in accordance with the above discussion about p_a):

$$\delta \cong \delta_c\left(1 - 2\xi^2 + \xi^4\right), \quad \text{with } \xi \equiv \frac{s}{s_l}, \tag{2.1.3.4}$$

in accordance with the assumption of a $\bar{\sigma}_a$ nearly constant along the arch (apart from perturbations near the abutments), i.e. if the arch dam is a circular-arch structure in a U-shaped valley or a parabolic-arch structure in a V-shaped valley; so the shape of the valley indirectly enters the formulation.

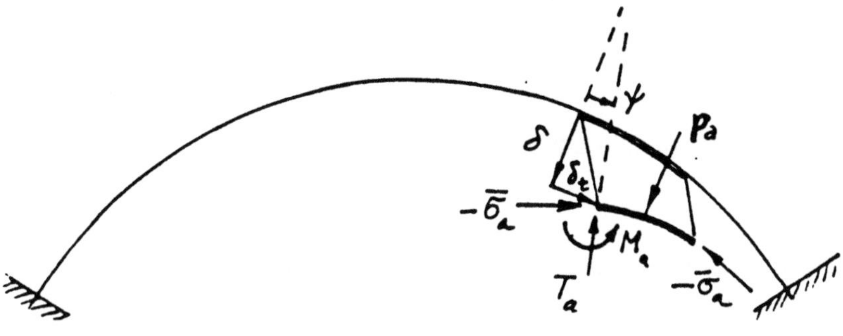

Dilatation (average) $\quad \bar{\varepsilon}_a = -\frac{\delta}{R} + \frac{\partial \delta_t}{R\partial\omega}$

Const. eq. $\quad \bar{\sigma}_a = E_c \bar{\varepsilon}_a$

Radial Equilibrium $\quad p_a R = -\bar{\sigma}_a h + \frac{\partial T_a}{R\partial\omega}$

Rotation $\quad \psi = -\frac{\delta_t}{R} + \frac{\partial \delta}{R\partial\omega}$

$$M_a = E_c J \frac{\partial \psi}{R\partial\omega}$$

Equilibrium to rotat. $\quad T_a = \frac{\partial M_a}{R\partial\omega}$

Tangent. Equilibrium $\quad \frac{\partial(\bar{\sigma}_a h)}{R\partial\omega} = -\frac{T_a}{R}$

Fig. 38
Displacements, deformations, equilibrium of an arch section under radial load

s_l being the arch half-length from crown to abutment [22] ; so that:

$$\varphi \cong \frac{\partial \delta}{\partial s} = \frac{1}{s_l}\frac{\partial \delta}{\partial \xi} = \frac{\delta_c}{s_l}\left(-4\xi + 4\xi^3\right), \qquad (2.1.3.5)$$

$$\frac{\partial \varphi}{\partial s} \cong -4\frac{\delta_c}{s_l^2} \quad \text{at crown } (\xi = 0), \qquad (2.1.3.6)$$

from which the contribution of torsional stiffness to central cantilever radial equiibrium can be taken approximately as:

$$p_t \cong -\frac{8}{12}E_c\frac{\partial}{\partial z}\left[h^3\frac{\partial}{\partial z}\left(\frac{\delta_c}{s_l^2}\right)\right] \qquad (2.1.3.7)$$

Taking into account that the contribution of cantilever bending to equilibrium (fig.39) is given by $p_m = \frac{\partial^2 M_z}{\partial z^2}$, with:

$$M_z = E_c\frac{h^3}{12}\frac{\partial^2 \delta_c}{\partial z^2} \quad , \qquad (2.1.3.8)$$

and denoting by $p_e(z)$ the local radial external pressure at crown, at last, the local radial equilibrium of the central cantilever is expressed by:

$$p_e(z) = p_a + p_m + p_t = -\frac{N_a}{R_c} + \frac{\partial^2 M_a}{\partial s^2} + \frac{\partial^2 M_z}{\partial z^2} + 2\frac{\partial^2 M_{sz}}{\partial s \partial z} =$$

$$\cong \frac{8}{15}E_c\frac{\delta_c h_c}{R_c^2} + \frac{E_c}{12}\frac{\partial^2}{\partial z^2}\left(h^3\frac{\partial^2 \delta_c}{\partial z^2}\right) - \frac{2}{3}E_c\frac{\partial}{\partial z}\left[h^3\frac{\partial}{\partial z}\left(\frac{\delta_c}{s_l^2}\right)\right]$$

(2.1.3.9) [23]

---

[22] The above is implicitly derived for the case of fixed (infinitely rigid: $E_c/E_r \cong 0$) abutments; for finite values of the ratio E_c/E_r , one can still apply the above formulation by adding to the "physical" half-lengths a "fictitious voussoir" length $\Delta s \cong \sqrt{(E_c/E_r)} \cdot h/2$.

[23] Note that in some versions of the Ritter method the torsional contrbution to equilibrium is neglected. In this case the last term of (2.1.3.9) is omitted. Such is the case of the Tölke method, in which, moreover, also influence of arch abutment constraints is neglected.

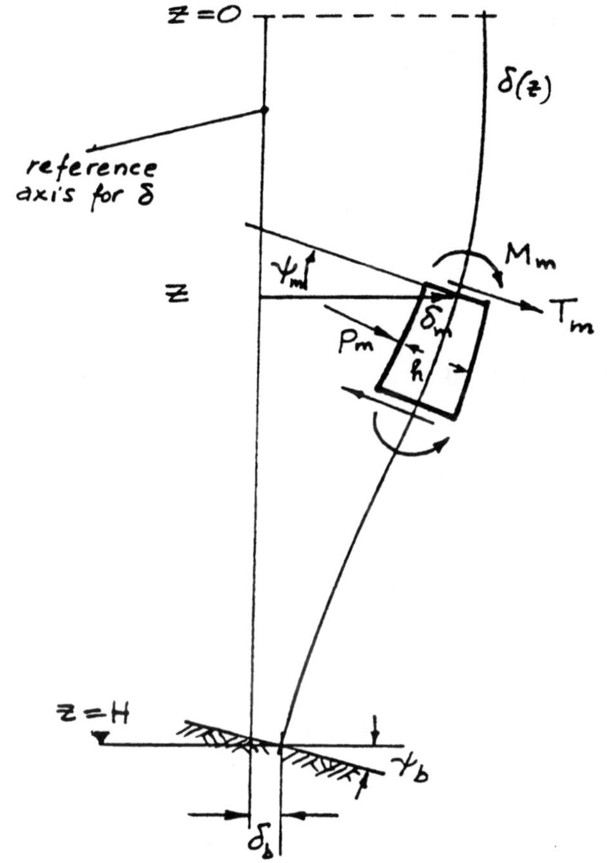

$$\psi_m = -\frac{\partial \delta_m}{\partial z}$$

$$M_m = -E_c J \frac{\partial \psi_m}{\partial z} = $$
$$= E_c J \frac{\partial^2 \delta_m}{\partial z^2}$$

$$T_m = \frac{\partial M_m}{\partial z}$$

$$P_m = \frac{\partial T_m}{\partial z} =$$
$$= \frac{\partial^2 M_m}{\partial z^2} =$$
$$= \frac{\partial^2}{\partial z^2}\left(E_c J \frac{\partial^2 \delta_m}{\partial z^2}\right)$$

$$\sigma_m = \frac{6}{h^2} M_m = E_c \frac{h}{2} \frac{\partial^2 \delta_m}{\partial z^2}$$

$$\tau_m = \frac{3}{2} \frac{T_m}{h}$$

(without torsional effects)

Fig. 39
The Ritter model: displacements, deformation, equilibrium of a cantilever section

under the restrictive simplifying assumptions mentioned in the development of the above discussion.

To the differential equation (2.1.3.9) above, one should add the B.C.'s at crest and at the base of the central cantilever.

At crest one has to impose the following two free-edge condtions:

i) M_z = bending moment $= 0$: since $M_z \sim \dfrac{\partial^2 \delta_c}{\partial z^2}$ this condition becomes:

$$\frac{\partial^2 \delta_c}{\partial z^2} = 0 \qquad \text{at crest ;} \qquad (2.1.3.10)$$

ii)- total "Kirchoff shear" $Q_{zx} = 0$:

$$Q_{zx} = \frac{\partial M_z}{\partial z} + \frac{\partial M_{sz}}{\partial s} = 0 \qquad \text{at crest ;} \qquad (2.1.3.11)$$

This last, making use of the previous results for M_z and $\dfrac{\partial M_{sz}}{\partial s}$, becomes:

$$\frac{E_c}{12}\frac{\partial}{\partial z}\left(h^3 \frac{\partial^2 \delta_c}{\partial z^2}\right) - \frac{1}{3}E_c h^3 \frac{\partial}{\partial z}\left(\frac{\delta_c}{s_l^2}\right) = 0 \quad \text{, or}$$

$$\frac{1}{4}\frac{\partial}{\partial z}\left(h^3 \frac{\partial^2 \delta_c}{\partial z^2}\right) - h^3 \frac{\partial}{\partial z}\left(\frac{\delta_c}{s_l^2}\right) = 0 \qquad \text{at crest} \qquad (2.1.3.12)$$

At the base of cantilever one should impose the two compatibility conditions between displacement and rotation of concrete and rock:

i) $$\delta_c(b) = C_{xx}\frac{Q_{zx}(b)}{E_r} + C_{\alpha x}\frac{M_z(b)}{E_r h_b} \text{ ,} \qquad (2.1.3.13)$$

ii) $$\left.\frac{\partial \delta_c}{\partial z}\right|_b = C_{\alpha x}\frac{Q_{zx}(b)}{E_r h_b} + C_{\alpha\alpha}\frac{M_z(b)}{E_r h_b^2} \text{ ,} \qquad (2.1.3.14)$$

i.e. equality of radial displacements and rotations (in the vertical plane) of the cantilever bottom section and of the underlying rocky foundation of Young modulus

E_r . The movements of the foundation are expressed in (2.1.3.13), (2.1.3.14) thanks to the so-called Vogt coefficients:

C_{xx} = radial displacements for $E_r =$ under unit shear per unit length (order of magnitude 2.6);

$C_{\alpha x}$ = radial displacements for $E_r =$ under unit bending moment per unit length, for unit thickness (ord. of m. 0.8);

$C_{x\alpha} = C_{\alpha x} =$ rotation in the vertical plane for $E_r =$ under unit shear per unit lenght, for unit thickness $h_b =$ (ord. of m. 0.8);

$C_{\alpha\alpha}$ = rotation in the vertical plane for $E_r =$ under unit bending moment per unit length, for unit thickness $h_b =$ (order of m. 5);

these non-dimensional coefficients have been derived by Vogt by assimilating the concrete-foundation contact surface to a uniformly loaded rectangular surface on an elastic half-space, thanks to Boussinesq'stheory. They are known functions of of the "aspect ratio" of the contact surface (or, better, of a suitably defined rectangle of the same area) and of the Poisson ratio, ν_r , of the rock (fig.40).

By using previously obtained results, the above base elastic constraint conditions become:

$$\text{i)}\quad \delta_c(b) = \frac{C_{xx}}{E_r}\frac{E_c}{12}\left[\frac{\partial}{\partial z}\left(h^3\frac{\partial^2\delta_c}{\partial z^2}\right) - 4h^3\frac{\partial}{\partial z}\left(\frac{\delta_c}{s_l^2}\right)\right] + \frac{C_{\alpha x}}{E_r}\frac{E_c}{12}h_b^2\left.\frac{\partial^2\delta_c}{\partial z^2}\right|_b \qquad (2.1.3.15)$$

$$\text{ii)}\quad -\left.\frac{\partial\delta_c}{\partial z}\right|_b = \frac{C_{\alpha x}}{E_r}\frac{E_c}{12h_b}\left[\frac{\partial}{\partial z}\left(h^3\frac{\partial^2\delta_c}{\partial z^2}\right) - 4h^3\frac{\partial}{\partial z}\left(\frac{\delta_c}{s_l^2}\right)\right] + \frac{C_{\alpha\alpha}}{E_r}\frac{E_c}{12}h_b\left.\frac{\partial^2\delta_c}{\partial z^2}\right|_b \qquad (2.1.3.16)$$ [24]

[24] If the torsional stiffness is neglected, the terms in (2.1.3.15), (2.1.3.16) containing derivative $\frac{\partial}{\partial z}\left(\frac{\delta_c}{s_l^2}\right)$ should be omitted. This, once again, leads to the Tölke approximation if one lets $\delta_c = -\frac{\bar{\sigma}_a}{E_c}R_c$ instead of $\delta_c = -\frac{15}{8}\frac{\bar{\sigma}_a}{E_c}R_c$, see (2.1.3.1)

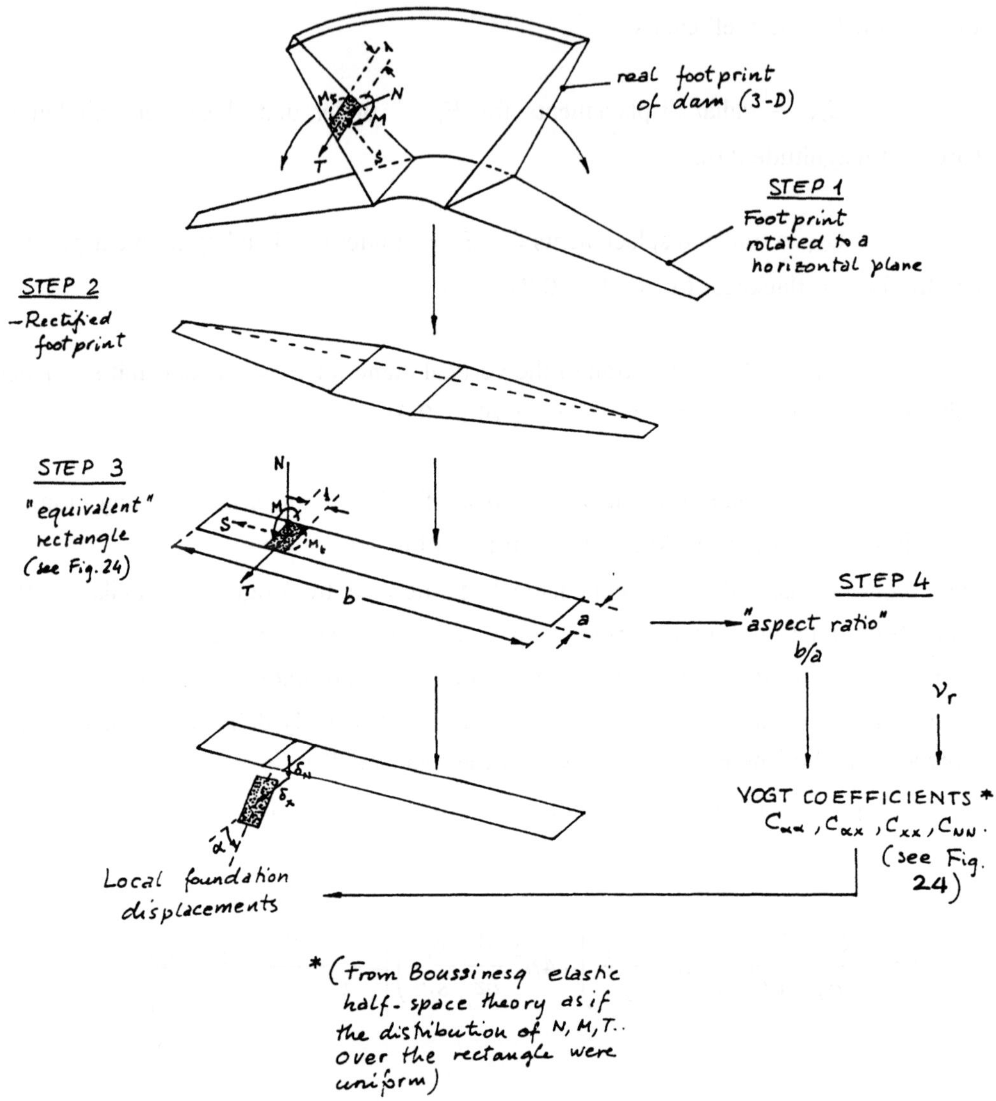

Fig. 40
The Ritter model: steps to represent foundation compliance

If one now substitutes the derivatives appearing in (2.1.1.8), (2.1.1.9), (2.1.3.17), (2.1.1.14), (2.1.1.15) with finite differences, one is reduced to an algebraic problem in which the only unknowns are given by the values of δ_c at the (equally spaced) chosen nodes along the z-axis.

The B.C.'s at crest and at the base necessitate the introduction of "fictitious nodes" above crest and below bottom. This can pose some numerical problems for the extrapolation of geometrical quantities, like h , R_c and s_l , which are physically defined only between base and crest.
However, at the end one is left with a linear system of as many equations as there are unknowns (the nodal values of δ_c). Once solved the system, application of (2.1.3.3), (2.1.3.5), (2.1.3.8), together with the obvious relations:

$$\sigma_a \cong \overline{\sigma}_a \pm \frac{h}{2} E_c \frac{\partial \varphi}{\partial s} \quad , \qquad \overline{\sigma}_a \cong -\frac{8}{15} E_c \frac{\delta_c}{R_c} \quad , \qquad (2.1.3.17)$$

$$\sigma_m \cong \pm \frac{6}{h^2} M_z = \pm \frac{h}{2} \frac{\partial^2 \delta_c}{\partial z^2} \quad , \qquad (2.1.3.18)$$

with the sign + valid for extrados and the sign - for intrados, allows to obtain the arch and cantilever stresses due to external pressures $p_e(z)$; for $\frac{\partial \varphi}{\partial s}$ see (2.1.3.5).
To these (in particular to the cantilever stresses) one should add the stresses given by vertical loads (dead weight) and by the thermal loads [25].
A drawback of the Ritter method is that it does not provide information about the bending state of cantilevers other than the central one, nor is it sensitive to local discontinuities (e.g. along the abutments).
However, good results can be obtained by judicious application of this method, as shown by comparison with more sophisticated methods or with experimental data (fig. 41 and 42). It can be extended to encompass also dynamic and non-linear problems; it is very economical in terms of SW development, number of variables and computer time, so that it is still used, especially for preliminary design.

---

[25] The treatment of thermal loads is similar, *mutatis mutandis*, to what already seen for the Tölke method. The vertical loads are usually taken as acting on the isolated cantilever, unless there are such pronounced overhangings that the free-standing cantilever would undergo excessive bending tensile stresses. In this last case particular tricks can be used (if one does not want to consider in an explicit way the vertical dispalcements).

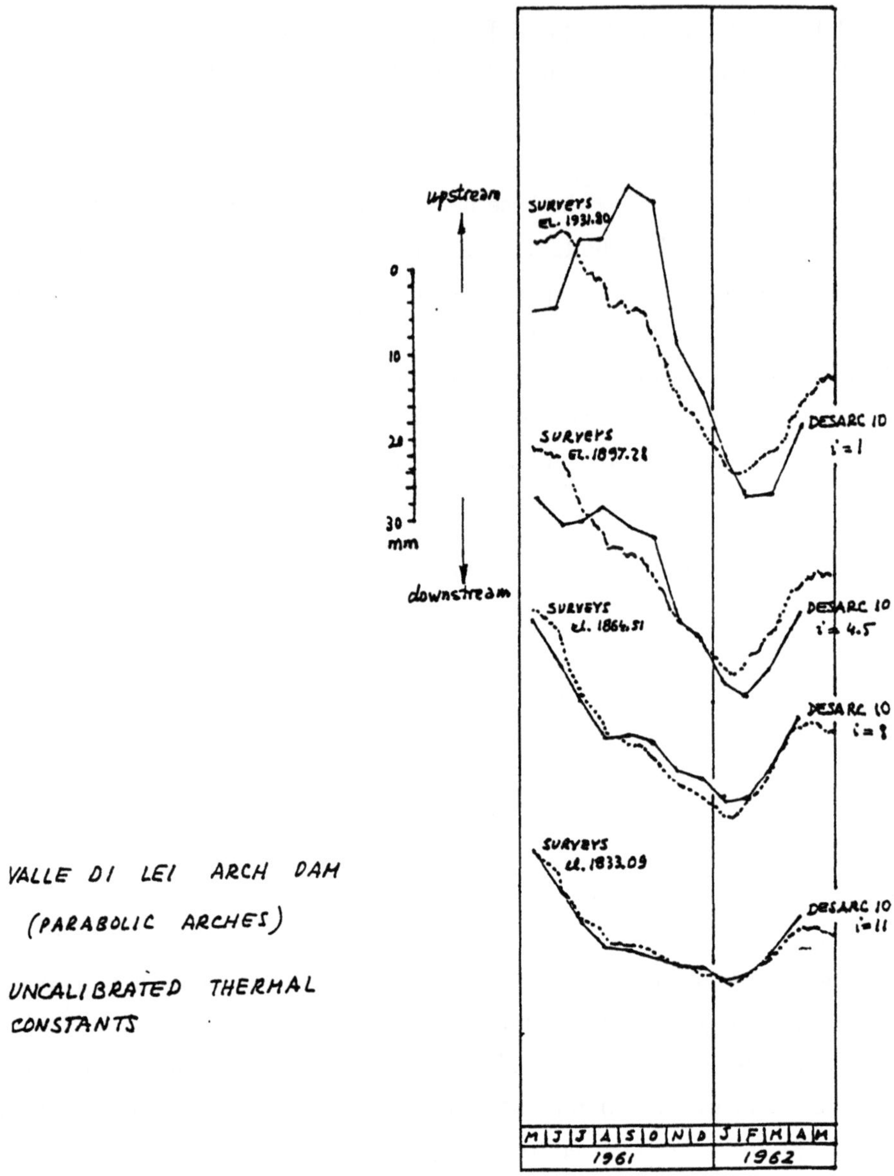

Fig. 41
Back-analysis of behaviour of Val di Lei dam with Ritter method

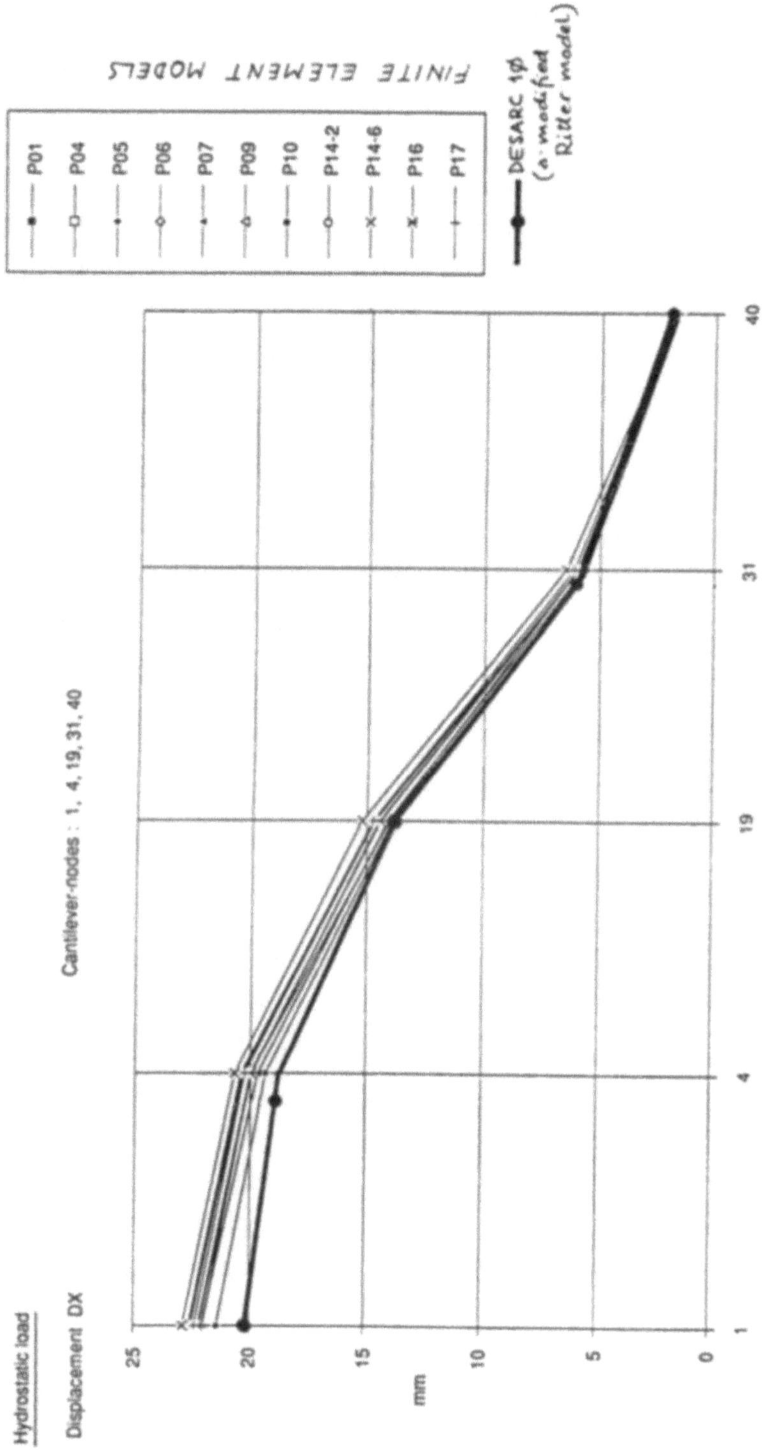

Fig. 42
The Ritter model: crown displacement of a circular arch dam as compared with results of different F.E. models

2.2 The "trial load method for analyzing arch dams" (T.L.M.)

2.2.1 Foreword

The simplifying, but restrictive assumptions (only one central cantilever, uniform load p_a on the generic arch) that lend the Ritter method its simplicity are relinquished in the T.L.M. in order to gain more complete information. Thus, a "grid" of many arches and many cantilevers is defined (fig.43), and compatibility of displacements, as well as equilibrium conditions, are imposed at each "node" of the grid. If the dam is geometrically and physically symmetric, and is subjected to symmetric loads, only half of it needs be analysed (by imposing the suitable symmetry conditions on displacement, and internal action, components at the nodes lying on the symmetry plane). In any other case, a complete grid needs be analysed, with all types of unknowns at each of the nodes.

Compatibility of displacements at the rock-concrete interface are expressed according to the technique of the already mentioned Vogt coefficents.

Compatibility conditions at the other grid nodes can be imposed at different levels of completeness of representation. In the simplest version of the method, only compatibility of radial displacements of arches and cantilevers is explicitly imposed; correspondingly, at each node we have this single compatibility equation and a single radial equlibrium equation, expressing that the volume element around a node is in equilibrium in the radial direction under the action of the external actions directly applied to it and of the internal actions transmitted to its vertical faces by the "arch" lying at the central elevation of the element, and to its horizontal faces by the "cantilever" passing through the element center.

In the more complete versions, more and more (n) compatibility equations between the components of displacement are written for each node, until the maximum number of $n = 6$ (for the three Cartesian components of the vector displacement of the element center and for the three orthogonal components of rotation); correspondingly, the same number of local equilibrium conditions must be expressed, always involving the directly applied external forces and the components of internal actions transmitted through the element faces.

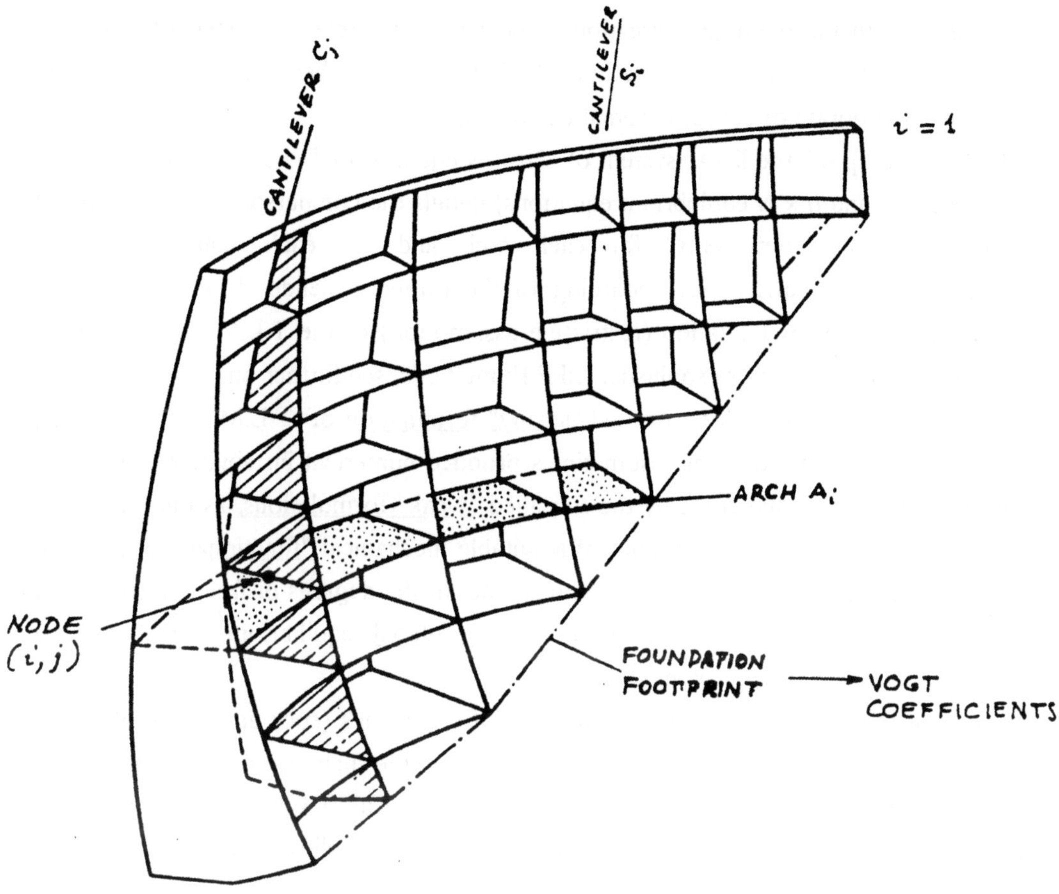

Fig. 43
The Trial Load Method: arch-cantilever grid.
(only half dam is shown for greater clarity)

These versions are known under the denomination of "n-components adjustments" after the fact that at the time of creation of the T.L.M. the relevant system of equations was solved by trial and error methods (although now, where used at all, T.L.M. versions incorporate of course automatic solvers).
The building up of the linear system of $2n \times N$ equations (if N are the nodes: $n \times N$ compatibility and $n \times N$ equilibrium equations) entails the cumbersome computation of many "influence coefficients" for each arch and for each cantilever. These computations are carried out according to the ordinary slender beam theory, and therein lies the major limitation (restrictive assumption) of the T.L.M., which can be overcome only by the more sophisticated F.E. methods, see further on.
It must also be said that this so-called "strong" formulation of T.L.M., imposing, i.e., equilibrium at every node, may sometimes produce numerical instabilities. This quite serious drawback is overcome by the so-called "weak" formulations, as the F.E. ones, in which the conditions of minimum of a suitable mechanical "functional" (usually the total potential energy) are expressed with respect to the degrees of freedom (d.o.f.) of the problem. This approach, while preserving global equilibrium, satisifies local equilibrium only in an approximate way.
In spite of these limitations, and of a number of other ones (such as the use of Vogt coefficients to represent the foundation compliance, the inability to accurately model local inhomogeneities or constructional details, etc.), the T.L.M. is still sometimes used, mainly where there is a large past experience in its application and interpretation of results. It is even mentioned or recommended in some Official Standards, for example in those of the People's Republic of China, as a reference analysis method for arch dams.
Many important structures (starting from the Boulder Canyon Project dam at the end of the 30's) were designed by using the T.L.M., and their subsequent in-service behaviour has proved to be completely satisfactory, so that there is a solid base of confidence in the general reliability of the method [26].
In the following we will briefly present the basis of the T.L.M., with special reference to the more simplified versions.

[26] The method was also "validated" by comparison of its results with in-situ measurements carried out on the experimental arch dam of Stevenson Creek in the 20's.

2.2.2 An outline of the T.L.M.

a) The radial adjustment

Let us define m cantilevers $C_j (1 \leq j \leq m)$, n arches $A_i (1 \leq i \leq n)$[27]. The "nodes" (i,j) will be the meeting point of the axis of arch i and of cantilever j.
If only the hydrostatic load is considered, at node (i,j) the external pressure $p_{ij} = \gamma_a z_{ij}$ will apply, with z_{ij} = depth of the node (i,j) under the free surface.
Let us now suppose, for simplicity sake, that only a radial adjustment is the aim of the exercise; the cantilever C_j will be subjected at its nodes $(1,j)$, $(2,j)$, ... (b_j,j)[28] to the unknown partial pressures:

$$p_m(1,j)\,;\ p_m(2,j)\,;.......p_m(b_j,j) \qquad (2.2.2.1)$$

and the arch A_i will be subjected at its nodes (i,S_i), (i,S_{i+1}), ... ,(i,D_i)[29] to the unknown partial pressures:

$$p_a(i,S_i);\ p_a(i,\ S_i+1);......\ p_a\ (i,\ D_i-1);\ p_a\ (i,\ D_i). \qquad (2.2.2.2)$$

Now let us compute, by the ordinary slender beam theory, the influence coefficients for each cantilever C_j:
Δ_j (i=k, i=l) radial displacement produced at node (k,l) by a "unit load" L^c_{ij} applied at node (l,j) see figure 44;

the $\quad \Delta_j$ (i=k, i=l) are known quantities; (2.2.2.3)

for each cantilever C_j (of unit thickness) the distribution of partial pressures p_m along its height can be decomposed into a linear combination of "unit loads", see fig. 45:

$$p_m = p_m(1,j)L^c_{1j} + p_m(2,j)L^c_{2j} +........+ p_m(b_j,j)L^c_{b_j,j} \qquad (2.2.2.4)$$

and the Δ_j = (k, l) will form a nearly symmetric matrix of known quantities.

---

27 Arch i=1 is supposed in the following to be at crest. The vertical distances $\Delta z = z_{i+1} - z_i$ are also supposed to be constant for simplicity.

28 b_j. being the lowermost arch meeting cantilever C_j.

29 S_i being the index of cantilever C_{Si} at the left abutment of arch A_i , and likewise D_i being the index ($j = D_i$) of cantilever C_{Di} at the right abutment of arch A_i (numbering of cantilevers is progressive from left to right abutment).

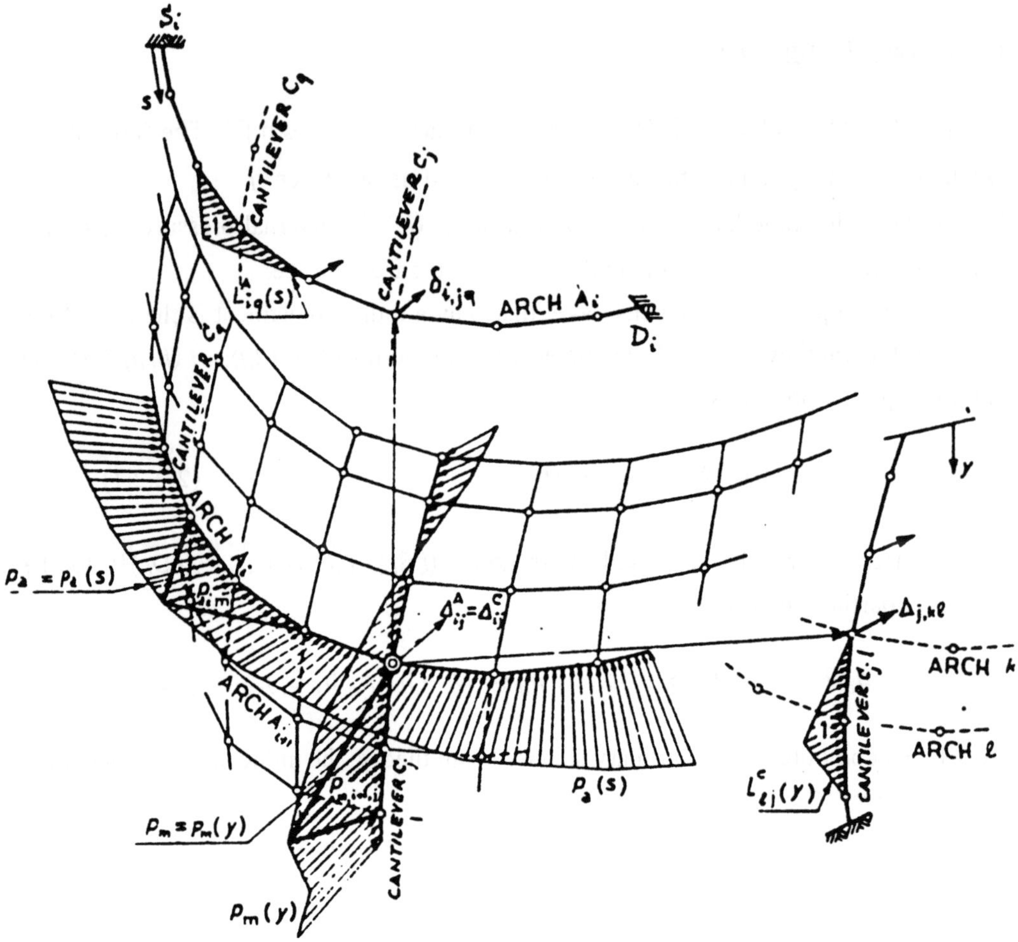

Fig. 44
T.L.M.: loads on arches and cantilevers; influence coefficients

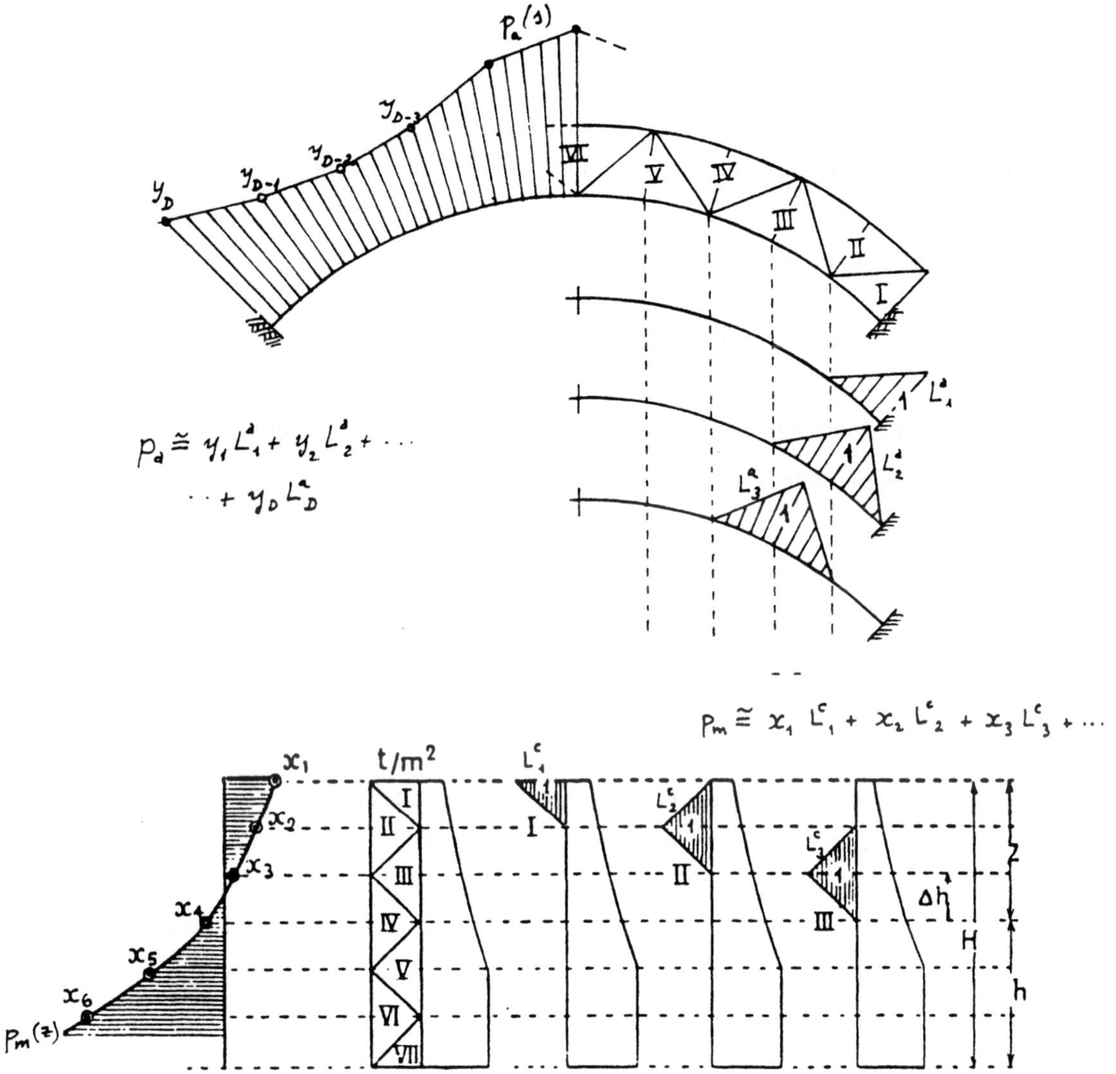

Fig. 45
The T.L.M. model:
radial arch & cantilever loads as linear combination of unit loads

Likewise, let us compute, by ordinary slender beam theory, the influence coefficients for each arch A_j:

$\delta_i\ (j=p, j=q)$ = radial displacement caused at node (i,p) by a "unit load" $L^A{}_{iq}$ applied at node (i,q) see figure 45. (2.2.2.5)

For each arch A_j of unit thickness, the distribution of partial pressures p_a along its upstream face can be decomposed into a linear combination of "unit loads" [30]:

$$p_a = p_a(i,\ S_i)L^A{}_{i,S_i} + p_a(i,\ S_i+1)L^A{}_{i,S_i+1} + \ldots \\ \ldots\ldots + p_a(i,\ D_i-1)L^A{}_{i,D_i-1} + p_a(i,\ D_i)L^A{}_{i,D_i} \quad (2.2.2.6)$$

and the $\delta_i(p,q)$ will form a nearly symmetric matrix of known quantities.

Now at the generic node (i,j) the following two equations must hold:

$$p_a(i,j) + p_m(i,j) = p_{ij} = \gamma_a z_{ij} + \ldots; \quad (2.2.2.7)$$

$$\sum_{l=1}^{l=b_j} \Delta_j\ (i,l)\ p_m(l,j) = \sum_{q=S_i}^{q=D_i} \delta_i\ (j,q)\ p_a(i,q) \quad (2.2.2.8)$$

Eqs. (2.2.2.7 et 8) form a linear system in the unknowns $p_a(i,j)$, $p_m(i,j)$ and can be solved without problems (barring numerical instabilities, see above).

The boundary conditions are implicitly satisfied if taken into account in the computation of the influence coefficients $\Delta_j\ (k,\ l)$, $\delta_i\ (p,\ q)$, e.g. by using the Vogt coefficients for the isolated cantilever C_j and for the isolated arch A_i, respectively. (In principle, more complicated conditions should be used to take into account the inclination of the abutments).

Once the partial pressures $p_a(i,j)$, $p_m(i,j)$ determined, the static régime of the cantilevers C_j and of the arches A_i can be solved by ordinary beam theory, applying the pressure distribution (2.2.2.4), (2.2.2.6) respectively; vertical and horizontal normal stresses, as well as shear stresses on surfaces normal to the facings, can then be determined.

30 The triangular distribution of unit loads in the T.L.M. is the first systematic application of the idea of "shape functions", see F.E.M.s

b) More complete adjustments

It will be observed, on one hand, that the consideration of external forces other than normally applied surface tractions (= pressures) is not possible in the preceding formulation without particular tricks.
[By contrast, consideration of thermal actions would be relatively easy; if $\Delta_T^{(i,j)}$ is the thermal displacement produced on the isolated cantilever C_j, and $\delta_T^{(i,j)}$ the one produced on the isolated arch A_i ,at node (i,j), eqs. (2.2.2.7 et 8) change into:

$$p_a(i,j) + p_m(i,j) = 0 \qquad (2.2.2.9)$$

$$\sum_{l=1}^{l=b_j} \Delta_j\,(i,l)\,p_m(l,j) + \Delta_T^{(i,j)} = \sum_{q=S_i}^{q=D_i} \delta_i\,(j,q)\,p_a(i,q) + \delta_T^{(i,j)} \qquad (2.2.2.10)$$

and the static régime produced in the cantilevers by pressures p_m, in the arches by pressures p_a, will have to be added to the thermal stresses caused in the isolated members].
On the other hand, it will be observed that in the isolated arches subjected to pressures not only radial displacements, but also tangential ones will be produced. In order to preserve the continuity of material, then, the cantilevers will have to deflect not only in their own (radial) planes, but also sideways (normally to the said planes) (fig.46). Thus mutual tangential actions will have to arise. Similar considerations can be made for vertical displacements and for rotations.

c) Concluding remarks

By way of conclusion, on the T.L.M. in general one can observe the following.
The scheme of interconnected plane structures (arches A_i and cantilevers C_j) is quite appealing to intuition and gives an engineering insight into the distribution of loads which undoubtely takes place in an arch dam. As such, it has great didactic and historic value.

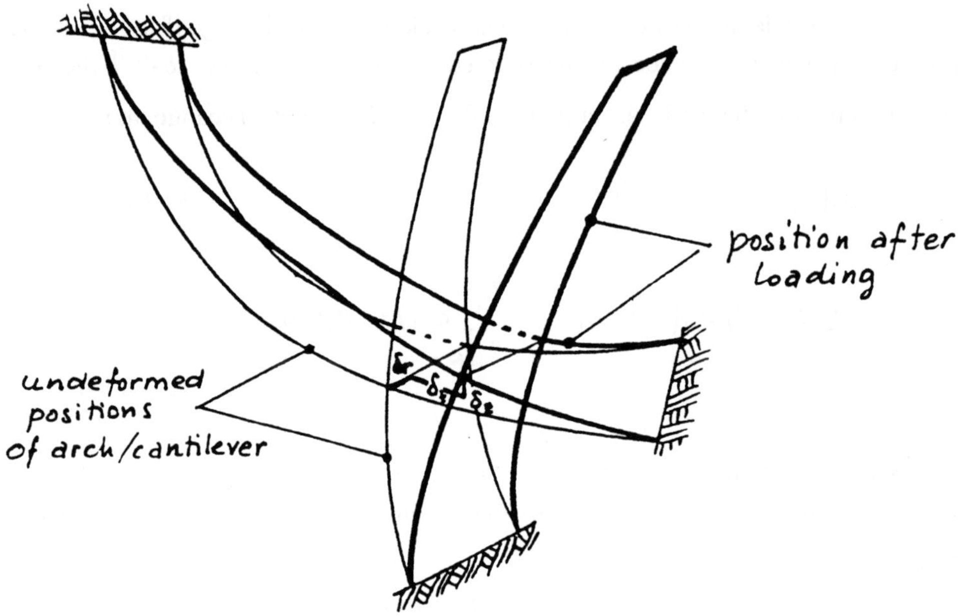

Fig. 46
The T.L.M. model: out-of-plane displacements of arch & cantilever.
Necessity of n-adjustements with n>1

However, it can be shown by simple considerations that "something" of the real behaviour is lost by adopting this grid-like scheme, at least if one sticks to the simpler scheme of <u>radial</u> adjustment.

By reference (fig.47) to a prism surrounding one of the grid crossing points (nodes) one can write, according to the 3-D eqs. of the Theory of Elasticity:

$$\frac{\partial \sigma_x}{\partial x} + \frac{\partial \tau_{xy}}{\partial y} + \frac{\partial \tau_{xz}}{\partial z} = 0$$

$$\frac{\partial \tau_{yx}}{\partial x} + \frac{\partial \sigma_y}{\partial y} + \frac{\partial \tau_{yz}}{\partial z} = 0 \qquad \text{where} \qquad \begin{cases} \tau_{xy} = \tau_{yx} \\ \tau_{xz} = \tau_{zx} \\ \tau_{yz} = \tau_{zy} \end{cases}$$

$$\frac{\partial \tau_{zx}}{\partial x} + \frac{\partial \tau_{zy}}{\partial y} + \frac{\partial \sigma_z}{\partial z} = \gamma_c \ , \qquad (2.2.2.11)$$

and by integrating with respect to y over the thickness h, i.e. from $y = -\frac{h}{2}$ to $y = +\frac{h}{2}$ keeping into account that:

$$\int_{-\frac{h}{2}}^{\frac{h}{2}} \frac{\partial \tau_{xy}}{\partial y} dy = \tau_{xy}\left(\frac{h}{2}\right) - \tau_{xy}\left(-\frac{h}{2}\right) = 0$$

$$\int_{-\frac{h}{2}}^{\frac{h}{2}} \frac{\partial \sigma_y}{\partial y} dy = \sigma_y\left(\frac{h}{2}\right) - \sigma_y\left(-\frac{h}{2}\right) = p_{ij} \ ,$$

$$\int_{-\frac{h}{2}}^{\frac{h}{2}} \frac{\partial \tau_{zy}}{\partial y} dy = \tau_{zy}\left(\frac{h}{2}\right) - \tau_{zy}\left(-\frac{h}{2}\right) = 0, \qquad (2.2.2.12)$$

eqs. (2.2.2.11) yield:

$$\int_{-\frac{h}{2}}^{\frac{h}{2}} \left(\frac{\partial \sigma_x}{\partial x} + \frac{\partial \tau_{xz}}{\partial z}\right) dy = 0,$$

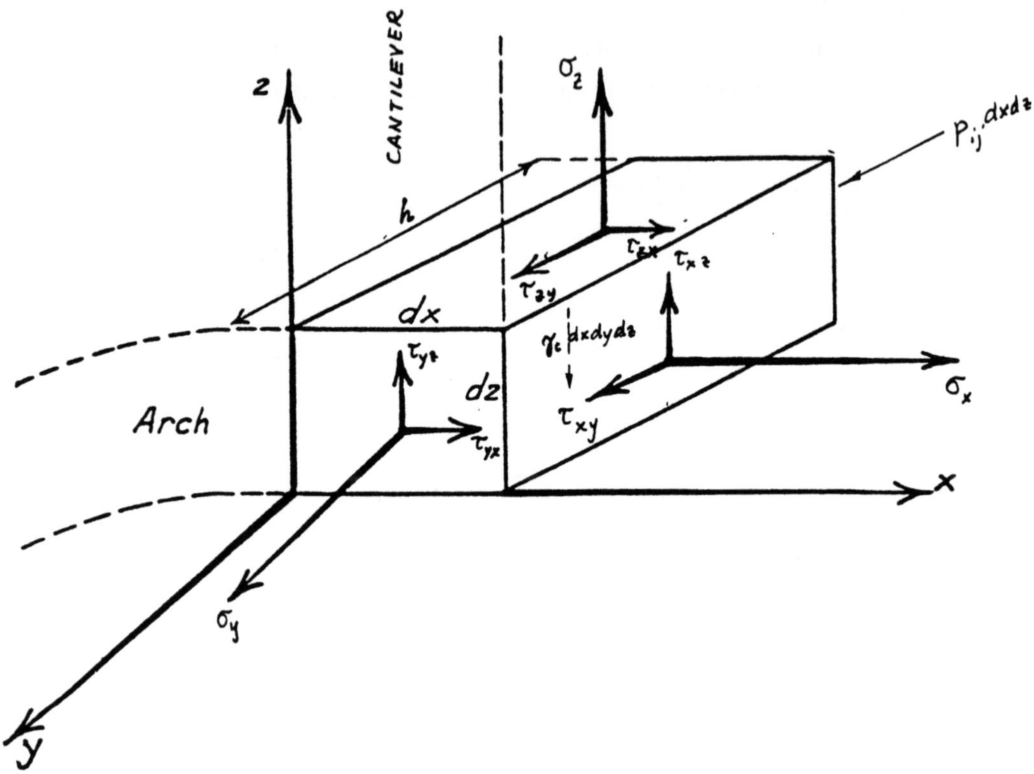

Fig. 47
T.L.M.: stresses, body forces, surface forces around node (i,j)

$$\int_{-\frac{h}{2}}^{\frac{h}{2}}\left(\frac{\partial\tau_{yz}}{\partial x}+\frac{\partial\tau_{yz}}{\partial z}\right)dy=-p_{ij},$$

$$\int_{-\frac{h}{2}}^{\frac{h}{2}}\left(\frac{\partial\tau_{zx}}{\partial x}+\frac{\partial\sigma_{z}}{\partial z}\right)dy=\gamma_c h. \tag{2.2.2.13}$$

By contrast, the assumption of "plane" behaviour of arches under pressures p_a would entail:

$$\sigma_z=\tau_{zx}=\tau_{zy}=0$$

the eqs. (2.2.2.13), keeping into account (2.2.2.12), then reduce to:

$$\int_{-\frac{h}{2}}^{\frac{h}{2}}\frac{\partial\sigma_x}{\partial x}dy=0,$$

$$\int_{-\frac{h}{2}}^{\frac{h}{2}}\frac{\partial\tau_{yx}}{\partial x}dy=-p_a(i,j); \tag{2.2.2.14}$$

similarly, the assumption of "plane" behaviour of cantilevers under pressures $p_m\,(i,j)$ would entail:

$$\sigma_x=\tau_{xz}=\tau_{xy}=0$$

eqs. (2.2.2.13) then would yield:

$$\int_{-\frac{h}{2}}^{\frac{h}{2}}\frac{\partial\tau_{yz}}{\partial z}dy=-p_m(i,j),\quad \text{where } p_m\,(i,j)=p_{ij}-p_a\,(i,j);$$

$$\int_{-\frac{h}{2}}^{\frac{h}{2}}\frac{\partial\sigma_z}{\partial z}dy=\gamma_c h. \tag{2.2.2.15}$$

By adding the second eq. of (2.2.2.14) to the first eq. of (2.2.2.15) one gets

$$\int_{-\frac{h}{2}}^{\frac{h}{2}}\left(\frac{\partial\tau_{yx}}{\partial x}+\frac{\partial\tau_{yz}}{\partial z}\right)dy=-\left[p_a(i,j)+p_m(i,j)\right]=-p_{ij} \tag{2.2.2.16}$$

which tallies with the second eq. of (2.2.2.13) derived from 3-D Theory of Elasticity; but the first eqs. of (2.2.2.13) and of (2.2.2.14) would require the following condition:

$$\int_{-\frac{h}{2}}^{\frac{h}{2}} \frac{\partial \tau_{xz}}{\partial z} dy = 0, \qquad (2.2.2.17)$$

while the third eq. of (2.2.2.13) and the second eq. of (2.2.2.15) would require the condition:

$$\int_{-\frac{h}{2}}^{\frac{h}{2}} \frac{\partial \tau_{zx}}{\partial x} dy = 0. \qquad (2.2.2.18)$$

Eq. (2.2.2.17), (2.2.2.18) express that the thickness-averaged value of the derivatives of τ_{xz} with respect to the local tangents to the shell mean surface be zero.
This hypothesis can generally be accepted - but only in an approximated way - in the case of a symmetric dam with no abrupt slope discontinuities at abutments (fig.48).
More "complete" adjustments can overcome this theoretical difficulty, but at a steep price in complication of numerical treatment and with the residual drawbacks already alluded to. (In particular the Vogt model for foundation compliance is unsatisfactory from a theoretical point of view and has little flexibility to represent inhomogeneities).
It is on the basis of these theoretical and practical shortcomings that the more modern F.E.M.s were evolved.

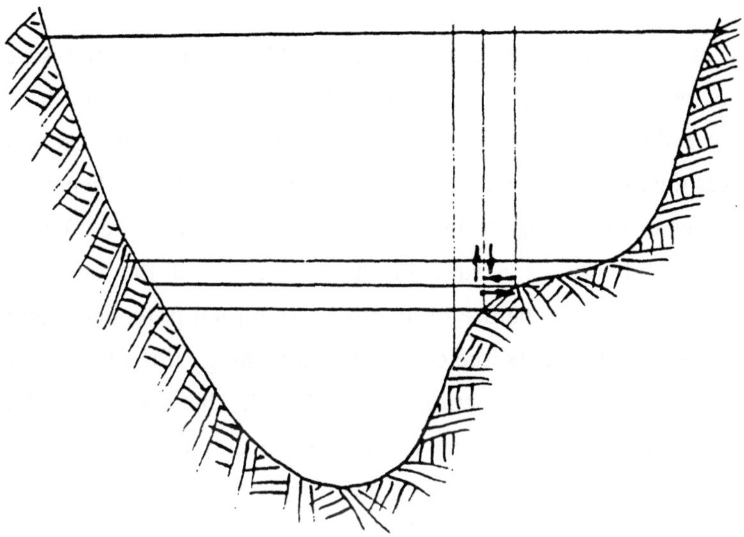

Fig. 48
Radial-adjustement T.L.M. is inadequate for abrupt slope changes at abutments

3. THE FINITE ELEMENT METHOD OF ANALYSIS

3.1 Introductory remarks

As already said in a passing remark, the F.E.M. belongs to the so-called "weak formulation" models of the problem of elastic equilibrium of a solid body. Instead of trying to solve directly the partial differential equations of the Theory of Elasticity, which express the local "proximity relationships" between the gradients of the variables (stresses, strains or displacements) one takes, as it were, a step back and tries to find the extremum conditions of a suitable "functional" [31] of the same variables. This functional has generally the nature of an energy (the total potential energy if the basic variables are the displacements, the complementary energy if the basic variables are the stresses, or the Airy stress functions) (fig. 49). If the classical differential Euler conditions for the extremum of the functional are derived, one comes back to the partial differential equations of the elastic equilibrium, so nothing would be gained this way. But, if the space distribution of basic variables -e.g. the displacements- are approximated as a superposition (linear combination) of a finite number of "suitable" elementary distributions, then the corresponding approximate expression of the functional becomes an algebraic quadratic function of the coefficients of the linear combination (the "degrees of freedom" of the model). Then the extremum conditions of the approximated functional reduce to the requirement that the partial derivatives of this approximated functional with respect to all these degrees of freedom be zero. These conditions will obviously be a finite number of linear equations having as unknowns the degrees of freedom[32].

Thus explained, the central idea of the numerical approximation methods based on the "weak formulation" seems very open to criticism. However, closer reflection, as well as practical results, show that this is not the case.

i)First of all, if one considers the elementary case of a discrete, single degree of freedom [33] system, the two approaches yield the same (exact) results. Take the case of

[31] A "functional" is a space integral (extended over the whole domain) of the problem variables and of their derivatives up to a certain order. As such, it depends on the infinitely main values taken by these variables in all points of the domain.

[32] Classical applications of this approach are the so-called Ritz-Rayleigh approximation methods, widely used e.g. in static and dynamic problems of beam and plate equilibrium.

[33] Or a discrete, finite-number of d.o.f. system.

$\delta = (u, v, w)$ = displacement field from which:

$$\varepsilon_x = \frac{\partial u}{\partial x}, \quad \varepsilon_y = \frac{\partial v}{\partial y}, \quad \varepsilon_z = \frac{\partial w}{\partial z},$$

$$\gamma_{yz} = \frac{\partial v}{\partial z} + \frac{\partial w}{\partial y},$$

$$\gamma_{zx} = \frac{\partial w}{\partial x} + \frac{\partial u}{\partial z},$$

$$\gamma_{xy} = \frac{\partial u}{\partial y} + \frac{\partial v}{\partial x};$$

P_0 = initial position of generic point

P_1 = displaced position at equilibrium

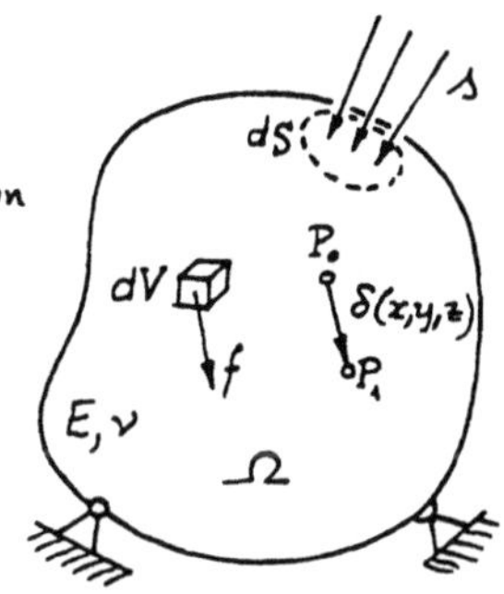

$$\sigma_x = E\left(\frac{\varepsilon_x}{1+\nu} + \frac{\nu(\varepsilon_x+\varepsilon_y+\varepsilon_z)}{1-\nu-2\nu^2}\right); \quad \tau_{yz} = \frac{E}{2(1+\nu)}\gamma_{yz};$$

$$\sigma_y = E\left(\frac{\varepsilon_y}{1+\nu} + \frac{\nu(\varepsilon_x+\varepsilon_y+\varepsilon_z)}{1-\nu-2\nu^2}\right); \quad \tau_{zx} = \frac{E}{2(1+\nu)}\gamma_{zx};$$

$$\sigma_z = E\left(\frac{\varepsilon_z}{1+\nu} + \frac{\nu(\varepsilon_x+\varepsilon_y+\varepsilon_z)}{1-\nu-2\nu^2}\right); \quad \tau_{xy} = \frac{E}{2(1+\nu)}\gamma_{xy}.$$

Total potential energy:

$$\mathcal{E}_{pt} = -\iint_S \vec{s} \times \vec{\delta}\, dS - \iiint_\Omega \vec{f} \times \vec{\delta}\, dV + \frac{1}{2}\iiint_\Omega (\sigma_x\varepsilon_x + \sigma_y\varepsilon_y + \sigma_z\varepsilon_z + \tau_{yz}\gamma_{yz} + \tau_{zx}\gamma_{zx} + \tau_{xy}\gamma_{xy})\, dV =$$

= a functional of the displacement field $\delta(x,y,z)$.

$\mathcal{E}_{pt}$ is minimum for the equilibrium configuration compatible with the constraints.

Fig. 49
The principle of min. total potential energy for a solid elastic body

a spring (elastic constant κ) statically loaded by a force F (fig. 50). The equation of equilibrium is

a) $F = \kappa x$,

x being the displacement from rest position of the loaded end of the spring.
In this case the total potential energy of the system is

b) $\mathcal{E} = -Fx + \frac{\kappa x^2}{2}$, and the condition of minimum of $\mathcal{E}$ with respect to x reduces to:

c) $\frac{d\mathcal{E}}{dx} = -F + \kappa x = 0$, which is identical to a).

ii) In the case of a continuum, it can be shown that one would also get the exact solution if a suitable linear combination of the chosen approximation functions is able to fit exactly the actual (unknown) spatial pattern of the problem variables. (See the classical Ritz-Rayleigh methods for vibration problems of beams and plates, fig. 51).
Of course, since the actual distribution is unknown, one cannot be sure "a priori" of having chosen the "right" approximation functions; but experience, analogy and intuition can give reasonable guidance in choosing approximation functions such that a linear combination thereof will give a "close enough" fit to the real distribution.
iii) So, even if one is in the dark about the qualitative "shape" of the actual solution, intuition suggests that, by choosing approximation functions that are in a "sufficient number" and are also sufficiently "indifferent" with respect to the shape of the actual solution, the weak formulation will yield coefficients of combination such that the corresponding linear combination of the approximation functions will come "close enough" to the actual solution.
iv) In any case, the approximate version of the weak formulation guarantees global equilibrium of the system. Local equilibrium is not generally assured (if the variables are displacements [34]), and this could be seen, again, as a serious drawback of the weak formulation approach. But, one may ask, is it really so important to demand that local equilibrium be assured at all points of the continuum?

[34] In this case local compatibility is usually automatically assured. (If the variables are stresses or stress functions, local equilibrium is generally assured but local compatibility conditions can be violated).

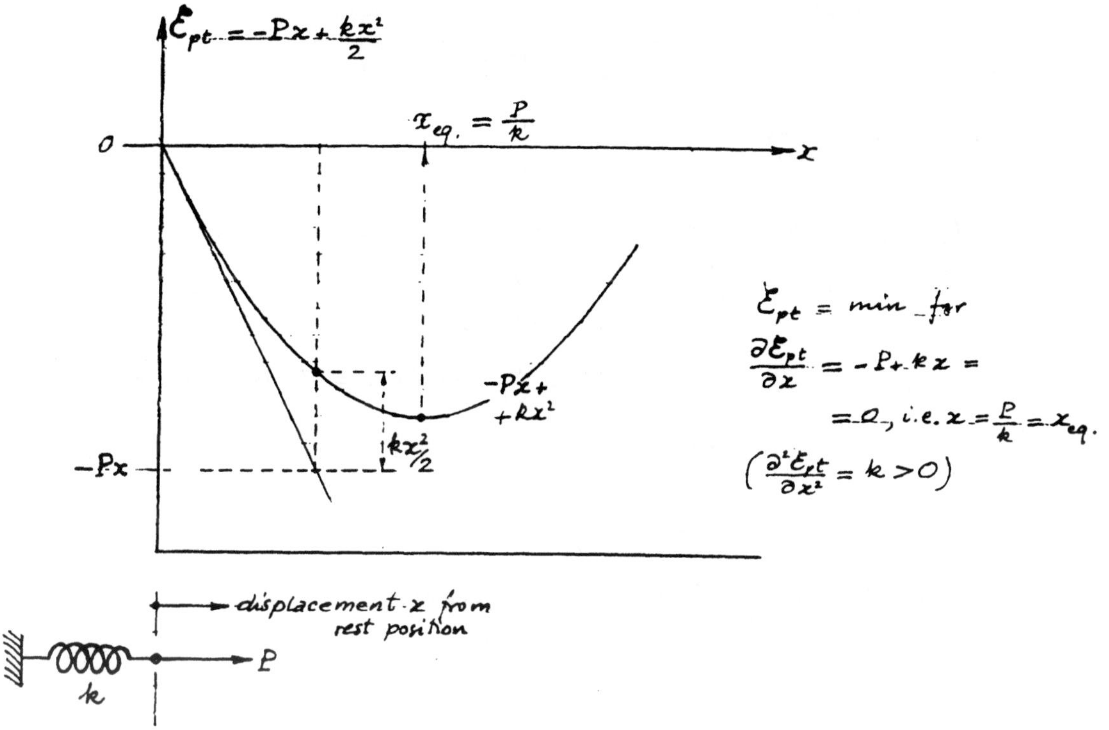

Fig. 50
The principle of min. of total potential energy: 1 d.a.f. system

PROBLEM:

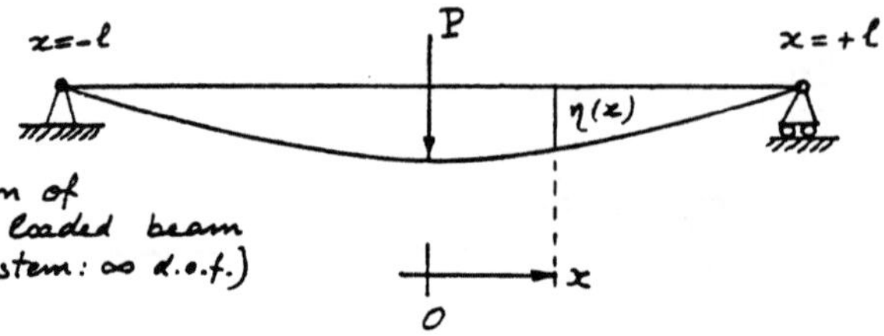

Equilibrium of
a centrally loaded beam
(continuous system: ∞ d.o.f.)

Ritz-Rayleigh method: with the A_{2n+1} as unknowns one assumes:

$$\eta \cong \sum_{n=0}^{j} A_{2n+1} \cos\frac{(2n+1)\pi x}{2\ell}$$

– Note that η complies with the constraints:

$$\eta(x=-\ell) = \eta(x=+\ell) = 0;$$

$$\frac{\partial^2\eta}{\partial x^2} = \frac{M}{EJ} = 0 \text{ for } x=-\ell \text{ and for } x=+\ell.$$

$$\mathcal{E}_{pt} = -P\cdot\eta(0) + \frac{EJ}{2}\int_{-\ell}^{+\ell}\left(\frac{\partial^2\eta}{\partial x^2}\right)^2 dx \cong -P\cdot\sum_m A_m + \frac{EJ}{2}\int_{-\ell}^{+\ell}\left(\sum_m \frac{m^2\pi^2}{4\ell^2} A_m \cos\frac{m\pi x}{2\ell}\right)^2 dx =$$

$$= -P\sum_m A_m + \frac{EJ}{2}\sum_m \frac{m^4\pi^4}{16\ell^4} A_m^2 \int_{-\ell}^{+\ell}\cos^2\frac{m\pi x}{2\ell}\, d\ell = -P\cdot\sum_m A_m + \frac{EJ}{2}\sum_m \frac{m^4\pi^4}{16\ell^4} A_m^2 \times \frac{1}{2} \times 2\ell =$$

$$= -P\sum_m A_m + EJ\frac{\pi^4}{32\ell^3}\sum_m m^4 A_m \qquad (m = 2n+1).$$

Minimum condition for $\mathcal{E}_{pt}$:

$$\frac{\partial\mathcal{E}_{pt}}{\partial A_m} = -P + \frac{EJ\pi^4}{16\ell^3} m^4 A_m = 0 \quad \text{or} \quad A_m = \frac{16P\ell^3}{m^4\pi^4 EJ};$$

$$\eta \cong \sum_{n=0}^{j} \frac{16P\ell^3}{(2n+1)^4\pi^4 EJ}\cos\frac{(2n+1)\pi x}{2\ell} \text{ from which } M(\eta) = EJ\frac{\partial^2\eta}{\partial x^2} \ldots$$

Fig. 51
The Ritz-Rayleigh method:
a functional is approximated by an algebraic function of a finite number of d.o.f.

Engineering practice suggests that in many cases this mathematically rigorous approach would, indeed, define an "ideal" solution, but this accuracy is not generally corresponding to the physical reality. Local inhomogeneities, or plastic adaptations, or micro-cracking, etc. etc., can make the local behaviour of the real thing depart quite appreciably from "exact" solution. What interests the engineer is rather the general picture of the strain/stress distribution and, from that, an indication (more qualitative, in many cases, than quantitative) of the zones where some critical condition may arise. These indications can usually be obtained in a reliable way by the approximated methods we are now discussing.
A big avantage of the F.E.M. is that all the operations involving the geometry of the contimuum, its mechanical properties, the loads etc. are carried out (thanks (fig. 52) to the definition of elementary "shape functions", see further on) preliminarly to, and independently of, the solution of the problem. Derivations (of a lower order than that involved in the partial differential equations of the Theory of Elasticity) are carried out on the shape functions, which are predefined; the integrations that appear in the mathematical expression of the "functional" are carried out element by element, also on the shape functions and their derivatives; at last, the "stiffness matrices" and "load vectors" thus obtained for each element are combined according to the topology of the element network (and to the space orientation of every single element). All along, matrix algebra is used, taking full advantage of its great power of synthetic expression. The geometric-physical background on which the elastic equilibrium is to take place is thus laid out and described in detail, giving a matrix representation of the necessary relationship between the elastic reactivity (stiffness) of the continuum, as "seen" from the nodes of the discretization, and the Lagrange-type generalized "nodal forces" (fig. 53). Great simplicity of formulation and of numerical operations is thus in principle achieved.
Of course, there is a price to pay. The first price consists in being prepared to build up, and solve, a system of many equations (hundreds, thousands or even tens of thousands of equations). Thanks to the great strides made in the techniques of numerical computation, in the power of HW and in the SW languages, this price is now not an insurmountable problem. There are other prices as well. One of them relates to the fact that, once solved the system equations, usually in terms of the nodal displacements, the next step is to obtain the stresses; these turn out to be mesh-dependent and, in any case, their accuracy is lower than that of the displacements. Nodal average of stresses

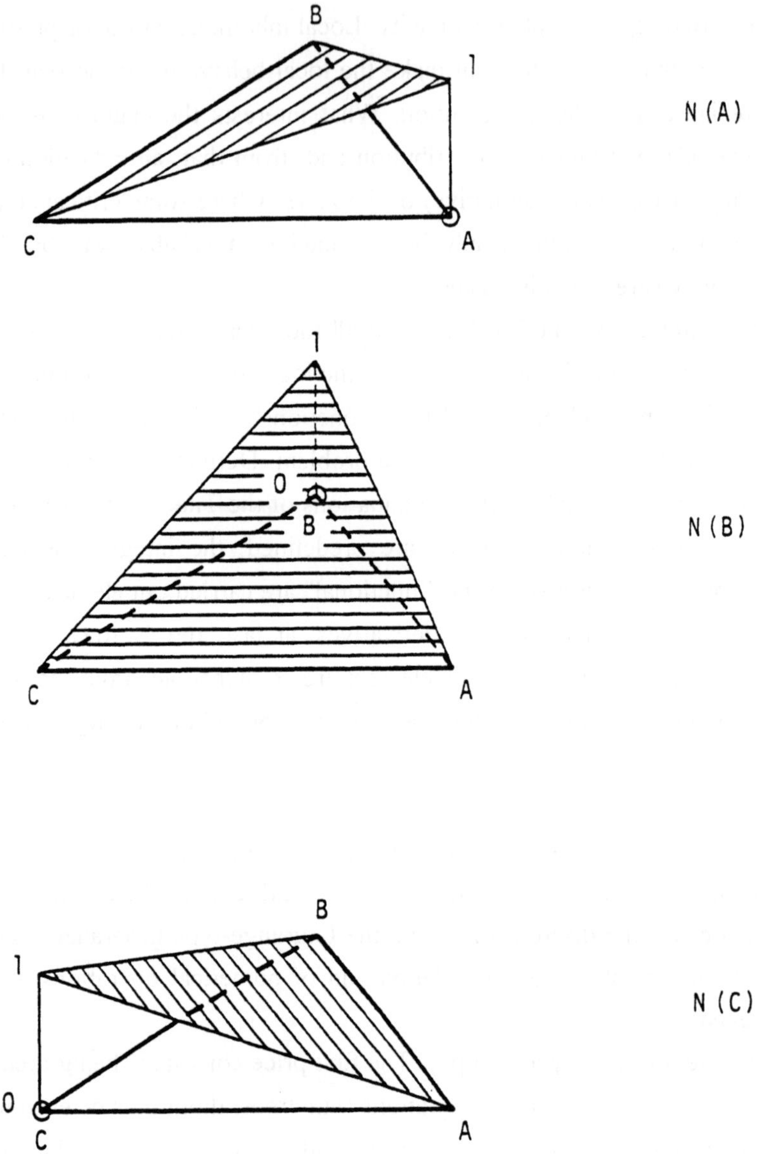

Fig. 52
Linear shape functions for 2-D elem.

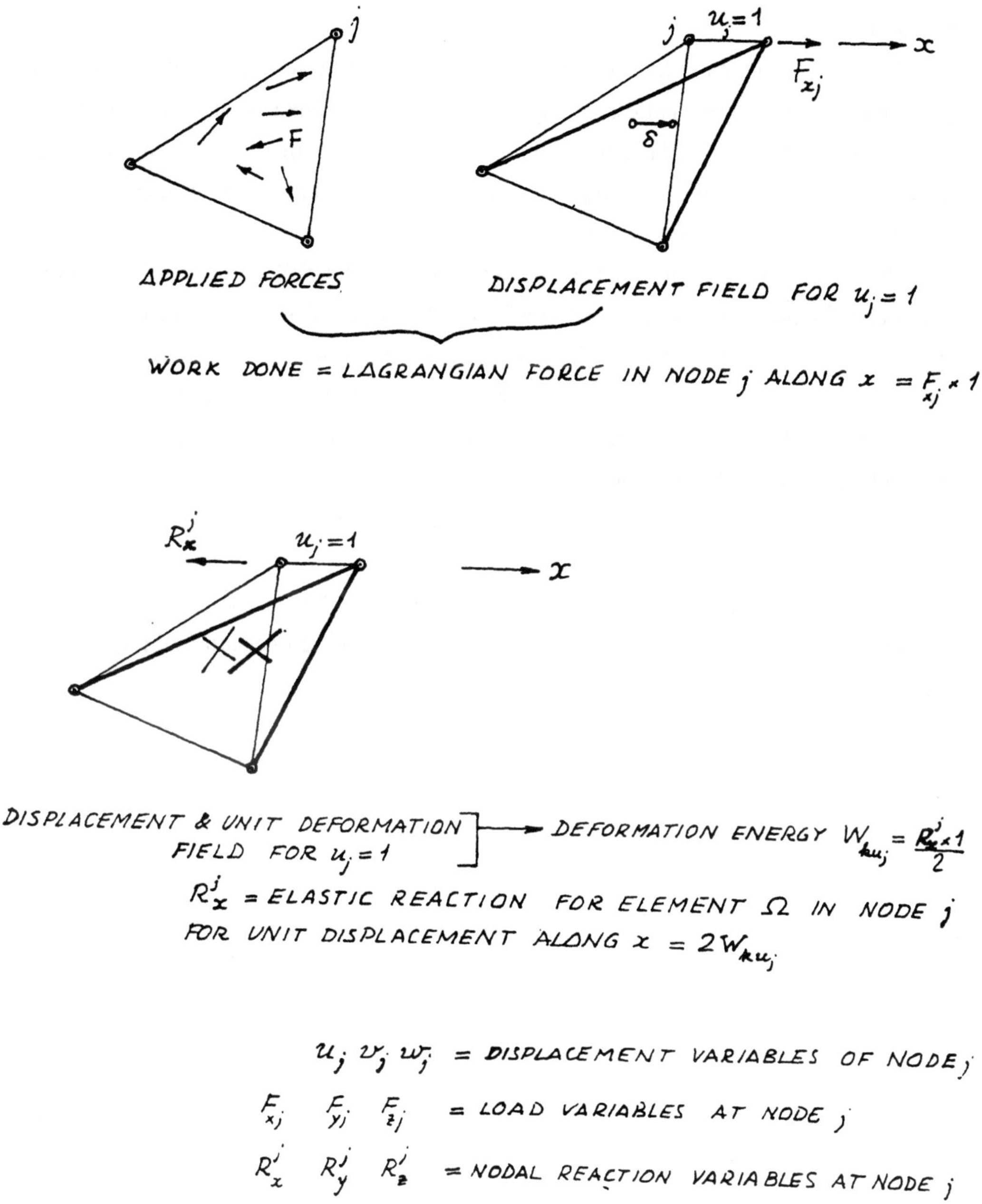

Fig. 53
The FEM: physical meaning of the nodal variables

in the concurring elements is necessary, or the Gauss-point stresses are used (see further on).

Finally, since it is essential to run problems by using large, involved computer programs, F.E.M.s are used essentially as "black boxes". The user has no control over the operations that take place from input of data to printout of the solution. Insight into the physical meaning of computations is lost; the potential for misuse of SW, or misinterpretations of results, is increased.

For all of these reasons, it is recommended that F.E.M.s SW be used by competent, experienced engineers. Independent checks on the order of magnitude of results, be it through experimental evidence, simpler analytical methods, or analogy with similar cases, should always be carried out. Parametric exploration of the sensitivity of results to variations of input data can give a feeling of the reliability of conclusions to be drawn from numerical computations.

In conclusion, F.E.M.s are now a necessary tool to solve many complex engineering problems, in particular those connected to arch dams; however, they should never be used lightly or carelessly, nor should the dam engineer be misled, by their apparent power, into thinking that they can suffice to provide a sound decision-making guidance indipendently from other sources of knowledge.

3.1.1 An outline of the 3-D linear finite element formulation

Let us suppose that the arch dam, and a convenient part of the foundation rock, have been geometrically defined and subdivided without gaps or overlappings into a great number of elements of simple shape (fig. 54). Let the overall geometric domain be $\Omega = \Omega_c + \Omega_r$, with Ω_c =partial domain corresponding to the concrete dam, Ω_r =partial domain corresponding to the rock. The elements will be denoted by Ω_k, $1 \le k \le M_e$.

Two adiacent elements will have a polygonal face in common; over each side (usually at the vertices of the face, sometimes also in intermediate points of the sides of the polygonal face) there will be some of the nodes of the discretization network, which will be numbered from 1 to M_n, using an index j : $1 \le j \le M_n$.

This will be the general numbering of nodes; in addition to that, inside each element the relevant nodes will have a local numbering established according to fixed rules of orientation along the sides.

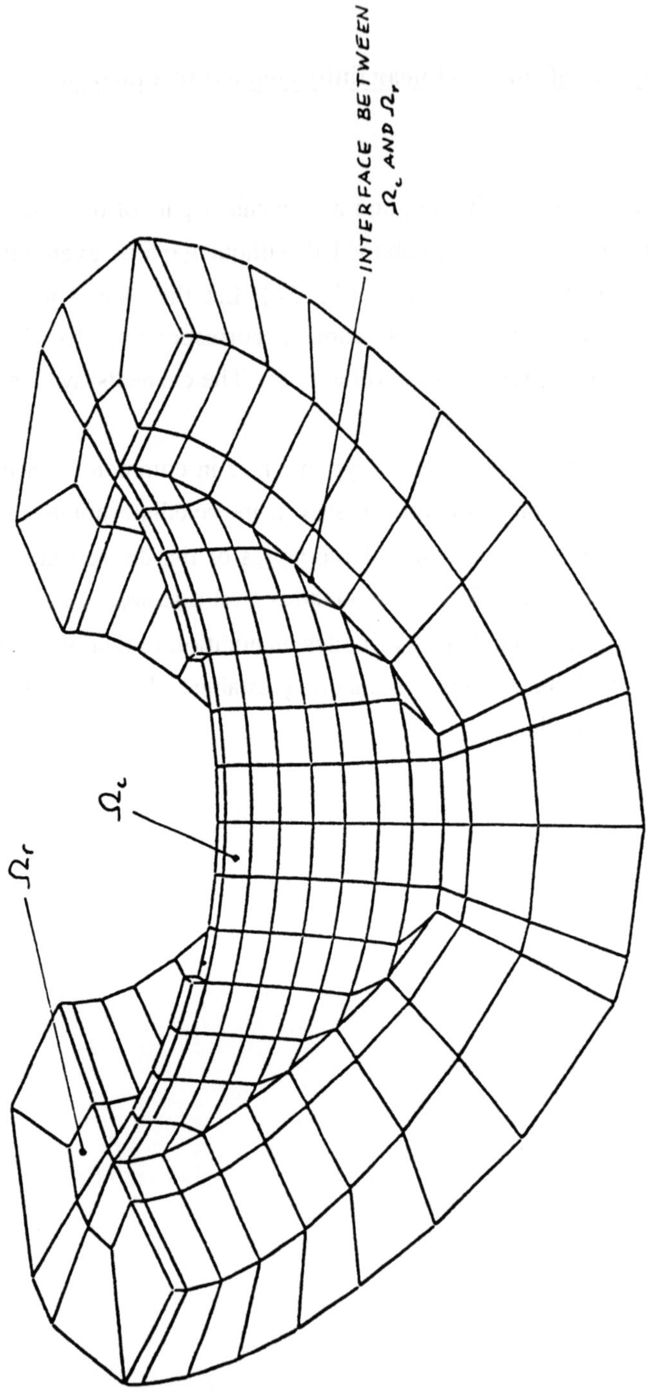

Fig. 54
Subdivision of the geometrical domain into concrete (Ω_c) and rock (Ω_r) parts

To each node j there will be in general <u>three</u> degrees of freeedom, corresponding to the three orthogonal components of displacement u, v, w along global axes x, y, z :

$$\begin{Bmatrix} u_i \\ v_i \\ w_i \end{Bmatrix} \qquad (3.1.1)$$

will then be the <u>nodal unknowns</u> of node j . Inside each element Ω_k containing nodes $j_1 \ldots j_n$ the <u>displacement field</u> will be defined as follows:

$$\begin{Bmatrix} u(x\,,\,y\,,\,z) \\ v(x\,,\,y\,,\,z) \\ w(x\,,\,y\,,\,z) \end{Bmatrix}_{\Omega_k} = \left[N_k(x\,,\,y\,,\,z)\right]\begin{Bmatrix} u_{j1} \\ v_{j1} \\ w_{j1} \\ \ldots \\ \ldots \\ \ldots \\ u_{jn} \\ v_{jn} \\ w_{jn} \end{Bmatrix}_k = [N]_k\{u\}_k \quad \text{for short,} \qquad (3.1.2)$$

where $[N_k] = [N_k(x,y,z)]$ [35] is a matrix of 3 rows and 3n columns, n being the number of nodes of the element, and $\{u\}_k$ is a vector of 3n rows containing the 3n displacements (unknown degrees of freedom) of the n nodes of the element Ω_k .
The generic element of $[N_k]$ is the "shape function" of the corresponding degree of freedom, giving the space distribution of the displacement field inside the element for a unit value of the degree of freedom in question (fig. 55). The shape functions can be

[35] The structure of $[N]_k$ is of the following type:

$$[N]_k = \begin{bmatrix} N_{j1} & 0 & 0 & \ldots & N_{jn} & 0 & 0 \\ 0 & N_{j1} & 0 & \ldots & 0 & N_{jn} & 0 \\ 0 & 0 & N_{j1} & \ldots & 0 & 0 & N_{jn} \end{bmatrix} \quad \text{with} \quad \begin{cases} N_j(x_j, y_j, z_j) = 1 \\ N_j(x_l, y_l, z_l) = 0 \ \text{for}\ j \neq l \\ \sum_{j1}^{jn} N_j(x, y, z) = 1 \ \text{for}\ (x, y, z) \in \Omega_k \end{cases}$$

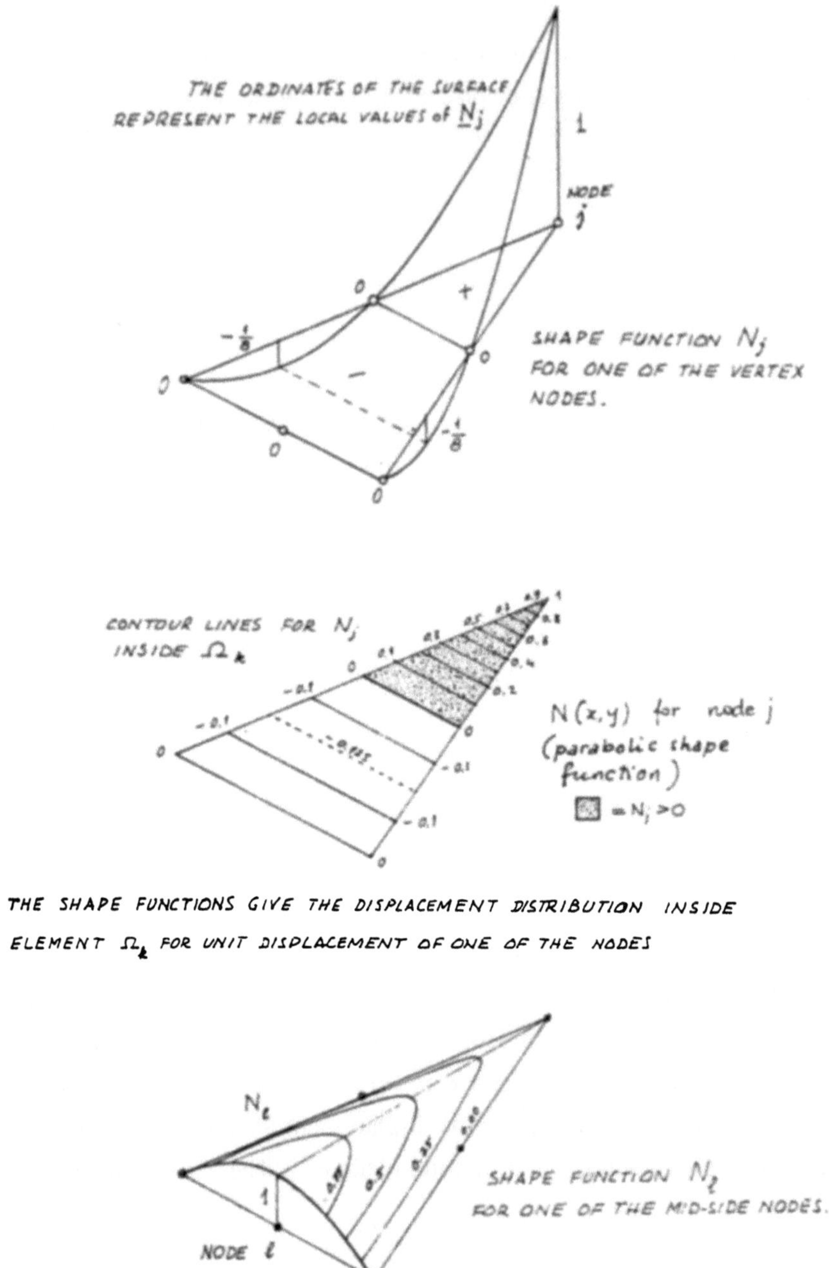

Fig. 55
The shape functions: continuous interpolation functions.
They allow to interpolate the δ-field inside Ω_k from nodal δ's. Second-order (parabolic) N are shown

chosen from a large spectrum of possibilities, according to the desired accuracy and other requirements[36].

It is then easy in principle to to build up the field of unit strains inside the element; in fact it is possible to write:

$$\{\varepsilon\}_k = \begin{Bmatrix} \varepsilon_x \\ \varepsilon_y \\ \varepsilon_z \\ \tau_{yz} \\ \tau_{zx} \\ \tau_{xy} \end{Bmatrix} = \begin{bmatrix} \frac{\partial}{\partial x} & 0 & 0 \\ 0 & \frac{\partial}{\partial y} & 0 \\ 0 & 0 & \frac{\partial}{\partial z} \\ 0 & \frac{\partial}{\partial z} & \frac{\partial}{\partial y} \\ \frac{\partial}{\partial z} & 0 & \frac{\partial}{\partial x} \\ \frac{\partial}{\partial y} & \frac{\partial}{\partial x} & 0 \end{bmatrix} \begin{Bmatrix} u \\ v \\ w \end{Bmatrix}_{\Omega_k} = [B(x,y,z)]_k \cdot \{u\}_k = [B]_k \cdot \{u\}_k, \tag{3.1.3}$$

by posing:

$$[B]_k = [B(x,y,z)]_k = \begin{bmatrix} \frac{\partial}{\partial x} & 0 & 0 \\ 0 & \frac{\partial}{\partial y} & 0 \\ 0 & 0 & \frac{\partial}{\partial z} \\ 0 & \frac{\partial}{\partial z} & \frac{\partial}{\partial y} \\ \frac{\partial}{\partial z} & 0 & \frac{\partial}{\partial x} \\ \frac{\partial}{\partial y} & \frac{\partial}{\partial x} & 0 \end{bmatrix} [N_k(x,y,z)]; \tag{3.1.4}$$

---

[36] Sufficient, but not necessary, conditions for convergence to the true solution as the size element goes to zero are the following:
- $N(x,y,z)$ = complete polynomial of degree G -1 if G is the maximum order of derivatives of N appearing in the functional;
- the N should be capable of representing rigid-body displacements;
- they should be capable of yielding a constant state of stress in the element;
- if not polynomial, they should have continuous derivatives up to order G -1 .
- they should provide across-element continuity ("conformity" requirements)

$[B(x,y,z)]_k$ is thus the 6x3*n* matrix of the distribution inside the element of the spatial gradients of the shape functions.The stress state is given inside the element by the constitutive law between stress components $\{\sigma\}_k$ and strain components; in the case of linear isotropic elasticity one has:

$$\{\sigma\}_k = \frac{E_k}{1+\nu_k}[D]_k\{\varepsilon\}_k = \frac{E_k}{1+\nu_k}[D]_k[B]_k\{u\}_k, \tag{3.1.5}$$

having defined $\{\sigma\}_k$ and $[D]_k$ as follows:

$$\{\sigma\}_K = \begin{Bmatrix} \sigma_x \\ \sigma_y \\ \sigma_z \\ \tau_{yz} \\ \tau_{zx} \\ \tau_{xy} \end{Bmatrix}, \tag{3.1.6}$$

$$[D]_k = \begin{bmatrix} \frac{1-\nu_k}{1-2\nu_k} & \frac{\nu_k}{1-2\nu_k} & \frac{\nu_k}{1-2\nu_k} & 0 & 0 & 0 \\ & \frac{1-\nu_k}{1-2\nu_k} & \frac{\nu_k}{1-2\nu_k} & 0 & 0 & 0 \\ & & \frac{1-\nu_k}{1-2\nu_k} & 0 & 0 & 0 \\ & & & \frac{1}{2} & 0 & 0 \\ & \textit{symmetric} & & & \frac{1}{2} & 0 \\ & & & & & \frac{1}{2} \end{bmatrix} \tag{3.1.7}$$

The elastic energy stored in the element as a consequence of the stress and strain state thus determined is then given by:

$$\begin{aligned} {}_k &= \frac{1}{2}\iiint_{\Omega_k}\{\sigma\}_k^T\{\varepsilon\}_k\,dxdydz = \frac{E_k}{2(1+\nu_k)}\iiint_{\Omega_k}\{\varepsilon\}_k^T[D]_k\{\varepsilon\}_k\,dxdydz = \\ &= \frac{E_k}{2(1+\nu_k)}\iiint_{\Omega_k}\{u\}_k^T[B]_k^T[D]_k[B]_k\{u\}_k\,dxdydzdxdydz \end{aligned} \tag{3.1.8}$$

If now on our element Ω_k there act body (volume) forces of density:

$$\{f\}_k = \begin{Bmatrix} f_x(x,y,z) \\ f_y(x,y,z) \\ f_z(x,y,z) \end{Bmatrix}_k \tag{3.1.9}$$

and surface forces of density:

$$\{\phi\}_k = \begin{Bmatrix} \phi_x(x,y,z) \\ \phi_y(x,y,z) \\ \phi_z(x,y,z) \end{Bmatrix}_k \tag{3.1.10}$$

[on the unit surface of director cosines $C_1(x,y,z)$, $C_2(x,y,z)$, $C_3(x,y,z)$].
then the potential energy of the external forces acting on element Ω_k will be expressed by:

$$P_k = -\iiint_{\Omega_k} \{f\}_k^T [N]_k \{u\}_k dxdydz - \iint_{S(\Omega_k)} \{\phi\}_k^T [N]_k \{u\}_k dS, \tag{3.1.11}$$

with $\qquad dS = C_1 dydz + C_2 dzdx + C_3 dxdy$. (fig.56)

The total potential energy of the element Ω_k will then be:

$$\mathcal{E}_k = P_k + W_k = -\iiint_{\Omega_k} \{f\}_k^T [N]_k \{u\}_k dxdydz + \iint_{S(\Omega_k)} \{\phi\}_k^T [N]_k \{u\}_k dS +$$

$$+ \frac{E_k}{2(1+\nu_k)} \iiint_{\Omega_k} \{u\}_k^T [B]_k^T [D]_k [B]_k \{u\}_k dxdydz \tag{3.1.12}$$

$$= -\{F\}_k^T \{u\}_k + \frac{1}{2} \{u\}_k^T [K]_k \{u\}_k$$

with: $\{F\}_k^T = \iiint_{\Omega_k} \{f\}_k^T [N]_k dxdydz + \iint_{S(\Omega_k)} \{\phi\}_k^T [N]_k dS$, $\qquad$ (3.1.13)

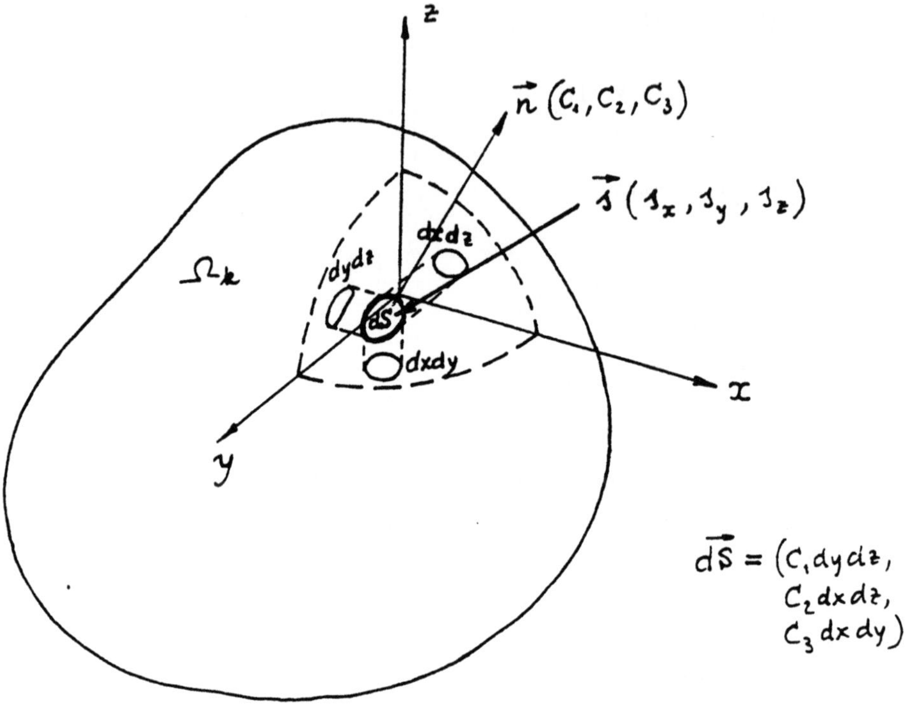

Fig. 56
Surface forces on element Ω_k

$\{F\}_k$ = vector of the generalized nodal forces for the element Ω_k ;

$$[K]_k = \frac{E_k}{1+\nu_k} \iiint_{\Omega_k} [B]_k^T [D]_k [B]_k \, dxdydz$$

$[K]_k$ = "stiffness matrix" of the element Ω_k ; $\{F\}_k$ and $[K]_k$ can be numerically computed for each element from the geometric and physical data; $[K]_k$ is definite positive.

$\mathcal{E}_k$ is thus a quadratic function of $\{u\}_k$.

The total potential energy of the whole system (domain Ω) will then be obtained by summing over all the elements Ω_k :

$$\mathcal{E} = \sum_k \mathcal{E}_k = -\sum_k \{F\}_k^T \{u\}_k + \frac{1}{2} \sum_k \{u\}_k^T [K]_k \{u\}_k \quad (3.1.14)$$

Keeping into account the identity of nodal displacements for the same node as belonging to different (adjacent) elements, one can rewrite (3.1.14) in terms of global forces and displacements vectors $\{F\}$, $\{u\}$ and and of a global stiffness matrix $[K]$: in fact it is possible to write (fig. 57):

$$\{u\}_k = [\,\mathrm{I}\,]_k \ \{u\} \quad (3.1.15)$$

with $[\,\mathrm{I}\,]_k$ = "incidence matrix" (with elements either 0 or 1) of $3n_k$ rows and $3M_n$ columns, of the n_k nodes of element Ω_k with respect to the general numbering of nodes, $\{u\}$ being the global nodal displacement vector of $3M_n$ rows; correspondingly:

$$\{F\} = \sum_k [\mathrm{I}]_k^T \{F\}_k \, , \quad (3.1.16)$$

so that

$$\mathcal{E} = -\{F\}^T \{u\} + \frac{1}{2} \{u\}^T [K] \{u\}, \quad (3.1.17)$$

with:

$$\{F\}^T = \sum_k \{F\}_k^T [\mathrm{I}]_k$$
$$[K] = \sum_k [\mathrm{I}]_k^T [K]_k [\mathrm{I}]_k \quad (3.1.18)$$

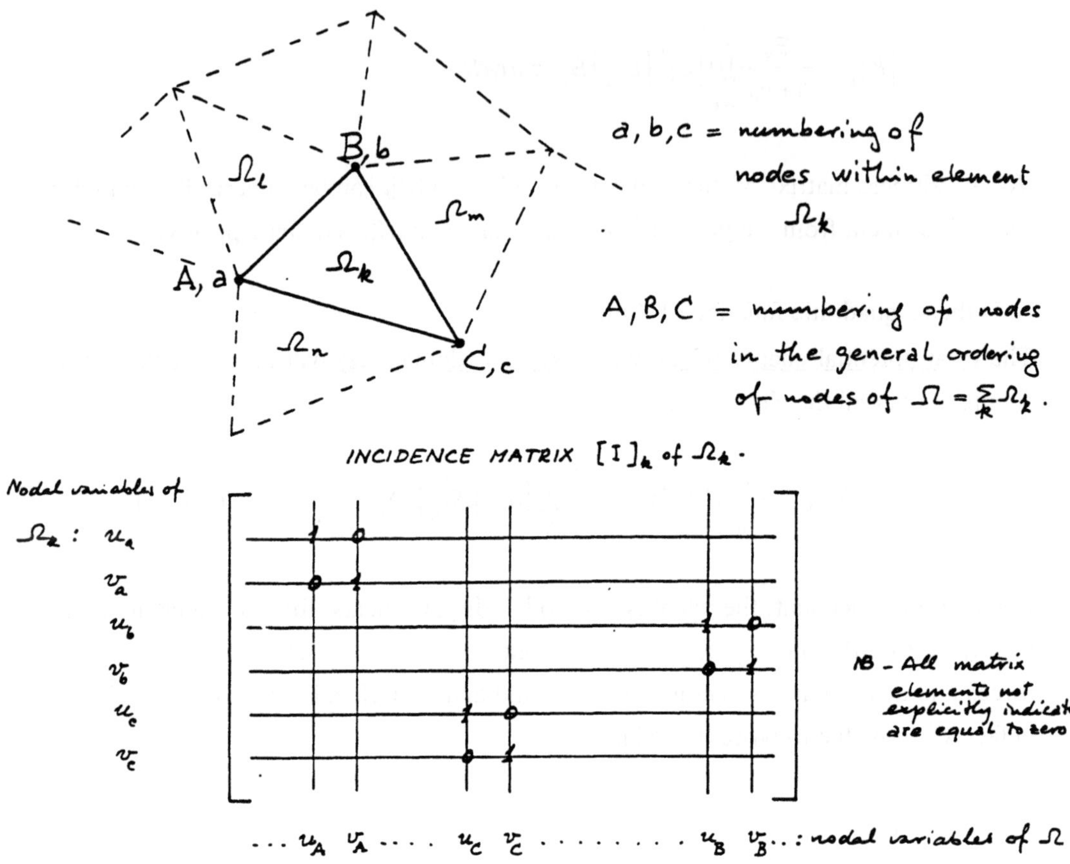

Fig. 57
Incidence matrix of nodal variables of el. Ω_k on the whole-domain (Ω) variables vector (2-D example)

Now it is easy, according to (3.1.17), to see that $\mathcal{E}$, total potential energy of the system (domain $\Omega = \Omega_c + \Omega_r$), is a quadratic function of the nodal displacement vector $\{u\}$; to express the minimum conditions of $\mathcal{E}$ with respect to all the possible values of nodal displacements $\{u\}$ it will be necessary to write that all the partial derivatives of ε with respect to each of the $\{u\}$ composing $\{u\}$ will go to zero:

$$\left\{\frac{\partial \mathcal{E}}{\partial u_j}\right\} = -\{F\} + [K]\{u\} = \{0\} \qquad (3.1.19)$$

or:

$$[K]\{u\} = \{F\} \qquad (3.1.20)$$

i.e. a linear system of equations in the unknowns $\{u\}$.

Once this sysem is solved, (3.1.5) allows computation of stresses inside each element [using (3.1.15) to derive the nodal displacements of the element from the solution vector].

Remarks (boundary conditions - numerical integration - stress evaluation - coordinate transformations - local vs. global reference systems)

The broad outlines of the F.E.M. having been described in the preceding pages, there remain a number of technical details worth at least a short mention.

a) Boundary conditions

(external constraints) are imposed, in 3-D F.E. models, simply by discretizing -with a suitable F.E. network matching, at the concrete/foundation interface, the dam own F.E. mesh - a "convenient" finite portion of the rock mass. The internal (deep-located) nodes of the rock mesh are usually rigidly constrained in all of their d.o.f.; i.e., if n_r is one of the nodes in question, the relevant constraints are:

$$u(n_r) = v(n_r) = w(n_r) = 0 \qquad (3.1.21)^{37}$$

A critical aspect of this rather empirical procedure is the choice of the size of the rock mass to be represented in the model. Usually one requires that the constrained boundary be distant from the concrete/foundation interface, in all directions, by lengths comparable with the dam dimensions (in particular with its height). See Fig. 58.
Of course, in reality small displacements (instead of null ones) will take place at the deep boundary under loading, so that by imposing conditions (3.1.21) one makes the rock mass stiffer than it should be. Sometimes one tries to compensate empirically for this effect by attributing to the rock (Ω_r) portion of the domain an elasticity modulus E_r lower than the real value E_r^* ($e.g.$ $E_r \cong \frac{2}{3} E_r^*$).
Another question of importance concerns the loads directly applied to Ω_r . Dead weight of rock masses is pre-existing to any subsequent application of loads and thus should not be represented in the model; but should one apply hydrostatic load to the upstream portion of the rock mesh external surface? There is no general consensus on the guidelines to follow in that respect, and case-by-case discussion should take place with the help of previous experience.

b) Numerical integration

In the building up of the system of equations (3.1.20) it is necessary to perform, element by element, space or surface integration, as indicated by (3.1.13). Only for

[37] This corresponds to eliminating from the system (3.1.20) those rows and columns that pertain to the d.o.f. in question.

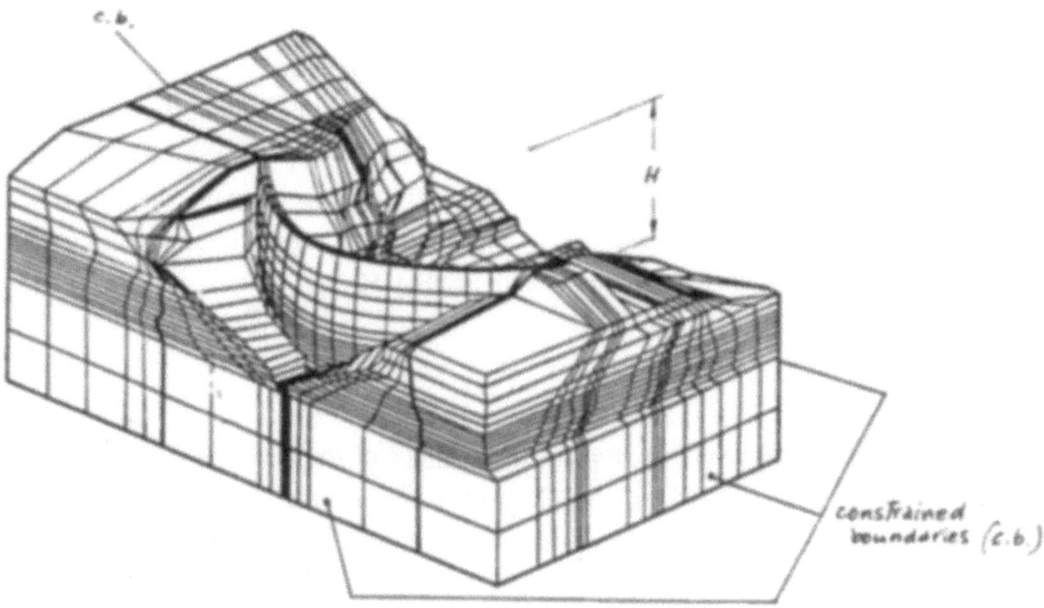

FE mesh of a complex foundation-arch dam system

Fig. 58
Distance of foundation constrained boundaries from dam

very simple shapes of the elements and very simple definition of the "shape functions" is it possible to carry out these integrations in closed form; most of the times it is necessary to use procedures of numerical integration. There are several different options to carry out this operation, but the widely preferred one (by reason of economy) is the so-called GAUSS integration. It consists in computing the integrand function in a finite (quite limited) number of points (Gauss points) suitably located inside the element (their number and location varying according to the order of accuracy one wants to achieve) and then multiplying the average of these values by the volume (or surface) of the integration domain. See Fig. 59.

c) Evaluation of stresses

It would seem natural to derive stresses at the nodes of the mesh, by using (3.1.5). However, if one does that for the several adjacent elements that share the same node, one will in general get different numerical values (the F.E.M.s ensuring continuity of displacements, but not of stresses, between adjacent elements).
One way out of this result, which is contrary to physical reality, is to average at each node the different nodal values that can be obtained from concurrent elements. But simple (unweighted) averaging can give biased results, nor there is agreement on different (weighted) criteria of averaging. It seems, on theoretical grounds, more correct to evaluate the stresses in the Gauss points (coordinates x_G , y_G , z_G) inside each element:

$$\{\sigma_G\}_k = \frac{E_k}{1+\nu_k}[D]_k\{\varepsilon_G\}_k = \frac{E_k}{1+\nu_k}[D]_k[B(x_G, y_G, z_G)]_k\{u\}_k , \quad (3.1.22)$$

where:

$$[B(x_G, y_G, z_G)] = \begin{bmatrix} \frac{\partial}{\partial x} & 0 & 0 & 0 & \frac{\partial}{\partial z} & \frac{\partial}{\partial y} \\ 0 & \frac{\partial}{\partial y} & 0 & \frac{\partial}{\partial z} & 0 & \frac{\partial}{\partial x} \\ 0 & 0 & \frac{\partial}{\partial z} & \frac{\partial}{\partial y} & \frac{\partial}{\partial x} & 0 \end{bmatrix}_{(x_G, y_G, z_G)} \cdot [N_k(x,y,z)]$$

and $\{u\}_k$ is now of course known from the solution of system (3.1.20).

<table>
<tr><th>Integration order</th><th>Degree of precision</th><th>Location of integration points</th></tr>
<tr><td>2 X 2</td><td>3</td><td>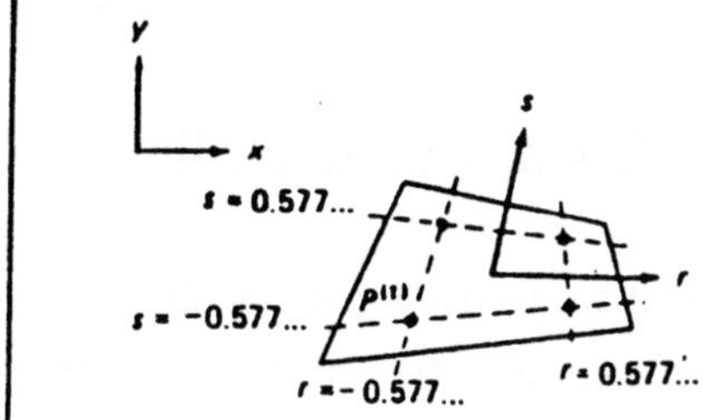
</td></tr>
<tr><td>3 X 3</td><td>5</td><td>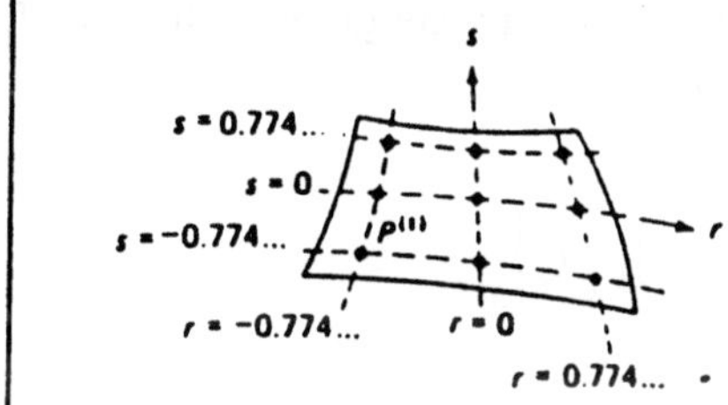
</td></tr>
<tr><td>4 X 4</td><td>7</td><td>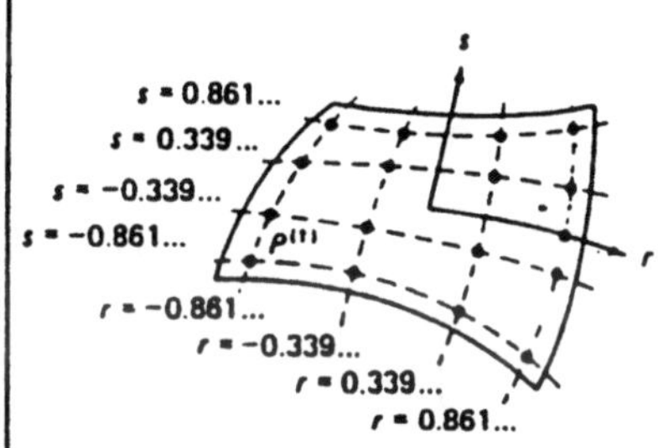
</td></tr>
</table>

Fig. 59
Numerical integration: the Gauss points (2-D)

$r = (\xi - 0.5) \times 2$
$s = (\eta - 0.5) \times 2$
(See Fig. 61)

If stresses are required at points on an interelement boundary, their value will be estimated by interpolating between the two nearest Gauss points belonging respectively to the two elements sharing the boundary in question.

d) Local vs. global reference system; coordinate transformation

Sometimes it is convenient to use local coordinate systems, different for each element, in order to facilitate the computation of $\{F\}_k$ and $[K]_k$, see (3.1.13).
Let the local reference Cartesian trihedron T_k be characterized by a matrix $[C_{Lg}]$ of director cosines with respect to the global trihedron T_g : it is then possible (fig. 60) to write for the element Ω_r :

$$\{u_L\}_k = [C]_k \{u_g\}_k \qquad (3.1.23)$$

where:

$$[C]_k = \begin{bmatrix} [C_{Lg}] & & & & \\ & [C_{Lg}] & & 0 & \\ & & [..] & & \\ & 0 & & [..] & \\ & & & & [C_{Lg}] \end{bmatrix}$$

(n_k blocks $[C_{Lg}]$ if Ω_r has n_k nodes)

with $[C_{Lg}]$ = matrix of director cosines of local trihedron

$$[C_{Lg}] = \begin{bmatrix} \cos x_L \hat{} x_g & \cos x_L \hat{} y_g & \cos x_L \hat{} z_g \\ \cos y_L \hat{} x_g & \cos y_L \hat{} y_g & \cos y_L \hat{} z_g \\ \cos z_L \hat{} x_g & \cos z_L \hat{} y_g & \cos z_L \hat{} z_g \end{bmatrix} \qquad (3.1.24)$$

In these cases it is necessary to effect all the coordinate transformation $T_k \rightarrow T_g$ before summing the $\{F\}_k$ and $[K]_k$, see (3.1.18).

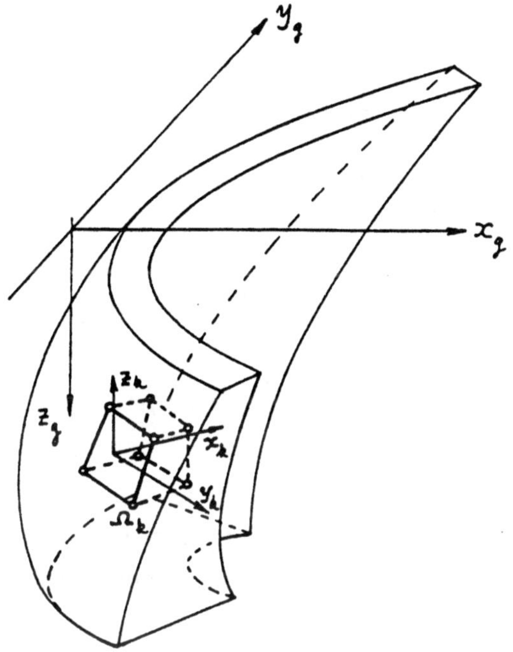

Fig. 60
Coordinate transform.: global (T_g) and local (T_k) reference Trihedra

It is easy to see that this can be done simply by pre -and post- multiplying the matrices by $[C]_k$: in place of (3.1.14) one has:

$$\begin{aligned}\mathcal{E} = \sum_k \mathcal{E}_k &= -\sum_k \{F_L\}_k^T \{u_L\}_k + \frac{1}{2}\sum_k \{u_L\}_k^T [K_L]_k \{u_L\}_k \\ &= -\sum_k \{F_L\}_k^T [C]_k \{u_g\}_k + \frac{1}{2}\sum_k \{u_g\}_k^T [C]_k^T [K_L]_k [C]_k \{u_g\}_k \\ &= -\sum_k \{F\}_k^T \{u_g\}_k + \frac{1}{2}\sum_k \{u_g\}_k^T [K]_k \{u_g\}_k\end{aligned} \qquad (3.1.25)$$

with $\quad \{F\}_k = [C]_k^T \{F_L\}_k \;,\quad [K]_k = [C]_k^T [K_L]_k [C]_k \;;$

after which one proceeds with (3.1.15).....(3.1.20), since $\{u_g\}_k$ is now identical with $\{u\}_k$ of former developments.

Another coordinate transformation that is often used for a numerical integration consists in expressing the physical (local or general) coordinates x, y, z of the generic point inside the element Ω_k as suitable functions of the Cartesian coordinates ξ, η, ζ of a point inside a "normalized" cubic element of unit sides (fig. 61).

This allows (see "isoparametric" elements), to build up in a very simple way the matrices of elements of irregular shape, even with curved faces if need be.

Let the transformations be defined by assigning the following functions (which are supposed to be continuous along with their first derivatives):

$$\begin{aligned} x &= f_1(\xi,\eta,\zeta) \\ y &= f_2(\xi,\eta,\zeta) \quad , \quad (x,y,z) \in \Omega_k \;, \\ z &= f_3(\xi,\eta,\zeta) \qquad 0 \le \xi,\eta,\zeta \le 1 \;; \end{aligned} \qquad (3.1.26)$$

now in order to effect the generic space integral in the space x, y, z over the domain Ω_k :

$$\mathrm{I} = \iiint_{\Omega_k} F(x,y,z)\,dxdydz = \iiint_{\Omega_k} F(x,y,z)\,dV \qquad (3.1.27)$$

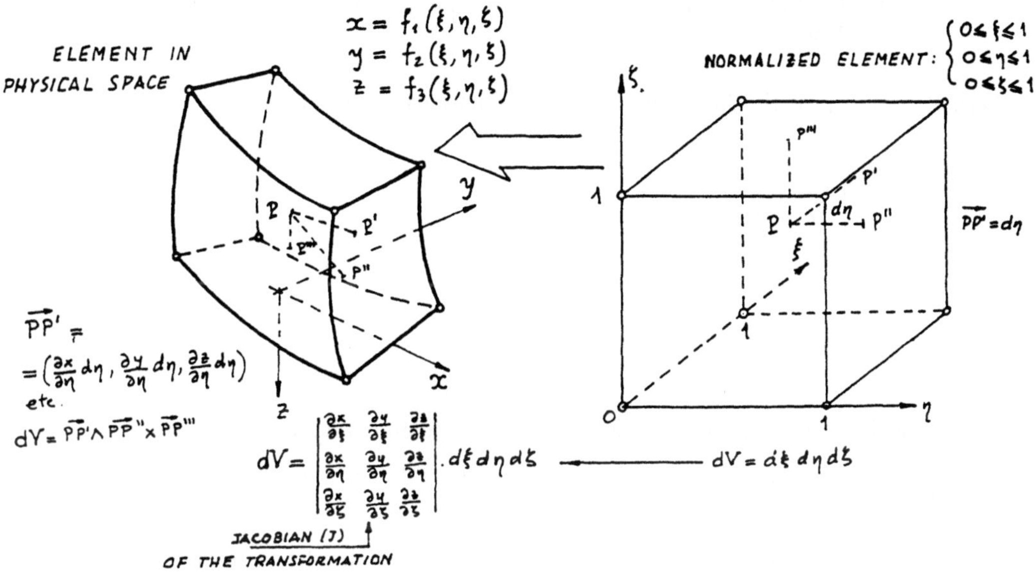

Fig. 61
Coordinate transform.: Ω_k derived from unit-sided "normalized" cubic element

one will instead carry out (numerically: Gauss points, see above) the following integration over the unit cubic domain in the space ξ, η, ζ :

$$I = \int_0^1\int_0^1\int_0^1 F\left[f_1(\xi,\eta,\zeta)\ ,f_2(\xi,\eta,\zeta)\ ,f_3(\xi,\eta,\zeta)\right] J(\xi,\eta,\zeta)\ d\xi d\eta d\zeta \qquad (3.1.28)$$

where $J(\xi, \eta, \zeta)$ is the Jacobian of the coordinate transformation:

$$J = \begin{bmatrix} \frac{\partial x}{\partial \xi} & \frac{\partial y}{\partial \xi} & \frac{\partial z}{\partial \xi} \\ \frac{\partial x}{\partial \eta} & \frac{\partial y}{\partial \eta} & \frac{\partial z}{\partial \eta} \\ \frac{\partial x}{\partial \zeta} & \frac{\partial y}{\partial \zeta} & \frac{\partial z}{\partial \zeta} \end{bmatrix} = J(\xi,\eta,\zeta) \qquad (3.1.29)$$

[In fact, for the elementary increments $d\xi, d\eta, d\zeta$ (elementary volume $d\xi d\eta d\zeta$ in the unit cube), the corresponding physical volume generated in the physical space of Ω_k is $dV = J\, d\xi d\eta d\zeta$].

While in the integrals for $\{F\}_k$ there appear the shape functions themselves, in the integrals for $[K]_k$, see (3.1.13), there appear the space derivatives of the space functions with respect to the physical coordinates x, y, z .

The shape functions and their derivatives are more easily expressed in terms of ξ, η, ζ, i.e. in the space of the normalized cubic element; if $[N_k(\xi, \eta, \zeta)]$ is the $3\times3n_k$ matrix of these shape functions, in place of (3.1.4) we shall then have to consider:

$$[B]_k = \begin{bmatrix} \frac{\partial\xi}{\partial x}\frac{\partial}{\partial\xi}+\frac{\partial\eta}{\partial x}\frac{\partial}{\partial\eta}+\frac{\partial\zeta}{\partial x}\frac{\partial}{\partial\zeta} & 0 & 0 \\ 0 & \frac{\partial\xi}{\partial y}\frac{\partial}{\partial\xi}+\frac{\partial\eta}{\partial y}\frac{\partial}{\partial\eta}+\frac{\partial\zeta}{\partial y}\frac{\partial}{\partial\zeta} & 0 \\ 0 & 0 & \frac{\partial\xi}{\partial z}\frac{\partial}{\partial\xi}+\frac{\partial\eta}{\partial z}\frac{\partial}{\partial\eta}+\frac{\partial\zeta}{\partial z}\frac{\partial}{\partial\zeta} \\ 0 & \frac{\partial\xi}{\partial z}\frac{\partial}{\partial\xi}+\frac{\partial\eta}{\partial z}\frac{\partial}{\partial\eta}+\frac{\partial\zeta}{\partial z}\frac{\partial}{\partial\zeta} & \frac{\partial\xi}{\partial y}\frac{\partial}{\partial\xi}+\frac{\partial\eta}{\partial y}\frac{\partial}{\partial\eta}+\frac{\partial\zeta}{\partial y}\frac{\partial}{\partial\zeta} \\ \frac{\partial\xi}{\partial z}\frac{\partial}{\partial\xi}+\frac{\partial\eta}{\partial z}\frac{\partial}{\partial\eta}+\frac{\partial\zeta}{\partial z}\frac{\partial}{\partial\zeta} & 0 & \frac{\partial\xi}{\partial x}\frac{\partial}{\partial\xi}+\frac{\partial\eta}{\partial x}\frac{\partial}{\partial\eta}+\frac{\partial\zeta}{\partial x}\frac{\partial}{\partial\zeta} \\ \frac{\partial\xi}{\partial y}\frac{\partial}{\partial\xi}+\frac{\partial\eta}{\partial y}\frac{\partial}{\partial\eta}+\frac{\partial\zeta}{\partial y}\frac{\partial}{\partial\zeta} & \frac{\partial\xi}{\partial x}\frac{\partial}{\partial\xi}+\frac{\partial\eta}{\partial x}\frac{\partial}{\partial\eta}+\frac{\partial\zeta}{\partial x}\frac{\partial}{\partial\zeta} & 0 \end{bmatrix} \left[N_k(\xi,\eta,\zeta)\right]$$

(3.1.30)

where at each Gauss point the nine derivatives $\dfrac{\partial(\xi,\eta,\zeta)}{\partial(x,y,z)}$ are readily computed; in fact:

$$\begin{bmatrix} \dfrac{\partial\xi}{\partial x} & \dfrac{\partial\eta}{\partial x} & \dfrac{\partial\zeta}{\partial x} \\ \dfrac{\partial\xi}{\partial y} & \dfrac{\partial\eta}{\partial y} & \dfrac{\partial\zeta}{\partial y} \\ \dfrac{\partial\xi}{\partial z} & \dfrac{\partial\eta}{\partial z} & \dfrac{\partial\zeta}{\partial z} \end{bmatrix} = \begin{bmatrix} \dfrac{\partial x}{\partial\xi} & \dfrac{\partial y}{\partial\xi} & \dfrac{\partial z}{\partial\xi} \\ \dfrac{\partial x}{\partial\eta} & \dfrac{\partial y}{\partial\eta} & \dfrac{\partial z}{\partial\eta} \\ \dfrac{\partial x}{\partial\zeta} & \dfrac{\partial y}{\partial\zeta} & \dfrac{\partial z}{\partial\zeta} \end{bmatrix}^{-1} = \begin{bmatrix} \dfrac{\partial f_1}{\partial\xi} & \dfrac{\partial f_2}{\partial\xi} & \dfrac{\partial f_3}{\partial\xi} \\ \dfrac{\partial f_1}{\partial\eta} & \dfrac{\partial f_2}{\partial\eta} & \dfrac{\partial f_3}{\partial\eta} \\ \dfrac{\partial f_1}{\partial\zeta} & \dfrac{\partial f_2}{\partial\zeta} & \dfrac{\partial f_3}{\partial\zeta} \end{bmatrix}^{-} \quad (3.1.31)$$

the R.H.S. being given by the coordinate transformation:

$$x = f_1(\xi,\eta,\zeta)$$

$$y = f_2(\xi,\eta,\zeta), \text{ see (3.1.26).}$$

$$z = f_3(\xi,\eta,\zeta)$$

3.1.2 A discussion of some aspects of the application of F.E.M. to arch dams

a) Discretization effects.

As already pointed out, the results obtained by application of the F.E.M. are approximated, and their degree of approximation is dependent upon the mesh fineness, the element geometric definition, the type of the shape functions (e.g. their degree if they are polynomial functions). Better and better approximation can be obtained by increasing the number of elements (decreasing their size h) or by increasing the order p of the polynomial shape functions; in the first case one speaks of h-convergence, in the second one of p-convergence.

The question arises, how should one interpret the local or global safety of the dam? To impose a pre-determined shape function (in place of the "exact"one) generally corresponds to introducing internal constraints, so the F.E. mathematical model will be stiffer than the real structure. For this reason, displacements are underestimated and convergence to the true solution takes place "from below" (fig. 62). (The opposite

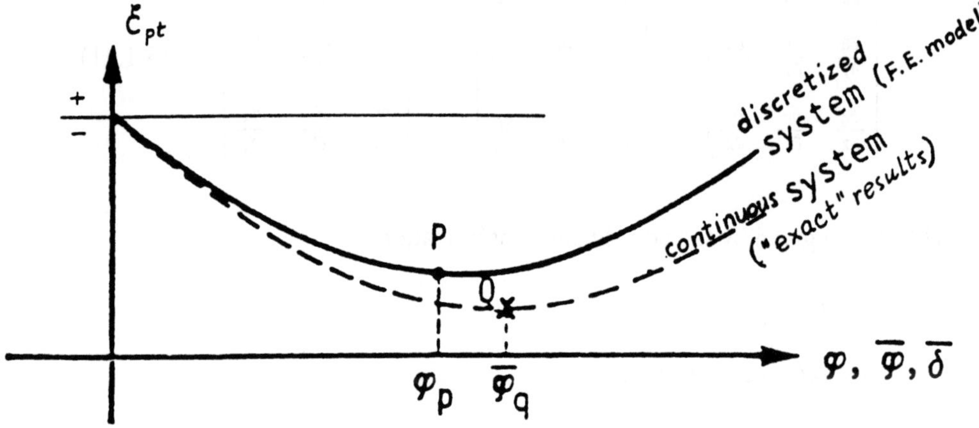

Fig. 62
Convergence "from below" to exact results for "deformation" F.E.M.

would happen with the less used "stress function" approach, in which local equilibrium, but not local compatibility, is preserved, and for which the functional is the "complementary energy"). So, in principle, stresses, too, are underestimated. Variable-degree polynomial F.E.M. with automatic local adjustment of degree p in order to achieve prescribed accuracy have been developed.

The consideration of discretization effects is particularly important in very thick sections of the dam near re-entrant corners of the domain (fig. 63) (e.g. near the concrete/rock interface section). In fact, in such points the theoretical solution is known to give edge stresses tending to infinity: as finer and finer meshes are used, larger and larger edge stresses are obtained, see fig. 64 (which on the other hand are not physically acceptable beyond a certain limit)[38]. The interpretation of the stress field given by application of F.E.M. is thus particularly delicate near such points; great experience is required from the user. By contrast, if one uses a very simple discretization (one element in the thickness, linear or parabolic shape functions) the results are grossly approximated, giving a linear distribution of stresses in these sections, which is certainly unrealistic; but the interpretation of the meaning of the global internal actions "normal force" and "bending moment" (that can readily be derived from the linear stress distribution) is much easier for the practising dam engineer, resting on a solid body of past experience in using just such simplified concepts.

In other words, quite often a gain in appreciation of the "real" behaviour does not lie in the direction of finer and finer discretization, but can rather be achieved by a change of model: instead of a linear elastic model, a non-linear one can be chosen according to the type of local situation and to the aim of the computation. Examples of such non-linear models are:

- the no-tension model. The material is supposed incapable of transmitting tensile stresses: the local principal tensile stresses are removed by the iterative "stress-transfer" process);
- the elastic-plastic models, of which there are several types;
- the "non-linear fracture mechanics" approaches (also in this case several options are available); etc (see APPENDIX).

The trouble with these non-linear models, besides their greater complexity and cost, is just that -contrary to the linear elastic model- there is not one universal, all-purpose

[38] In fact, plastic flow or localized cracking will take place if the stresses exceed the local resistance of the material. So the pursuit of an "exact" solution would not reflect the "real" behaviour.

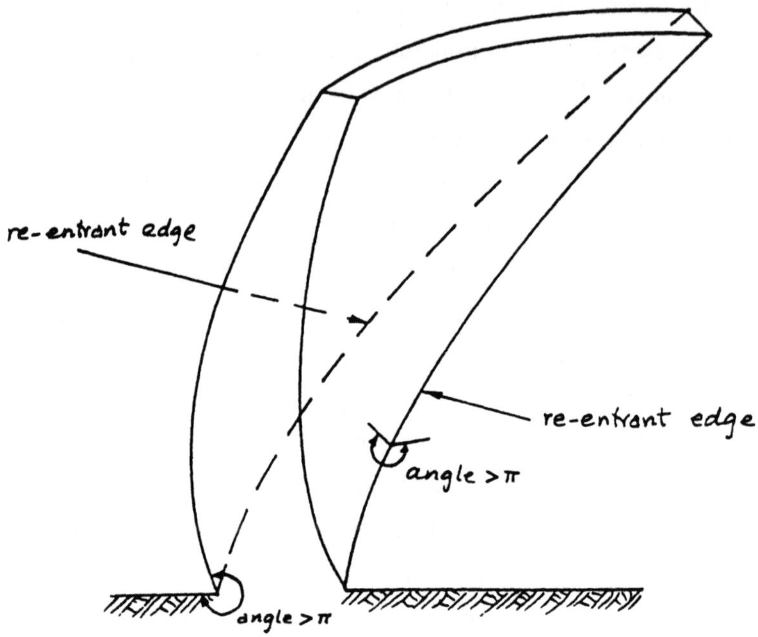

Fig. 63
Re-entrant corners: lines along which theoretical stresses go to ∞

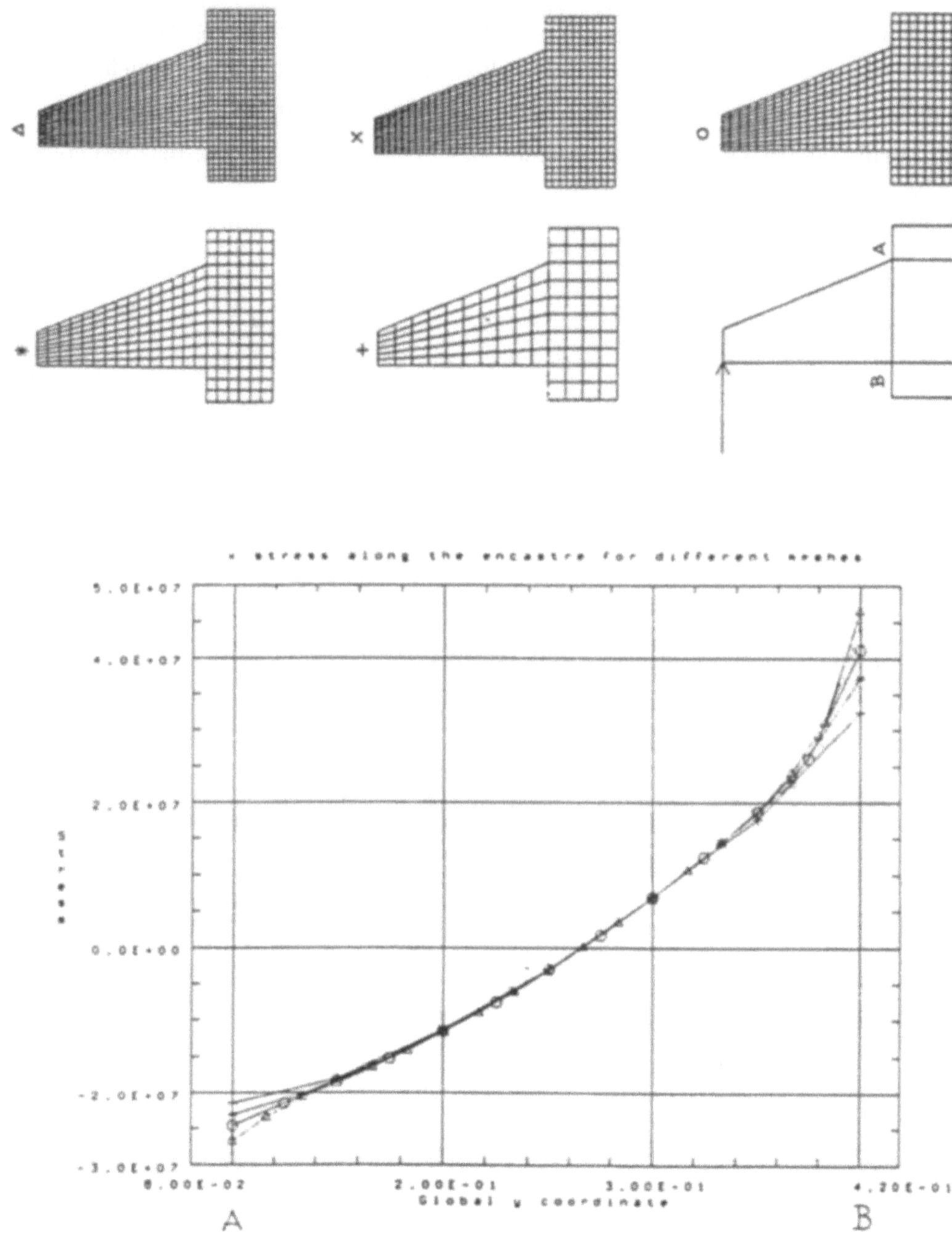

Fig. 64
Mesh-dependence of F.E.-computed stresses near re-entrant corners

model which can guide us beyond the frontier of the elastic domain, but first of all we are obliged to choose the "most appropriate" path from among the many existing options. So, in the first place, we need qualitative guidelines to identify which of the available models is closer to the probable material behaviour in the particular case in hand. Again, only experience -in the present state of the art- can provide reliable choice criteria.

b) Thermal loads

These can be very important for arch dams, both in terms of displacements and of surface stresses. The thermal components of these quantities, indeed, can be of the same order of magnitude as those caused by live loads. In the context of F.E.M. we can suppose that inside our domain Ω there is a known temperature distribution $\vartheta(x,y,z)$ (obtained, e.g., by a separate thermal F.E. diffusion analysis). It is convenient to suppose that inside element Ω_k the space distribution $\vartheta(x, y, z)$, $(x, y, z) \in \Omega_k$, can be approximated from the nodal values $\{\vartheta\}_k$ by the same interpolating functions (shape functions) that are used for displacements, see (3.1.2):

$$\vartheta(x,y,z) \cong \sum_{j_l}^{l=1...n} N_{j_l}(x,y,z)\vartheta_{j_l} = \left[N_{j_l}(x,y,z)\right]_k^T \left\{\vartheta_{j_l}\right\}_k \qquad (3.1.32)$$
$$= \{N\}_k^T \{\vartheta\}_k \quad \text{for short}$$

Now the unit dilatations $\{\varepsilon\}_k$ inside Ω_k, see (3.1.3), will be given by:

$$\{\varepsilon\}_k = [B]_k \{u\}_k = \{\varepsilon_s\} + \alpha_k \begin{Bmatrix} \vartheta(x,y,z) \\ 0 \end{Bmatrix}_k, \qquad (3.1.33)$$

where $\{\varepsilon_s\}$ are the elastic dilatations associated with stress components (α_k being the thermal dilatation coefficient), so that:

$$\{\sigma\}_k = \frac{E_k}{1+\nu_k}[D]_k \{\varepsilon_s\}_k, \quad \text{see (3.1.5)}, \qquad (3.1.34)$$

and

$$\begin{Bmatrix} \vartheta(x,y,z) \\ 0 \end{Bmatrix}_k = \begin{Bmatrix} \vartheta(x,y,z) \\ \vartheta(x,y,z) \\ \vartheta(x,y,z) \\ 0 \\ 0 \\ 0 \end{Bmatrix} \cong \left[N^*(x,y,z)\right]_k \{\vartheta\}_k \tag{3.1.35}$$

with

$$\left[N^*(x,y,z)\right]_k = \begin{bmatrix} N_{j1}(x,y,z) & N_{j2}(x,y,z) & \ldots\ldots & N_{jn}(x,y,z) \\ N_{j1}(x,y,z) & N_{j2}(x,y,z) & \ldots\ldots & N_{jn}(x,y,z) \\ N_{j1}(x,y,z) & N_{j2}(x,y,z) & \ldots\ldots & N_{jn}(x,y,z) \\ 0 & 0 & \ldots\ldots & 0 \\ 0 & 0 & \ldots\ldots & 0 \\ 0 & 0 & \ldots\ldots & 0 \end{bmatrix} \tag{3.1.36}$$

(for short $\left[N^*\right]_k$)

From (3.1.33), (3.1.35) we get:

$$\{\varepsilon\}_k = \{\varepsilon_S\}_k + \alpha_k\left[N^*\right]_k\{\vartheta\}_k \ , \tag{3.1.37}$$

and from(3.1.34), (3.1.37) together with (3.1.33)

$$\begin{aligned} \{\sigma\}_k &= \frac{E_k}{1+\nu_k}[D]_k\left(\{\varepsilon\}_k - \alpha_k\left[N^*\right]_k\{\vartheta\}_k\right) \\ &= \frac{E_k}{1+\nu_k}[D]_k\left([B]_k\{u\}_k - \alpha_k\left[N^*\right]_k\{\vartheta\}_k\right) \end{aligned} \tag{3.1.38}$$

in place of (3.1.5); $\{\vartheta\}_k$ being known, the problem consist in determining $\{u\}_k$.
The density of elastic energy (energy stored in the unit volume) inside element Ω_k is:

$$e = \frac{1}{2}\{\sigma\}_k^T\{\varepsilon_s\}_k$$

$$= \frac{E_k}{2(1+\nu_k)}\left(\{u\}_k^T[B]_k^T - \alpha_\kappa\{\vartheta\}_k^T[N^*]_k^T\right)[D]_k\left([B]_k\{u\}_k - \alpha_k[N^*]_k\{\vartheta\}_k\right)$$

(3.1.39)

in more explicit form,

$$e = \frac{E_k}{2(1-2\nu_k)}\left[\frac{(1-\nu_k)\left(\varepsilon_x^2+\varepsilon_y^2+\varepsilon_z^2\right)+2\nu_k\left(\varepsilon_z\varepsilon_y+\varepsilon_z\varepsilon_x+\varepsilon_x\varepsilon_y\right)}{1+\nu_k} - 2\alpha_k\vartheta_k\left(\varepsilon_x+\varepsilon_y+\varepsilon_z\right)+\right.$$

$$\left.+\frac{\alpha^2\vartheta^2}{1-2\nu_k}+\frac{1-2\nu_k}{1+\nu_k}\left(\gamma_{yz}^2+\gamma_{zx}^2+\gamma_{xy}^2\right)\right]$$

After some manipulations, one gets, by integration over the volume, the elastic energy stored in the element :

$$W_k = \frac{E_k}{2(1+\nu_k)}\iiint_{\Omega_k}\{u\}_k^T[B]_k^T[D]_k[B]_k\{u\}_k dV +$$

$$-\frac{E_k\alpha_k}{1-2\nu_k}\iiint_{\Omega_k}\{u\}_k^T[B^*]_k^T[N^{**}]_k\{\vartheta\}_k dV + W_k(\vartheta) \quad , \qquad (3.1.40)$$

where: $[B^*]_k = \left[\frac{\partial N_{j1}}{\partial x}\frac{\partial N_{j1}}{\partial y}\frac{\partial N_{j1}}{\partial z}\cdots\cdots\cdots\frac{\partial N_{jn}}{\partial x}\frac{\partial N_{jn}}{\partial y}\frac{\partial N_{jn}}{\partial z}\right]$,

$$[N^{**}]_k = \left[N_{j1}\ N_{j2}\ \cdots\cdots\cdots N_{jn}\right] \quad , \qquad (3.1.41)$$

$W_k(\vartheta)$ function only of $\{\vartheta\}_k$ and not of $\{u\}_k$.

Therefore, applying the Green-Gauss-Ostrogradsky theorem to the second space integral of (3.1.40), one gets:

$$W_k = \frac{E_k}{2(1+\nu_k)} \iiint_{\Omega_k} \{u\}_k^T [B]_k^T [D]_k [B]_k \{u\}_k dV +$$

$$-\frac{E_k \alpha_k}{1-2\nu_k} \left[\iint_{S(\Omega_k)} \{u\}_k^T [N]_k^T \{C\}_k [N^{**}]_k \{\vartheta\}_k dS + \right. \tag{3.1.42}$$

$$\left. -\iiint_{\Omega_k} \{u\}_k^T [N]_k^T [B^{**}]_k \{\vartheta\}_k dV \right] + W_k(\vartheta)$$

where $\{C\}_k$ is the vector of director cosines (fig. 65) of the local external normal to the surface $S(\Omega_k)$, and $[B^{**}]_k$ is the $3\times 3n$ matrix of space derivatives of the shape functions:

$$\{C\}_k = \begin{Bmatrix} C_{nx} \\ C_{ny} \\ C_{nz} \end{Bmatrix}_k = \begin{Bmatrix} \cos \hat{nx} \\ \cos \hat{ny} \\ \cos \hat{nz} \end{Bmatrix}_k ;$$

$$[B^{**}]_k = \begin{Bmatrix} \frac{\partial}{\partial x} \\ \frac{\partial}{\partial y} \\ \frac{\partial}{\partial z} \end{Bmatrix} [N^{**}]_k = \begin{bmatrix} \frac{\partial N_{j1}}{\partial x} & \frac{\partial N_{j2}}{\partial x} & \cdots & \frac{\partial N_{jn}}{\partial x} \\ \frac{\partial N_{j1}}{\partial y} & \frac{\partial N_{j2}}{\partial y} & \cdots & \frac{\partial N_{jn}}{\partial y} \\ \frac{\partial N_{j1}}{\partial z} & \frac{\partial N_{j2}}{\partial z} & \cdots & \frac{\partial N_{jn}}{\partial z} \end{bmatrix} \tag{3.1.43}$$

To form the expression of the total potential energy for Ω_k , to W_k one has to add the potential energy P_k of the live loads (body forces $\{f(x, y, z)\}_k$ and surface loads $\{\phi(x, y, z)\}_k$):

$$P_k = -\iiint_{\Omega_k} \{f\}_k^T [N]_k \{u\}_k dV - \iint_{S(\Omega_k)} \{\phi\}_k^T [N]_k \{u\}_k dS, \qquad \text{see (3.1.12).}$$

So it is: $\mathcal{E} = W_k + P_k = \frac{1}{2} \{u\}_k^T [K]_k \{u\}_k - \{F\}_k^T \{u\}_k + W_k(\vartheta)$

identical with (3.1.12) apart from the last term, provided we define in a different way $\{F\}_k$, i.e. the vector of nodal forces:

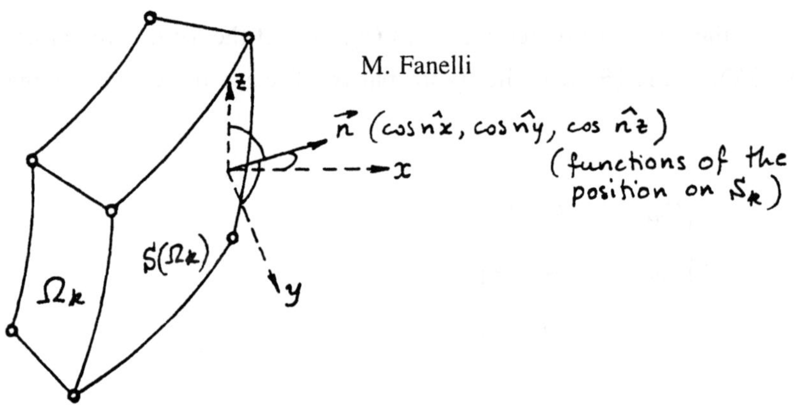

Fig. 65
Local external normal to S_k

$$\{F\}_k = \iiint_{\Omega_k} [N]_k^T \{f\}_k dV + \iint_{S(\Omega_k)} [N]_k^T \{\phi\}_k dS +$$

$$+\frac{E_k \alpha_k}{1-2\nu_k}\left[-\iiint_{\Omega_k} [N]_k^T [B^{**}]_k \{\vartheta\}_k dV + \iint_{S(\Omega_k)} [N]_k^T \{C\}_k [N^{**}] \{\vartheta\}_k dS \right] \tag{3.1.44}$$

The first two terms on the R.H.S. of (3.1.44) originate from live loads and are identical with (3.1.12); the last term, with the factor $E_k\ \alpha_k$, comes from the existence of the thermal field.

By summing $\mathcal{E}_k$ over all the elements Ω_k contained in Ω, the surface integrals containing $\{\vartheta\}_k$ cancel out for the internal faces belonging to two adjacent elements. In fact, they are taken once with a certain sign for one of the two adjacent elements, and a second time with the opposite sign (because the external normal is oriented in the opposite direction) for the other element. Of these integrals, thus, only those pertaining to the external surface of Ω will remain in the expression of the total potential energy for the whole domain.

With the new definition (3.1.44) of nodal forces, the expression of the total potential energy for the whole domain Ω takes on the same formal expression as in (3.1.17), except for one last term $\mathcal{E}(\vartheta)$ which does not depend on the d.o.f. :

$$\mathcal{E} = \frac{1}{2}\{u\}^T [K]\{u\} - \{F\}^T \{u\} + \mathcal{E}(\vartheta).$$

Note that $[K]$, global stiffness matrix, is defined and formed exactly as in the case of no thermal loads. The conditions for the minimum of $\mathcal{E}$ with respect to all the possible values of the d.o.f. $\{u\}$ are expressed likewise by the usual formula:

$$\left\{\frac{\partial \mathcal{E}}{\partial u}\right\} = [K]\{u\} - \{F\} = \{0\}, \qquad \text{as in (3.1.20).}$$

It is seen, thus, that the presence of a thermal field $\vartheta(x,y,z)$ produces, on the solutions $\{u\}$ of (3.1.20), i.e. on displacements, the same effect as the presence of a system of additional surface forces $\{\phi^*\}$ and body forces $\{f^*\}$ given respectively by:

$$\text{a)}\qquad \{s^*\} = \sum_k [\mathrm{I}]_k^T \frac{E_k\alpha_k}{1-2\nu_k} \{C\}_k [N^{**}]_k \{\vartheta\}_k,$$

$$\text{b)}\qquad \{f^*\} = -\sum_k [\mathrm{I}]_k^T \frac{E_k\alpha_k}{1-2\nu_k} [B^{**}]_k \{\vartheta\}_k. \qquad (3.1.45)$$

It can be shown that these "thermal nodal forces" have a simple physical meaning. In fact, let us imagine that the thermal field is applied first, without the live loads, but with internal (distributed) constraints such that the thermal dilatations are completely eliminated. This can be done by applying locally isotropic stresses given by :

$$-\frac{E_k\alpha_k\vartheta(x,y,z)}{1-2\nu_k}.$$

The external displacements $\{u\}_1$ are evidently zero in this stage: $\{u\}_1 = \{0\}$.
In stage 2 we apply on the constrained structure the live loads. These will produce the same displacements $\{u\}_2$ as would be caused by those same live loads acting by themselves on the unloaded structure. In stage 3, to reach a situation that reproduces real conditions, we imagine to remove the internal constraints introduced in stage 1; this last step is effected by applying to the structure the distributed isotropic stress of stage 1 with the reverse sign, i.e. $\frac{E_k\alpha_k\vartheta(x,y,z)}{1-2\nu_k}$ (fig.66). Now this is readily seen to be equivalent to applying surface tractions $\frac{E_k\alpha_k\vartheta(x,y,z)}{1-2\nu_k}$ to each unit of area of the external surfaces of the domain Ω, and body forces $-\frac{E_k\alpha_k \,\mathrm{grad}\,\vec{\vartheta}}{1-2\nu_k}$ to each unit of volume inside Ω. These distributed surface and body forces give rise, in the F.E.M. formalism, to the nodal forces (3.1.45) a) and b) respectively. After the last step the displacements $\{u\}$ of the system will evidently be:

$$\{u\} = \{u\}_1 + \{u\}_2 + \{u\}_3 = \{0\} + \{u\}_2 + \{u\}_3 = \{u\}_2 + \{u\}_3$$

i.e. the sum of those caused by the live loads in step 2 and those caused by the fictitious body forces (3.1.45) of step 3.
Of course, after solution of (3.1.20) for $\{u\}$, stresses will be given by (3.1.38) instead of by (3.1.5).
The above treatment shows that F.E.M. can handle the effects of a known thermal field $\vartheta(x,y,z)$ under the assumption of linear elastic behaviour. The determination of $\vartheta(x,y,z)$ from boundary and initial conditions is a separate problem of significant

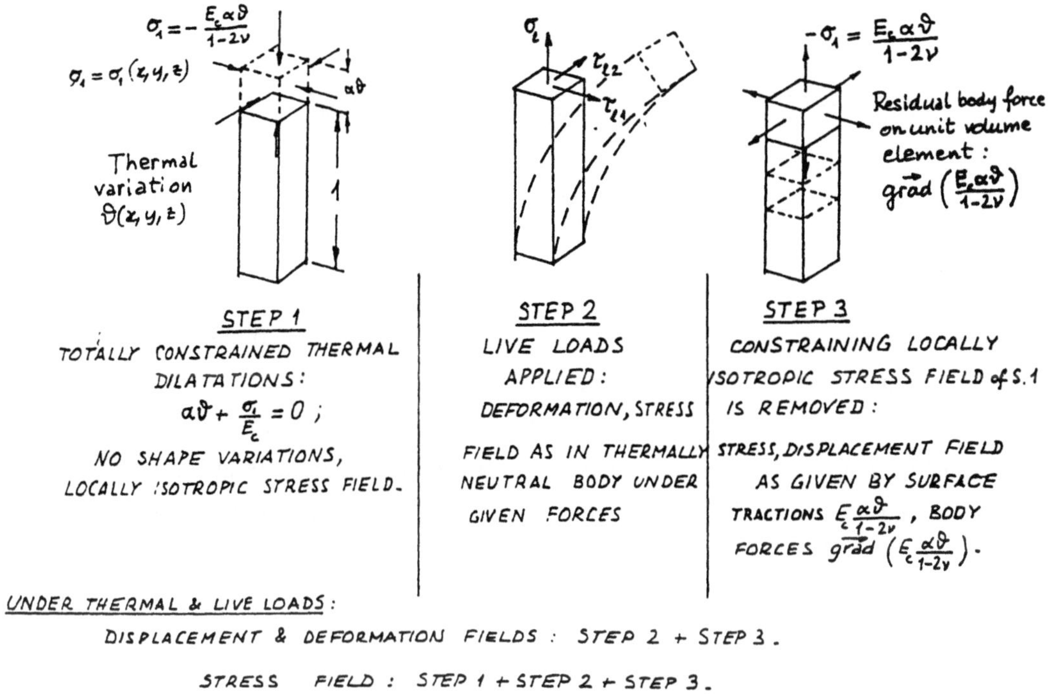

Fig. 66
Thermal loads: "equivalent" surface & body forces

importance for arch dams, and can be solved as well by F.E.M. based on the discretization of FOURIER's partial differential equation for heat diffusion:

$$\frac{\partial}{\partial x}\left(\kappa_x \frac{\partial \vartheta}{\partial x}\right)+\frac{\partial}{\partial y}\left(\kappa_y \frac{\partial \vartheta}{\partial y}\right)+\frac{\partial}{\partial z}\left(\kappa_z \frac{\partial \vartheta}{\partial z}\right)+q=c\frac{\partial \vartheta}{\partial t}, \qquad (3.1.46)$$

with κ_x, κ_y, κ_z = orthotropic heat transmissivity coefficients, q = rate of internal heat generation for unit volume, c = thermal capacity for unit volume (see APPENDIX).

The above treatment, it can be remarked, is valid in general for thermoelastic problems of any 3-D elastic body. What can be said specifically for arch dams?

From a qualitative point of view, it is easy to see that the highest thermal stresses (for "reasonable" temperature distributions) occur at the arch abutments in the horiziontal direction and at the base of the central cantilever in the vertical one. For a generalized temperature rise $\overline{\tau} > 0$ and a concurring thermal jump $\Delta\tau > 0$ (see. fig. 67) (upstream face warmer than downstream face) there will be, tendentially, tensile stresses in a horizontal direction along the downstream face of arch abutments, and tensile vertical stresses, also along the downstream face, at the basis of central cantilevers. For $\overline{\tau} < 0$, $\Delta\tau < 0$ the tensile stresses will tend to occur at the upstream face (the horizontal tensile stresses at upstream arch abutments being somewhat greater than those on the downstream side of the preceding case).

Keeping into account the qualitative stress distribution caused by hydrostatic presssure, it can be said that the worst conditions will tend to occurr in winter with full reservoir (or in summer, with upstream-directed seismic forces, for empty reservoir). As a last remark, it can be said that the application of F.E.M. to the thermal behaviour of arch dams requires -as it is always the case with "sophisticated" methods- a quite complete information about the temperature distribution inside the dam.

c) Anelastic foundation settlements

If these are known, e.g. let us consider a single component δ_{ajx} at node j along the foundation, the corresponding row of the system of equations (3.1.20) is omitted, and the corresponding column elements of matrix $[K]$ are brought to the R.H.S. of the system, with change of sign, after multiplication by δ_{ajx} .After this is done for all affected nodes and for all Cartesian components of their settlements, the system is

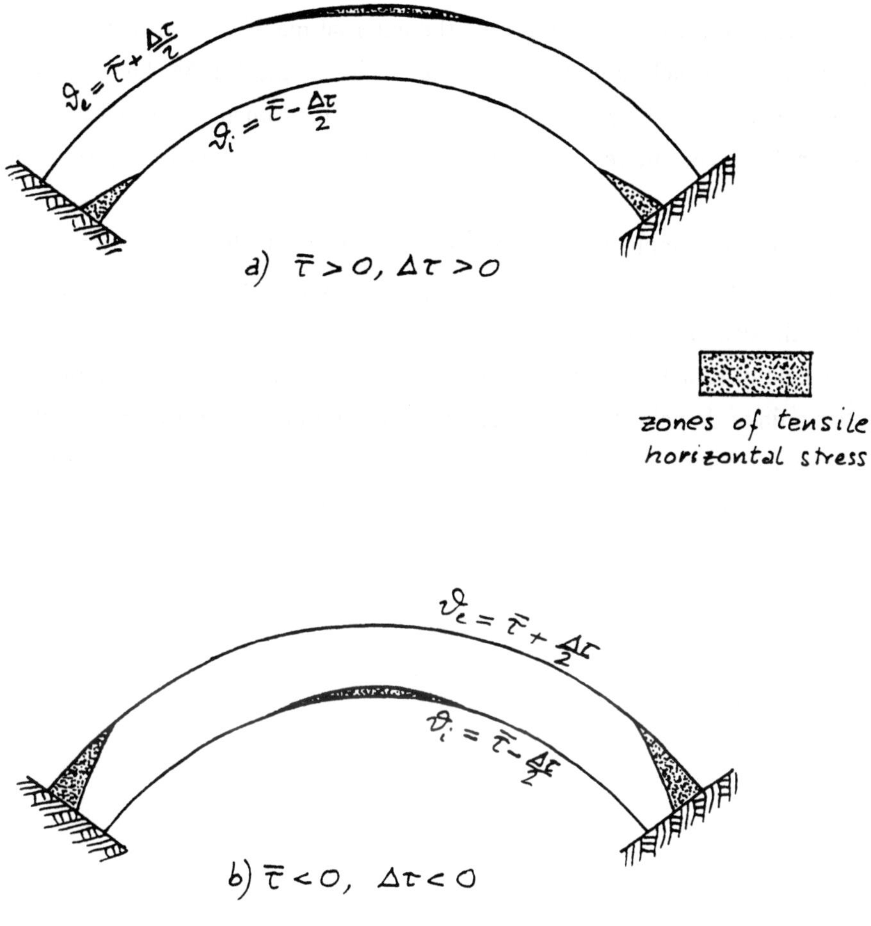

Fig. 67
Thermal variations in an arch dam: zones with tensile thermal stresses

solved and then stresses are obtained using (3.1.5) with the complete $\{u\}$ vector, i.e. including the known anelastic components. This procedure, it is to be noted, will give reliable results for the arch dam, but not for the foundation itself near the settled zone (the anelastic settlement being only the external manifestation of an unknown internal phenomenon which is certainly not elastic-linear, the F.E.M. linear analysis does not model in a correct way the local condition for the domain Ω_r). Away from the settled region, however, the results of the F.E. analysis can give useful information also inside Ω_r .

Compare the simplicity of the F.E. treatment of this important effect with the complications one would incur if trying to solve the problem in question by one of the more classical methods.

Speaking of foundations, it should be observed that the F.E.M., by representing - albeit over a finite domain Ω_r - the real geometry and physiçal properties of the rock mass, allows in principle a far better appreciation of the interaction between the dam and its abutments than, e.g., the Vogt coefficients in the "classical" methods illustrated in the preceding chapters.

3.2 Linear dynamic analysis via the F.E.M.

The dynamic behaviour of arch dams (dynamic intrinsic characteristics, response to external dynamic excitations) can be investigated not only by F.E.M., but also using the "classical" simplified models illustrated in earlier chapters, e.g. by the Ritter method. In fact, quite acceptable results (i.e. close agreement with experimental in-situ investigations) have been obtained in this way (fig. 68).

However, for the sake of brevity it is expedient to limit the present exposé to the more modern, general, powerful F.E. formulation.

We will consider first the behaviour of an arch dam with empty reservoir; determination of eigenfrequencies, eigenshapes and response to external excitation (in particular seismic input through "ground shaking") will be outlined. The influence of structural damping will then be considered. The important subject of fluid/structure interaction, which brings about important changes in the dynamic behaviour in the case of full (or partially full) reservoir, will next be treated in steps, beginning with the classical WESTERGAARD approach and proceeding to illustrate the more modern formulations. Some critical considerations on the present "state of the art" will conclude the chapter.

a) Undamped system, empty reservoir

To represent the interplay of dynamic effects, the approximate $3M_n$ d.o.f. displacement field will have to be made time-dependent:

$$\begin{Bmatrix} u(x,y,z,t) \\ v(x,y,z,t) \\ w(x,y,z,t) \end{Bmatrix} = [N]\ \{u(t)\} \qquad (3.2.1)$$

where $[N]=3\times 3M_n$ matrix of shape functions, $\{u(t)\}=3M_n$ - vector of nodal displacements. The logical choice is to keep the shape functions $[N(x,y,z)]$ independent of time, and confine the time-variability to the $3\text{-}M_n$ nodal displacement vector:

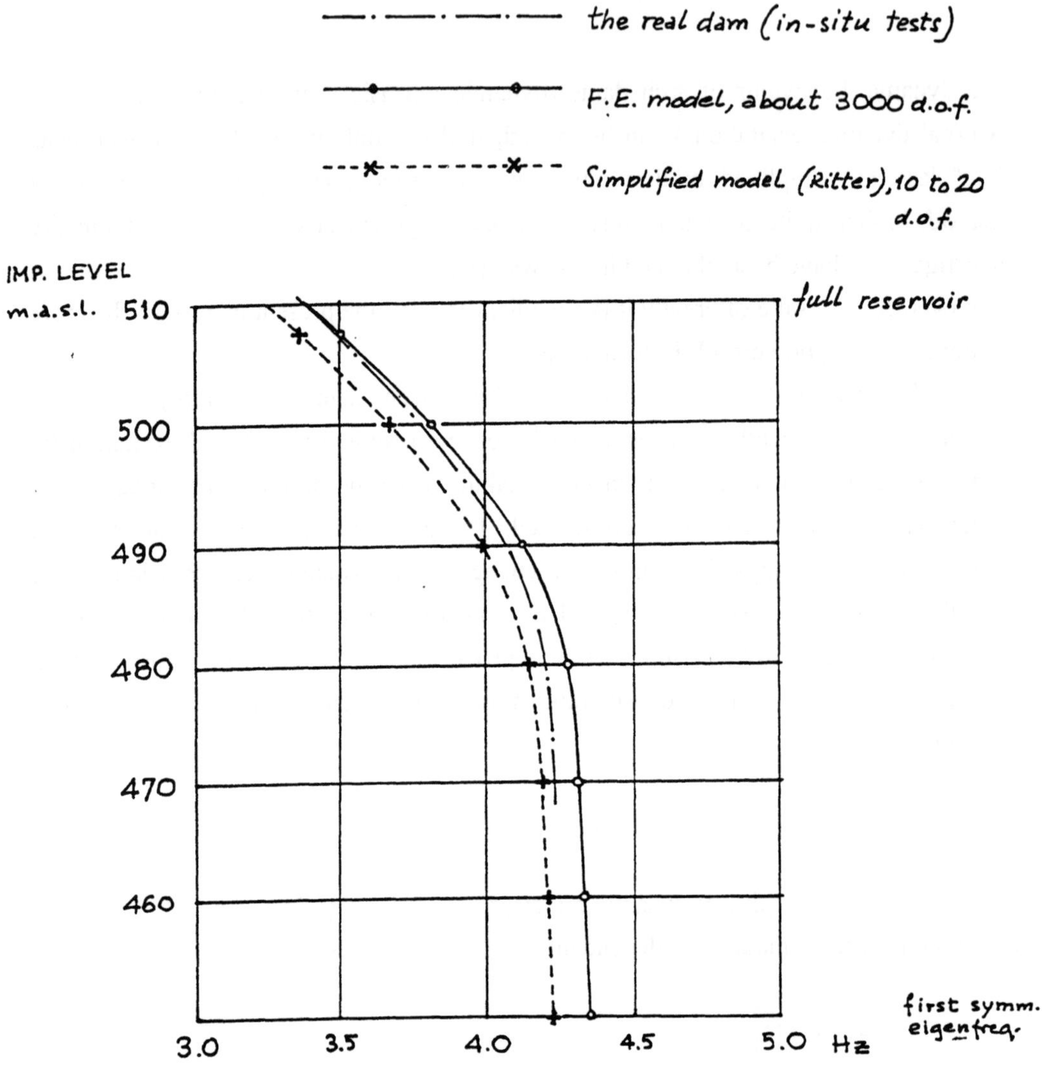

Fig. 68
Variation of the First Symmetric Eigenfrequency of Arch Dam with the Impoundment Level (a fluid-structure interaction problem)

$$\{u\} = \{u(t)\} \quad \text{, i.e.}$$
$$u_j = u_j(t) \quad , \quad 1 \le j \le 3M_n . \qquad (3.2.2)$$

The "weak" formulation of the dynamic problem, so well suited to discretization methods like F.E.M., will now entail consideration not only of the potential energy (as in the static case), but also of the kinetic energy of the system.

As for the potential energy, apart from the time-variability of $\{u\}$ and of the external forces (or externally imposed displacemetns as in the case of "ground shaking") its expression will remain formally the same as before: thus, e.g. the elastic deformation energy W stored in the system at any time instant t will be, with the same stiffness matrix of the static case:

$$= \frac{1}{2}\{u\}^T[K]\{u\} = \frac{1}{2}\{u(t)\}^T[K]\{u(t)\} = W(t). \qquad (3.2.3)$$

The kinetic energy $\mathcal{E}_c$ stored in the system is defined as:

$$\mathcal{E}_c = \frac{1}{2}\iiint_\Omega \rho\,(\dot{u}^2 + \dot{v}^2 + \dot{w}^2)dV \qquad (3.2.4)$$

where

$$\dot{u} \cong \frac{\partial u}{\partial t} \quad , \quad \dot{v} \cong \frac{\partial v}{\partial t} \quad , \quad \dot{w} \cong \frac{\partial w}{\partial t} \qquad (3.2.5)$$

in the linear approximation.

By using the interpolation of $\begin{Bmatrix} u \\ v \\ w \end{Bmatrix}$ within each element Ω_k form the nodal values through the shape functions, see (3.2.1), the approximate value of $\mathcal{E}_c$ in our discretization scheme will be (for each element Ω_k), assuming ρ to be constant in the element (ρ_k) :

$$\mathcal{E}_{ck} \cong \frac{\rho_k}{2}\iiint_\Omega \{\dot{u}\}_k^T[N]_k^T[N]_k\{\dot{u}\}_k dV = \frac{1}{2}\{\dot{u}\}_k^T[M]_k\{u\}_k \qquad (3.2.6)$$

where $[M]_k$ = "mass matrix" of the element Ω_k is defined as:

$$[M]_k = \rho_k \iiint_{\Omega_k} [N]_k^T[N]_k\, dV \qquad (3.2.7)$$

Note that $[M]_k$ can easily be computed for each element Ω_k, e.g. through Gauss numerical integration.

By use of the incidence matrix $[\ \mathrm{I}\ ]_k$ of the nodes of element Ω_k on the general numbering of nodes for the whole discretization domain Ω we can bring each $[M]_k$ to the $3M_n \times 3M_n$ order of the global matrix which is then obtained by summation over $k(1 \leq k \leq$ number of elements) :

$$\mathcal{E}_c = \sum_k \mathcal{E}_{ck} = \frac{1}{2}\{\dot{u}\}^T [M]\{\dot{u}\}, \text{ discretized form of (3.2.4),} \quad (3.2.8)$$

where $$[M] = \sum_k [\mathrm{I}]_k^T [M]_k [\mathrm{I}]_k;$$

$[M]$ = <u>mass matrix</u> of the whole discretized domain Ω , and $[\ \mathrm{I}\ ]_k$ = incidence matrix of element Ω_k . Note that $[M]$ is a symmetric, definite-positive, $3M_n \times 3M_n$ matrix.

The potential energy of external forces, if any, will have the same formal expression as in the static case, apart from the time-variability of both the external actions and the displacements:

$$\mathcal{E}_p = -\{F(t)\}^T \{u(t)\}, \quad (3.2.9)$$

where for the definition of $\{F\}$ the reader is referred back to (3.1.9).......(3.1.13).

A different variational principle has now to be used in place of the "minimum of the total potential energy" that applies in the static case. In fact, the Lagrange equations for a dynamic, undamped, multi-degree-of-freedom system with generalized coordinates $q(t)$ are:

$$\frac{\partial}{\partial t}\left(\frac{\partial H}{\partial \dot{q}}\right) = -\frac{\partial H}{\partial q} \quad (3.2.10)$$

where

$$H = \mathcal{E}_p + W + \mathcal{E}_c, \text{ and } \{q\} = \{u\}, \{\dot{q}\} = \{\dot{u}\}.$$

By (3.2.10), (3.2.11) together with (3.2.3) for W , (3.2.8) for $\mathcal{E}_c$, (3.2.9) for $\mathcal{E}_p$, we now obtain the following matrix equation:

$$[M]\{\ddot{u}\} = \{F(t)\} - [K]\{u\}, \quad (3.2.11)$$

or $$[K]\{u\} + [M]\{\ddot{u}\} = \{F\}, \text{ as usually written.}$$

This matrix equation could be solved (= integrated in time) directly, once the external actions and initial ($\{u(0)\}$, $\{\dot{u}(0)\}$) and boundary conditions are defined. However, usually $\{F(t)\}$ is not known in a deterministic way, but through its statistical properties (Fourier or response spectra, see further on); such is the important case, e.g., of earthquake excitation. In this case, it becomes necessary to solve the problem in steps, beginning with the determination of the dynamic properties intrinsic to the mechanical system: frequencies and shapes of the free-vibration "modes" of the dam (eigenfrequencies, eigenshapes).

a$_1$) Eigenfrequencies and eigenshapes at empty reservoir

The free vibration modes are defined as those dynamic configurations of motions which, in a conservative (undamped) system, could be sustained indefinitely, with only periodic changes at a well-defined single frequency, without any external excitation. Each of the modes behaves as an independent, single-degree-of-freedom linear oscillator, as it will be illustrated further on. Thus we are looking for particular solutions of (3.2.11) under the following conditions:

$$\{F\}=\{0\},\quad \{u\}=\{U\}e^{i\omega t} \tag{3.2.12}$$

where $\{U\}=3M_n$ - vector of nodal amplitudes (unknown) and $\omega/2\pi$ = modal frequency (unknown as well). With these assumptions (3.2.11) yields the equation for the modal characteristics $\{U\}$, ω:

$$[K]\{U\}=\omega^2[M]\{U\}, \tag{3.2.13}$$

i.e. a classical eigenvalue-eigenvector problem. Non-trivial ($\{U\}\neq\{0\}$) solutions of (3.2.13) exist only for the discrete set of values of $\omega(\omega_1, \omega_2, \ldots\omega_{3M_n})$ which make $\mathrm{Det}([K]-\omega^2[M])=0$.

In a continuous system (like an arch dam) i.e. a system with an infinite number of d.o.f., there would be infinitely many eigenvalues ω_m, to each of which there would correspondingly exist an eigenshape $\{U\}_m$.

By contrast, in our F.E. discretization, at most $3M_n$ values for ω_m, and correspondingly $3M_n$ vectors for $\{U\}_m$, could be determined $(1 \le m \le 3M_n)$[39]
In practice, a limited number of eigenvalues and eigenvectors are of interest[40], namely those corresponding to the lower values of ω_m.
In any case, the values of ω_m (and the corresponding modal shapes $\{U\}_m$) would become grossly approximated as m increases to values comparable with M_n. The spatial "wavelenght" of the modes, indeed, becomes then comparable with the element size (fig. 69), and the shape functions are not usually able, as a consequence, to represent the distortions taking place inside the elements.
The solutions of (3.2.13), i.e. the search for eigenvalues $\omega_1, \omega_2, \ldots\omega_m$ (with $m << M_n$, see above) and for the corresponding eigenvectors $\{U\}_1, \{U\}_2, \ldots, \{U\}_m$ can be carried out by well-known numerical techniques (e.g. the "iterated subspace technique", see APPENDIX). In the following it will be assumed that:
i) $[K]$ and $[M]$ have been computed for the F.E. mesh covering Ω;
ii) $\omega_1, \omega_2, \ldots,\omega_m$ and $\{U\}_1, \{U\}_1, \{U\}_2, \ldots,\{U\}_m$ have been determined numerically from (3.2.13); note that $\{U\}_1, \{U\}_1, \{U\}_2, \ldots,\{U\}_m$ are defined apart from arbitrary numerical factors $\alpha_1, \alpha_2, \ldots,\alpha_m$ unless some "normalization criteria" are used (see further on).
It can be shown that for any two distinct modes j and l, characterized by:

$$\begin{aligned} &\omega_j \text{ and } \{U\}_j, \\ &\omega_l \text{ and } \{U\}_l, \text{ with } j \neq l;\ \omega_j \neq \omega_l \end{aligned} \tag{3.2.14}$$

the following orthogonality condition holds:

$$\{U\}_j^T[M]\{U\}_l = 0 \tag{3.2.15}$$ [41]

---

[39] We establish to put $m = 1$ for the lowest eigenvalues, and to proceed in an orderly way through successively increasing values: $\omega_1 < \omega_2 < \omega_3 \ldots$

[40] This "limited number" could be as low as 5 to 10, according to some authors and to current practice. Of course, if the response to seismic input is to be determined, all the modes corresponding to frequencies $\omega_m/2\pi$ relevant to the earthquake spectrum energy content should be included in the number in question.

[41] In fact we have: $[M]\omega^2_j\,\{U\}_j = [K]\,\{U\}_j$, $[M]\omega^2_l\,\{U\}_l = [K]\,\{U\}_l$ Premultiplying the first of these matrix equations by $\{U\}^T_l$ and the second one by $\{U\}^T_j$ we have:
$\{U\}^T_l\,[M]\omega^2_j\,\{U\}_j = \{U\}^T_l\,[K]\,\{U\}_j$, $\{U\}^T_j\,[M]\omega^2_l\,\{U\}_l = \{U\}^T_j\,[K]\,\{U\}_l$; taking the transpose of L.H.S. and R.H.S. of the last one gets $\{U\}^T_l\,[M]\omega^2_l\,\{U\}_j = \{U\}^T_l\,[K]\,\{U\}_j$; comparing with the last-but-one matrix equation we see that the two R.H.S. are the same; so the two L.H.S. must be equal. This entails: $(\omega^2_j - \omega^2_l)\cdot(\{U\}^T_l\,[M]\,\{U\}_j) = 0$ and, since by (3.2.14) $(\omega^2_j - \omega^2_l) \neq 0$, the orthogonality condition (3.2.15) is deduced. It is easy to show that also $\{U\}^T_j\,[K]\,\{U\}_l = 0$.

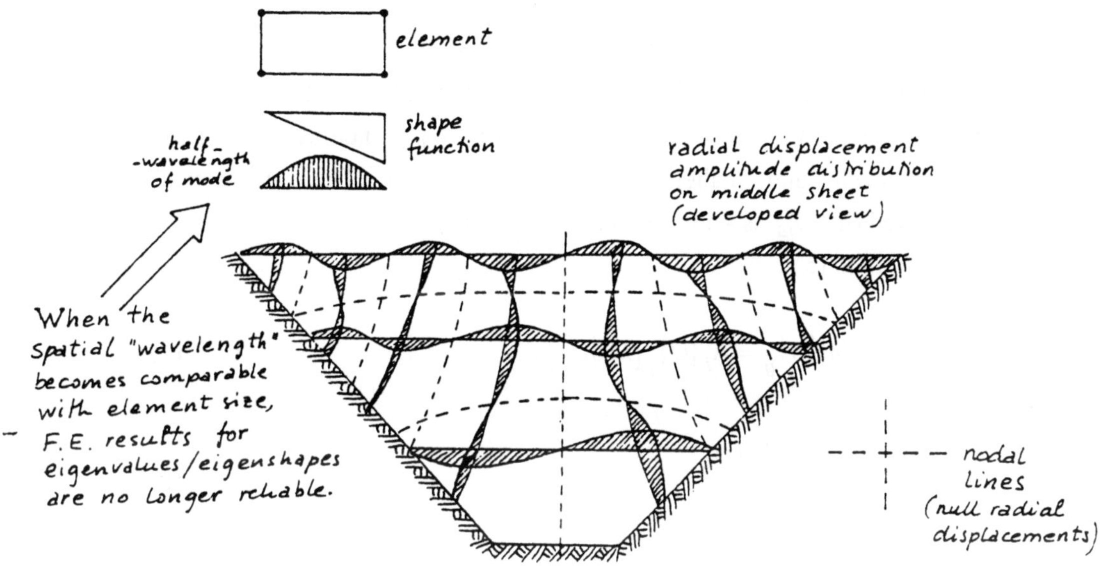

Fig. 69
High-order vibration modes: m comparable with M_n

Another useful property is the following:
Any given arbitrary $3M_n$ distribution of nodal amplitudes $\{U\}_g$ can be developed as a univocally defined linear combination of the modal shapes:

$$\{U\}_g = A_1\{U\}_1 + A_2\{U\}_2 + \ldots\ldots + A_m\{U\}_m + \ldots\ldots\ldots \quad (3.2.16)$$

where in principle the sum should be extended to the complete set of $3M_n$ modes. In practice (3.2.16) can be truncated at a value $m << M_n$, since usually the coefficients A_m become smaller and smaller for "large" values of m. These coefficients can be easily computed (if $\{U\}_g$ is known) thanks to (3.2.15):
from (3.2.16), (3.2.15)

$$\{U\}_j^T[M]\{U\}_g = A_j\{U\}_j^T[M]\{U\}_j, \quad (3.2.17)$$

so that

$$A_j = \frac{\{U\}_j^T[M]\{U\}_g}{\{U\}_j^T[M]\{U\}_j} \quad (3.2.18)$$

Note that if we change $\{U\}_j$ by an arbitrary factor αj, A_j will change by the factor $\frac{1}{\alpha_j}$ and the product $A_j\ \{U\}_j$ in sum (3.2.16) will remain unaltered. For reasons that will be apparent further on, we choose to remove the arbitrariness in the "scale" of each eigenvector $\{U\}_m$ by adopting the following normalization criterion: given a vector $\{1\}$ defined as a $3M_n$ vector of unit displacement in a given direction $\vec{g}$, repeated uniformly at all M_n nodes, the <u>normalized eigenvector</u> $\{\psi\}_m$ is defined as:

$$\{\psi\}_m = \alpha_m\{U\}_m, \quad (3.2.19)$$

with
$$\alpha_m = \frac{\{1\}^T[M]\{U\}_m}{\{U\}_m^T[M]\{U\}_m}$$

It is easy to show that $\{\psi\}_m$ is independent of the arbitrary "scale" initially chosen for $\{U\}_m$; $\{\psi\}_m$ is also a <u>dimensionless</u> vector, having the dimension of a length only $\{U\}_m$.. However, its "scale" will change with the direction $\vec{g}$ chosen; for practical use (ground-shaking analysis, see further on) in principle it would be convenient to effect three distinct normalizations for each $\{U\}_m$: one with $\vec{g}$ along the vertical and the other two with $\vec{g}$ along two perpendicular horizontal directions (for instance one with $\vec{g}$ along the N - S alignment and the other with $\vec{g}$ along the E - W alignment).

a$_2$) **Modal response**

The orthogonality condition (3.2.15) holds of course also for the normalized modes:

$$\{\psi\}_j^T [M] \{\psi\}_l = 0 \qquad \text{for } j \neq l\,;\ \omega_j \neq \omega_l \tag{3.2.20}$$

If for every mode we define the "modal mass" μ_m along the direction chosen for vector $\vec{g}$ as:

$$\mu_m = \{\psi\}_m^T [M] \{\psi\}_m \tag{3.2.21}$$

(μ_m = a scalar quantity) it can be seen that (for any direction that can be chosen for $\vec{g}$):
$\sum_m \mu_m = \sum_k \rho_k V_k$ = total mass of the system.

Also a modal stiffness χ_m along direction of $\vec{g}$ can be defined:

$$\chi_m = \{\psi\}_m^T [K] \{\psi\}_m\,, \tag{3.2.22}$$

(χ_m is also a scalar);

and since by (3.2.13) it is:

$$[K]\{\psi\}_m = \omega_m{}^2 [M] \{\psi\}_m\,, \tag{3.2.23}$$

if we premultiply both sides of (3.2.23) by $\{\psi\}_m^T$ we obtain:

$$\chi_m = \omega_m{}^2 \mu_m\,, \tag{3.2.23bis}$$

as for a single d.o.f. linear oscillator of mass μ_m and spring stiffness χ_m.
Now let us consider the stationary response of our arch dam, i.e. $\{u\} = \{U\}\exp(i\omega t)$, $\{U\}$ unknown, to an external sinusoidal excitation, of arbitrary frequency $\omega/2\pi$, defined by a force-vector $\{F\} = \{\phi\}\exp(i\omega t)$.
From (3.2.11) one can write, after the initial transient has died out:

$$[K]\{U\}e^{i\omega t} - \omega^2[M]\{U\}e^{i\omega t} = \{\phi\}e^{i\omega t}, \qquad (3.2.24)$$

where the common factor $e^{i\omega t}$ can be dropped from all terms of the equation.
Now after (3.2.16) we can express $\{U\}$ as a linear combination (with yet unknown coefficients x_m) of the normalized eigenvectors $\{\psi\}_m$:

$$\{U\} = x_1\{\psi\}_1 + x_2\{\psi\}_2 + \ldots\ldots + x_m\{\psi\}_m + \ldots \qquad (3.2.25)$$

Note that the "modal amplification factors" x_m are homogeneous with a length.
Premultiplying both sides of (3.2.24) (without the factor $e^{i\omega t}$) by $\{\psi\}_j^T$ and keeping into account (3.2.25) together with (3.2.20) it comes:

$$x_j\chi_j - \omega^2 x_j\mu_j = \{\psi\}_j^T\{\phi\}; \qquad (3.2.26)$$

since by (3.2.23bis) $\mu_j = \dfrac{\chi_j}{\omega_j^2}$ it comes:

$$x_j = \frac{\{\psi\}_j^T\{\phi\}}{\chi_j\left(1 - \dfrac{\omega^2}{\omega_j^2}\right)} \qquad (3.2.27)\,^{42}$$

The formula is valid for negligible damping and $\omega \neq \omega_j$. Undamped response would tend to $\pm\infty$ for $\omega \to \omega_j \mp \varepsilon$ (with $\varepsilon \to 0$); so it is necessary to consider a finite "modal damping factor" $\nu_j \neq 0$. Using complex notation:

$$x_j = \frac{\{\psi\}_j^T\{\phi\}}{\chi_j\left(1 - \dfrac{\omega^2}{\omega_j^2} + 2i\nu_j\dfrac{\omega}{\omega_j}\right)}$$

After all the modal amplitudes x_j ($1 \leq j \leq m$) have been thus obtained, (3.2.25) gives the nodal amplitudes $\{U\}$.
Eq. (3.2.27) shows that the amplification factor x_j of mode $\{\psi\}_j$ has the same value as the response amplitude of a linear, single d.o.f. oscillator of frequency $\omega/2\pi$, subjected to the force:

---

[42] Note that the $\{\psi\}_m$ are dependent upon the direction $\vec{g}$ chosen for the vector {1}; but suppose we change this direction. Then a particular mode $\{\psi\}_j$ may change by a factor β ; χ_j , by (3.2.22), will change by a factor β^2; $\{\psi\}^T_j$ $\{\phi\}$ will change by a factor β. Then x_j , by (3.2.27), changes by a factor $\beta/\beta^2 = 1/\beta$, and the product $x_j\{\psi\}_j$ in (3.2.25) will not change ($\beta \times 1/\beta = 1$): the response of a mode to a given external excitation is univocally defined by (3.2.27), unless $\mu_j = \chi_j = 0$ for chosen $\vec{g}$.

$$f_j = \{\psi\}_j^T \{\phi\} e^{i\omega t} \tag{3.2.28}$$

In this connection, remark that the "base" of the normalized modes $\{\psi\}_m$ can be considered as defining a multi-dimensional "generalized projection space" in which the $\{\psi\}_m$ are the directions of the reference "axes". The x_1 , x_2 , ...x_m are then the "coordinates" of the displacement vector $\{U\}$, i.e. the projections of $\{U\}$ onto these "axes".

By the same token, f_j is the Lagrangian component of the system of external forces - defined (in the Cartesian space) by $\{\phi\}$ - onto the "coordinate axis" j of the "generalized projection space"; see also the meaning of f_j in the context of the virtual work principle.

The above result for the stationary modal response to a sine excitation can be extended also to the perturbations caused by initial conditions.

In this way the statement that each eigenmode behaves (indipendently of the others) as a single d.o.f. linear oscillator holds for all the frequencies of excitation and for all conditions. The overall response of the system will consist of the superposition (separately for every degree of freeedom u_r) of the responses of all modes along that d.o.f. to the sine excitation., according to (3.2.25), (3.2.28).

As a further step, let us look for the response to an arbitrarily time-varying excitation $\{F(t)\}$. This can always be developed as a Fourier integral:

$$\begin{aligned} \{F(t)\} &= \left\{ \int_0^\infty A(\omega)\cos\omega t \, d\omega + \int_0^\infty B(\omega)\sin\omega t \, d\omega \right\} \\ &= \int_0^\infty \{A(\omega)\}\cos\omega t \, d\omega + \int_0^\infty \{B(\omega)\}\sin\omega t \, d\omega \end{aligned} , \tag{3.2.29}$$

and indicating for brevity $\{F(t)\}$ as :

$$\{F(t)\} = \left\{ \int_0^\infty A(\omega)\, e^{i\omega t} \, d\omega \right\}, \quad A(\omega) \text{ complex}$$

after (3.2.27) it is easy to conclude that the response of the system will be:

$$\{u(t)\} = \sum_j \int_0^{\infty} \frac{\{\psi\}_j^T \{A(\omega)\}\{\psi\}_j \, e^{i\omega t}}{\chi_j \left(1 - \frac{\omega^2}{\omega_j^2} + 2i\nu_j \frac{\omega}{\omega_j}\right)} d\omega \qquad (3.2.30)^{43}$$

Again, the overall response (3.2.30) is the superposition (node by node and d.o.f. by d.o.f. of our F.E. mesh) of the responses of each single mode (ω_j , $\{\psi\}_j$) computed indipendently of each other as those of single, one d.o.f. linear oscillators of frequency $\omega_j /2\pi$.

As to the number of (j_{max}) of modes that need be included in the Σ_j in order to get a sufficiently accurate response representation, opinions differ. Some authors, as already said, consider j_{max} = 5 to 10 to be sufficient, while others claim that it must be $j_{max} \geq 30$ to 40.

a$_3$) Ground-shaking response

The more interesting type of dynamic excitation for dams is that of ground shaking, which is the typical input a dam receives e.g. during an earthquake. This input can be defined (in the "time domain") by an accelerogram (ground accelerations $a_x(t)$, $a_y(t)$, $a_z(t)$, versus time), or by Fourier spectra along the three Cartesian directions, or by the so-called (Housner) "response spectra" (see further on). The last two types of representation are in the "frequency domain".

We will for the moment consider the simple case of uniform ground shaking with an assigned law:

$$u_g(t) \qquad (3.2.32)$$

---

43 With $\{G(\omega)\} = \{[A(\omega)A^*(\omega)]^{1/2}\}$ that is the "spectral density" of the excitation function, the spectral density of the response is: $\{R(\omega)_r\} = \left\{\left[\sum_j R_j(\omega)_r \cdot \sum_j R_j^*(\omega)_r\right]^{\frac{1}{2}}\right\}$, (3.2.31)

where $R_j(\omega)_r = \dfrac{\{\psi\}_j^T \{A(\omega)\} \cdot \psi_{jr}}{\chi_j (1 - \frac{\omega^2}{\omega_j^2} + 2i\nu_j \frac{\omega}{\omega_j})}$

where r denotes the rth degree of freedom.

along a given direction $\vec{g}$; that motion is assumed to carry along rigidly the boundary nodes of the mesh that are internal to the foundation of rock mass. The motions $\{u(t)\}$ of the nodes of our discretized system (arch dam + foundation) can be written as:

$$\{u(t)\} = u_g(t)\{1\} + \{u_r(t)\}, \tag{3.2.33}$$

(see fig. 70). $u_g(t)\{1\}$ being short-hand notation for a vector of uniform (rigid-body) nodal displacements of value $u_g(t)$ along direction $\vec{g}$ and $\{u_r(t)\}$ being the vector of nodal displacements relative to the shaking ground. Since for rigid body motion no elastic strain is generated, $[K]u_g\{1\}=\{0\}$ for any time t and for arbitrary direction $\vec{g}$ of ground shaking.

Then application of (3.2.11) yields:

$$[K]\{u_r(t)\} + [M]\left(\ddot{u}_g(t)\{1\} + \{\ddot{u}_r(t)\}\right) = \{0\}, \tag{3.2.34}$$

or:

$$[K]\{u_r(t)\} + [M]\{\ddot{u}_r(t)\} = -[M]\ddot{u}_g(t)\{1\}, \tag{3.2.35}$$

Thus the motion of our system relative to the shaking ground is the same as if -with fixed ground, i.e. with $u_g(t)=0$ - the dam were acted upon by forces $\{F\}$ given by:

$$\{F\} = -[M]\{1\}\ddot{u}_g(t) \qquad (3.2.36)^{44}$$

These are seen to represent the generalized Lagrangian forces deriving from the D'Alembert inertial forces as a consequence of the acceleration $\ddot{u}_g$, along direction $\vec{g}$, of the reference frame fixed to the shaking ground, with respect to which the relative displacements $\{u_r\}$ are measured.

---

[44] As in (3.2.19), $\{1\}$ is a unit (dimensionless) vector along the given direction of ground shaking $\vec{g}$ repeated identically for all nodes:

$$\{1\} = \begin{Bmatrix} \cos \hat{gx} \\ \cos \hat{gy} \\ \cos \hat{gz} \\ \cdots\cdots \\ \cos \hat{gx} \\ \cos \hat{gy} \\ \cos \hat{gz} \end{Bmatrix},$$

($3M_n$ rows) - It is assumed for simplicity that $\vec{g}$ is the same direction chosen to compute the modal masses and stiffnesses in (3.2.19)..........(3.2.22) .

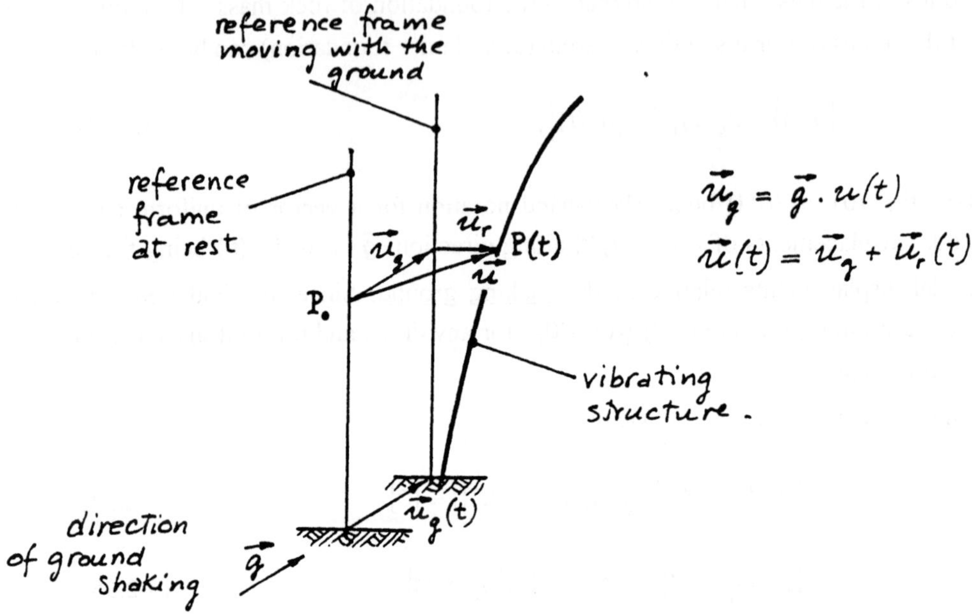

Fig. 70
Near-field ground shaking: absolute and relative dam displacement

In principle, we can now decompose $\{u_r(t)\}$ onto the base-space of the $\{\psi\}_m$, see (3.2.16) and (3.2.25):

$$\{u_r(t)\} = x_1(t)\{\psi\}_1 + x_2(t)\{\psi\}_2 + \ldots + x_m(t)\{\psi\}_m + \ldots \quad (3.2.37)$$

from which obviously

$$\{\ddot{u}_r(t)\} = \ddot{x}_1\{\psi\}_1 + \ddot{x}_2\{\psi\}_2 + \ldots + \ddot{x}_m\{\psi\}_m + \ldots \quad (3.2.38)$$

Premultiplying each side of (3.2.35) by $\{\psi\}^T_m$ and recalling the orthogonality condition (3.2.20) it comes:

$$\begin{aligned}\{\psi\}_j^T[K]\{\psi\}_j x_j + \{\psi\}_j^T[M]\{\psi\}_j \ddot{x}_j &= -\{\psi\}_j^T[M]\{1\}\ddot{u}_g \\ &= -\{1\}^T[M]\{\psi\}_j \ddot{u}_g\end{aligned} \quad (3.2.39)$$

which by virtue of (3.2.21), (3.2.22) becomes[45]:

$$\chi_j x_j + \mu_j \ddot{x}_j = -\mu_j \ddot{u}_g\,, \quad (3.2.40)$$

i.e. the time-varying amplitude $x_j(t)$ with which mode j participates in the motion ("participation factor of mode j"):

$$x_j(t) = \frac{\{\psi\}_j^T[M]\{u_r(t)\}}{\{\psi\}_j^T[M]\{\psi\}_j}, \text{ after (3.2.37), (3.2.18),} \quad (3.2.41)$$

obeys the same equation of motion as a linear, single d.o.f. oscillator of frequency $\omega/2\pi$:

$$\omega_j^2 = \frac{\chi_j}{\mu_j} \quad (3.2.42)$$

having a mass μ_j equal to the modal mass along the direction of ground shaking, when the support of the oscillator moves with the same assigned law $u_g = u_g(t)$ of ground shaking.

Another way of seeing this is the remark that modal mass $\mu_m = \{1\}^T[M]\{\psi\}_m$ is the "participation factor" of the unit ground motion along direction $\vec{g}$ to the mode $\{\psi\}_m$,

---

[45] It is immediate to show that, if in (3.2.19) we put $\{\psi\}_m$ in place of $\{U\}_m$, we get $\alpha=1$, so that $\{\psi\}^T_m[M]\{\psi\}_m = \{1\}^T[M]\{\psi\}_m$, i.e $\mu_m = \{1\}^T[M]\{\psi\}_m$ (unless $\{1\}^T[M]\{\psi\}_m = 0$).
Since evidently the product $\{\psi\}^T_m[M]\{\psi\}_m$ is equal to its transpose, it is also $\mu_m = \{\psi\}^T_m[M]\{1\}$, see (3.2.39). Note that only with the above chosen normalization criterion for $\{\psi\}_m$ is it possible to have on the R.H.S. of eq. (3.2.40) the actual ground accelaration $\ddot{u}_g(t)$.

i.e. the work done by the inertial forces generated by unit accelerations acting along $\vec{g}$ on virtual displacements $\{\psi\}_m$.

If $\ddot{u}_g(t)$ is decomposed into Fourier integrals:

$$\ddot{u}_g(t) = \int_0^\infty A_g(\omega)\cos\omega t \, d\omega + \int_0^\infty B_g(t)\sin\omega t \, d\omega,$$

in a way analogous to (3.2.29), it comes for the response $\{u_r(t)\}$ of the system, see (3.2.30):

if $\ddot{u}_g(t) = \int_0^\infty A_g(\omega)\, e^{i\omega t} \, d\omega$, A_g complex

$$\{u_r(t)\} = -\sum_j \int_0^\infty \frac{\mu_j A_g(\omega)\, e^{i\omega t}\{\psi\}_j}{\chi_j(1-\frac{\omega^2}{\omega_j^2}+2i\nu\frac{\omega}{\omega_j})} d\omega \qquad (3.2.43)$$

The spectral density of the response at frequency $\omega/2\pi$ is given, for the generic r^{th} d.o.f. $u_r(t)$, by:

$$\{R(\omega)_r\} = \left\{ \left[\sum_j R_j(\omega)_r \cdot \sum_j R_j^*(\omega)_r \right]^{\frac{1}{2}} \right\}, \qquad (3.2.44)$$

where $$R_j(\omega)_r = \frac{\mu_j A_g(\omega)\psi_{jr}}{\chi_j(1-\frac{\omega^2}{\omega_j^2}+2i\nu_j\frac{\omega}{\omega_j})}$$

Thus, in principle, the problem of the elastic response of an arch dam to uniform, but otherwise arbitrary, ground shaking is amenable to solution[46].

As a last step, let us see how to proceed if, instead of operating with a Fourier decomposition of the ground shaking motion, the Housner response spectrum technique is used.

The velocity response spectrum $V(\omega)$ of a given ground-shaking event is defined as the maximum relative velocity assumed by a single d.o.f.[47] linear oscillator of frequency $f = \omega/2\pi$ acted upon by the support motion $u_g(t)$ experienced during that event. Spectra of this kind are routinely produced for every important earthquake (as

---

[46] A note on the B.C.: evidently the nodes of the foundation rock mass that are assumed to move with law $u_g(t)$ will have zero relative displacements, as in the determination of the free vibrations modes $\{\psi\}_m$.

[47] The coordinate of the d.o.f. being coincident in direction with the direction $\vec{g}$ of the ground motion; thus usually three spectra, for the three Cartesian directions, need be known.

are the Fourier spectra), so that there is a "catalog" of Housner spectra which the dam engineer can use for his/her seismic analyses (fig. 71).

By developments analogous to the above ones we can derive, for the "modal velocities" $\dot{x}_j$ of the relative motion, see (3.2.37)....(3.2.40):

$$\max(\dot{x}_j) = V(\omega_j) \tag{3.2.45}$$

(remember that x_j is homogeneous with a length)

It is further known (Housner) that (3.2.45) implies:

$$\max(x_j) \cong \frac{1}{\omega_j} V(\omega_j), \tag{3.2.46}$$

so that the maximum relative response of <u>each mode</u> (in terms of relative displacements) can readily be determined.

The <u>global</u> response (summation over all the modes for each d.o.f.) is then obtained in an approximate way through the technique known as SRSS ("Square Root Sum of Squares"). Not only for nodal displacements $\{u\}$, but for every quantity Q of interest (strain, stress, etc.), we can indeed determine the partial max. response Q_{jmax} pertaining to mode j:

$$Q_{j\max} \cong \frac{1}{\omega_j} V(\omega_j) q_j, \quad \text{if } q_j \text{ be the value of } Q \tag{3.2.47}$$

for $x = 1$, i.e. for nodal displacements given by the normalized mode $\{\psi\}_j$; (q_j can be called the "<u>normalized modal effect</u>" for the quantity in question).

The global SRSS max. response for Q is then approximately given by:

$$Q_{\max} \cong \sqrt{Q^2_{1\max} + Q^2_{2\max} + \ldots + Q^2_{m\max} + \ldots}$$

$$= \sqrt{\sum_j Q^2_{j\max}} = \sqrt{\sum_j \frac{1}{\omega_j^2} V^2(\omega_j) q_j^2}, \tag{3.2.48}$$ [48]

which is much faster to compute -if the Housner response spectrum is known- than through the Fourier decomposition.

Sometimes the Housner response spectra are available in terms of max. relative accelarations $A(\omega)$ for the linear oscillator of frequency $\omega/2\pi$ (instead of in terms of

[48] The phase information for each modal component, which can in principle be preserved when using the Fourier technique if $A_g(\omega)$ and $B_g(\omega)$ are known separtely, is lost with the Housner approach. Considering the time-varying realtive phases of the modal motions with different frequencies $\omega/2\pi$ as pseudo-random variables, the "effective" (r.m.s) average of the global effect is given by (3.2.48), if the excitation covers an interval of time large in comparison with the period $\omega/2\pi$ of all the modes that are significantly excited.

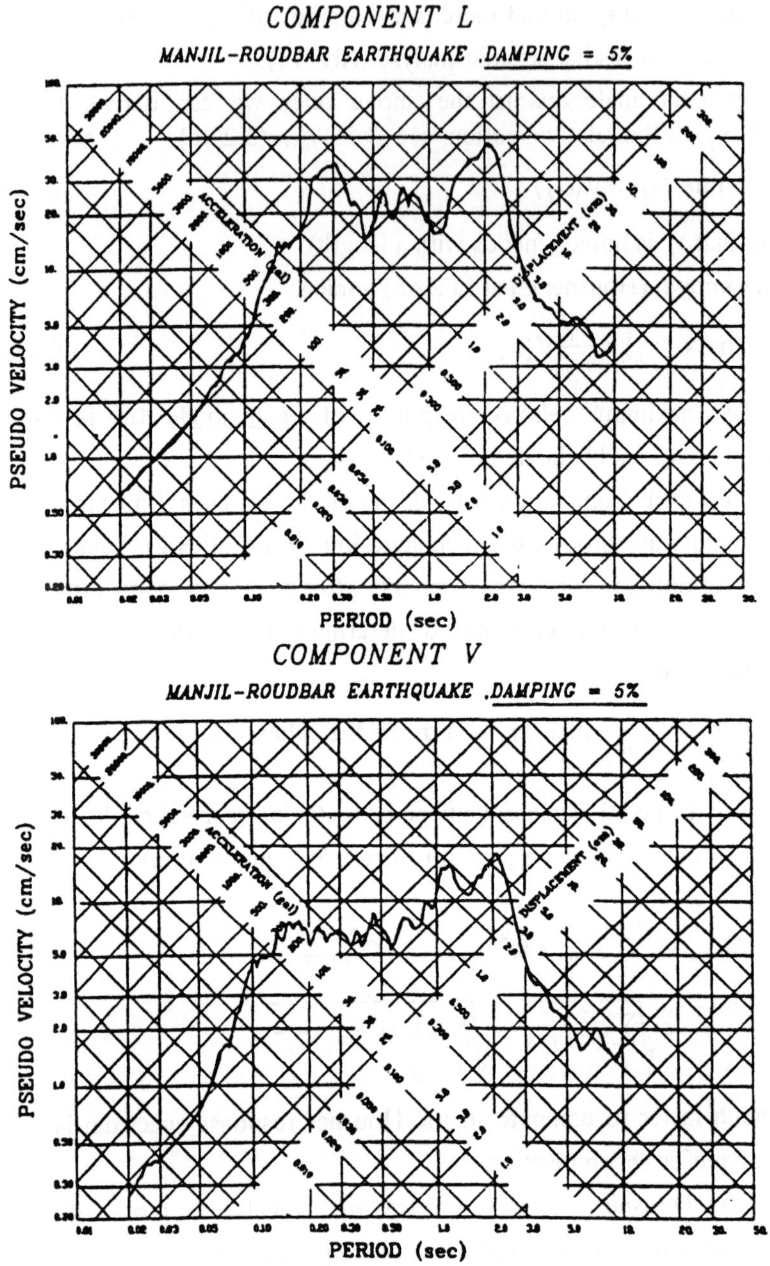

Fig. 71
Examples of Housner response spectra (υ=0.05)

max. relative velocity $V(\omega)$ as above). In that case, instead of (3.2.46) and (3.2.47), one uses the approximate relationships:

$$\begin{aligned} \max(\dot{x}_j) &= V(\omega_j) \cong \frac{A(\omega_j)}{\omega_j} , \\ \max(x_j) &\cong \frac{V(\omega_j)}{\omega_j} \cong \frac{A(\omega_j)}{\omega_j^2} , \\ Q_{j\max} &\cong \frac{A(\omega_j)}{\omega_j^2} q_j , \end{aligned} \tag{3.2.49}$$

so that instead of (3.2.48) the SRSS technique gives:

$$Q_{\max} \cong \sqrt{\sum_j Q^2{}_{j\max}} \cong \sqrt{\sum_j \frac{A^2(\omega_j)}{\omega_j^4} q_j{}^2} \tag{3.2.50}$$

Note that until now we have assumed <u>zero damping</u>. The Housner spectra are parametrized as a function of the relative (viscous) damping ν (ratio between the actual damping factor of the oscillator and its critical damping factor) (fig. 72). By choosing for each mode j the "right" damping, the effects of energy dissipations can be automatically included in the evaluation of max. response through (3.2.50). More about the effects of damping will be said in the next sections.

<u>Note on the mass of the foundation rock</u> - If, as usual, a certain volume Ω_r of the rocky foundation is discretized, it would seem natural to include it (with the proper specific mass ρ_r) in the determination of the mass matrix $[M]$, and, implicitly, in all the subsequent developments. However, by so doing one introduces, when computing eigenvalues ω_j and eigenshapes $\{\psi\}_j$, very-low-frequency modes which correspond mainly to "almost rigid-body" motions of the dam (e.g. translational or rocking modes) on a compliant foundation mass. These modes are not of interest as far as the stress-strain state of the dam is concerned, and the only appreciable effect of including them would be a considerable increase of the total number of modes to be considered, thus making the computation unnecessarily heavy and costly. A trick to avoid this inconvenient consists in including the rocky domain Ω_r in the computation of the stiffness matrix $[K]$, but <u>not in the computation of the mass matrix</u> $[M]$ (i.e. assuming $\rho_r = 0$). The only justification of this procedure is, admittedly, an empirical one: it has

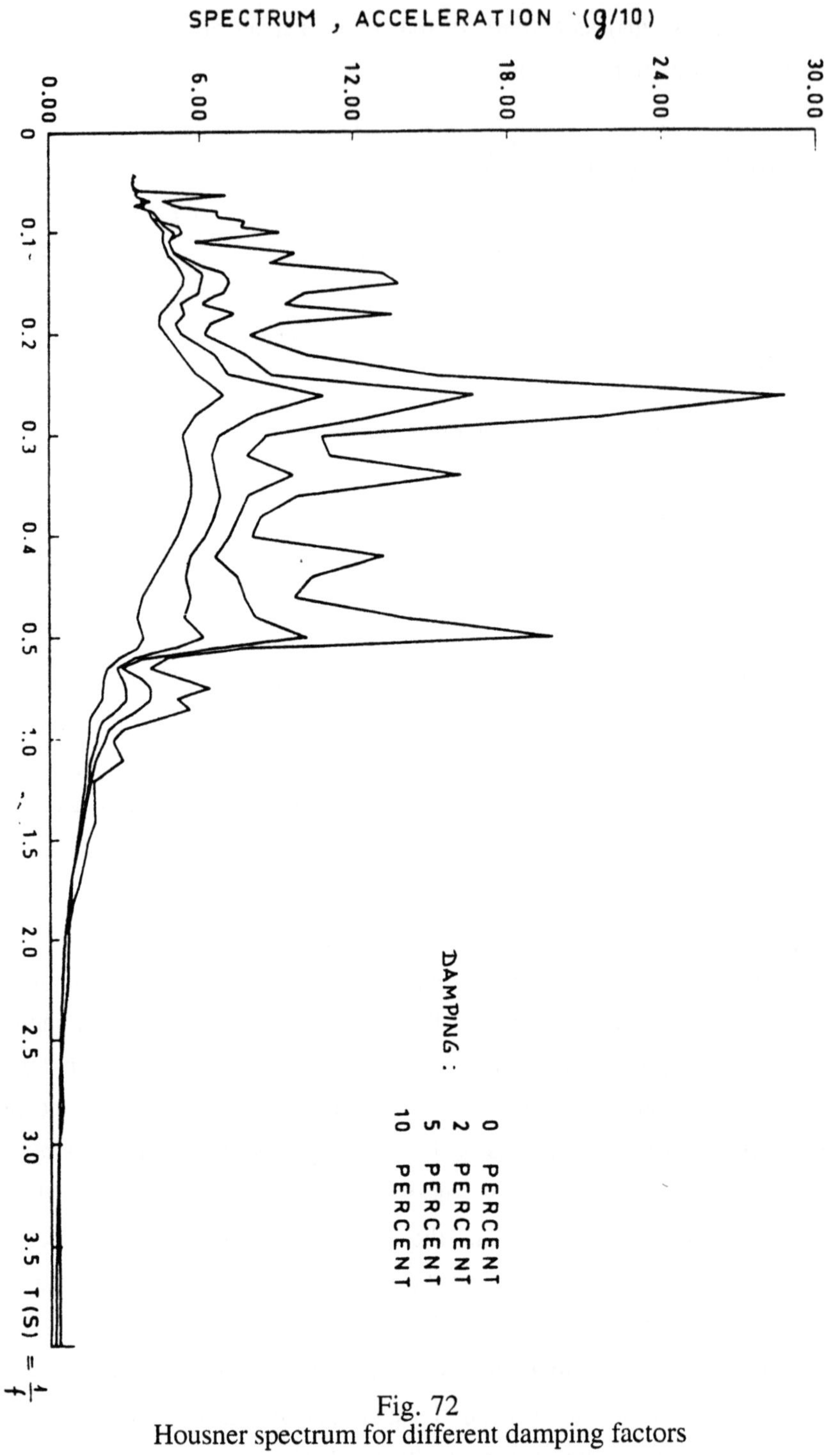

Fig. 72
Housner spectrum for different damping factors

been proved in practice that it gives acceptable results (see comparisons of computed vs. measured eigenfrequencies and eigenshapes, fig. 73).

On the other hand, it can be remarked that in the more "classical" methods (Ritter, T.L.M....) the rocky mass was represented (if at all) only through its compliance, e.g. via the Vogt coefficient, to the exclusion of any inertial effect; yet eigenfrequencies and eigenshapes determined on the basis of these assumptions were often more than acceptable. This can be seen as one more empirical "proof" that a compliant, but massless foundation can be used as a model in dynamic analyses of an arch dam.

b) Influence of structural damping (empty reservoir, rigid boundaries at the outermost nodes of foundation mesh)

When treating the problem of free vibrations (nodal forces $\{F\}=\{0\}$) for undamped systems (par. 3.2.a), the assumption (3.2.12), i.e.:

$$\{u\} = \{U\}e^{i\omega t}$$

implicitly based on a real value for ω, led to a set of real eigenmodes $\{U\}_m$ or, if normalized, $\{\psi\}_m$.

In other words, in the absence of damping all pure modal motions are harmonic functions of time having the same phase at all points of the system. This circumstance of exactly in-phase modal motions does not hold in general for damped systems.

From the energetic viewpoint, the existence of damping implies -for a linear system- the existence of a dissipation function giving the rate of the local density of energy dissipation and, by integration over the volume Ω, the global energy loss of the system per unit time. In the case of viscous isotropic damping in the material the dissipation function can assume, e.g., the following form:

$$\dot{\varepsilon}_d = -\frac{dH}{dt} = \iiint_{\Omega} f^2(x,y,x,\dot{q})dV, \tag{3.2.51}$$

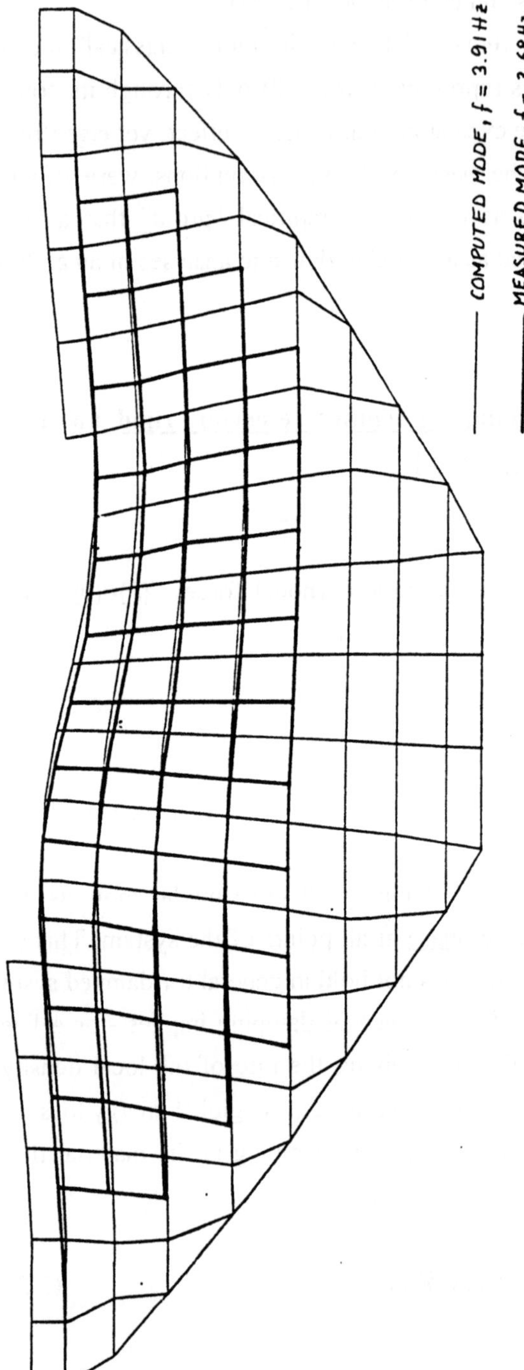

Fig. 73
Comparison between F.E. computed and in-situ measured modes, for $\rho_r=0$

q being the generalized Lagrangian coordinates describing the dynamic configuration of the system at instant t, and $\dot{q}=\frac{\partial q}{\partial t}$. If the material is characterized by a "viscosity" η, one can asssume the following relationships:

$$\dot{\mathcal{E}}_d = \iiint_\Omega \eta\{\dot{\gamma}\}^T\{\dot{\gamma}\}dV, \qquad (3.2.52)$$

with $\{\dot{\gamma}\}^T =< 2\dot{\varepsilon}_x \;\; 2\dot{\varepsilon}_y \;\; 2\dot{\varepsilon}_z \;\; \dot{\gamma}_{yz} \;\; \dot{\gamma}_{zx} \;\; \dot{\gamma}_{xy} >$,

and if we use the same shape functions $[N]_k$ of the static analysis for interpolating the displacement field inside the generic element from the nodal values, we obtain successively:

$$\{\dot{\gamma}\}^T = [\mathcal{D}]_k\{\dot{u}\}_k \;,\; \text{with} \;\; [\mathcal{D}]_k = [\partial]_k[N]_k$$

$$[\partial]_K = \begin{bmatrix} 2\frac{\partial}{\partial x} & 0 & 0 & 0 & \frac{\partial}{\partial z} & \frac{\partial}{\partial y} \\ 0 & 2\frac{\partial}{\partial y} & 0 & \frac{\partial}{\partial z} & 0 & \frac{\partial}{\partial x} \\ 0 & 0 & 2\frac{\partial}{\partial z} & \frac{\partial}{\partial y} & \frac{\partial}{\partial x} & 0 \end{bmatrix}; \text{ for } [N]_k \text{ see (3.1.2);} \qquad (3.2.53)$$

$$\dot{\mathcal{E}}_{dk} = \eta_k \iiint_{\Omega_k} \{\dot{u}\}_k^T[\mathcal{D}]_k^T[\mathcal{D}]_k\{\dot{u}\}_k\, dV = \frac{1}{2}\{\dot{u}\}_k^T[C]_k\{\dot{u}\}_k\,, \qquad (3.2.54)$$

with $[C]_k = 2\eta_k \iiint_{\Omega_k}[\mathcal{D}]_k^T[\mathcal{D}]_k\, dV$, under the assumption (3.2.52).

By using the incidence matrix $[\,\mathrm{I}\,]_k$ of element Ω_k we get for the whole domain Ω, remembering earlier results,

$$\dot{\mathcal{E}}_d = \frac{1}{2}\{\dot{u}\}^T[C]\{\dot{u}\}, \quad \text{with} \quad [C] = \sum_k [\mathrm{I}]_k^T[C]_k[\mathrm{I}]_k \qquad (3.2.55)$$

(Note that under assumption (3.2.52) the matrix $[C]$ is proportional to the stiffness matrix $[K_o]$ computed with Poisson ratios ν_k all set to zero: $\nu_k \equiv 0$ implies $[C]=\eta/E\,[K_o]$) ; $[C]$ is symmetric and positive-definite.

More generally, the form (3.2.55) for $[C]$ can be retained, but with definitions of $[C]_k$ that can differ from (3.2.54).

The total energy H of the isolated system (i.e. without any external force):

$$H = W + \mathcal{E}_c = \frac{1}{2}\{u\}^T[K]\{u\} + \frac{1}{2}\{\dot{u}\}^T[M]\{\dot{u}\} , \tag{3.2.56}$$

H = sum of the deformation energy and the kinetic energy, will now no longer be a constant, since (see 3.2.51):

$$\dot{H} = \frac{dH}{dt} = -\dot{\mathcal{E}}_d = -\frac{1}{2}\{\dot{u}\}^T[C]\{\dot{u}\} \tag{3.2.57}$$

The Lagrangian equations of motion will take the form:

$$\frac{\partial}{\partial t}\left(\frac{\partial H}{\partial \dot{q}}\right) = -\frac{\partial H}{\partial q} - \frac{\partial \dot{H}}{\partial \dot{q}} , \tag{3.2.58}$$

with q and $\dot{q}$ = any of the $3M_n$ d.o.f $\{u\}$ and its time derivative, respectively. In our discretized system, thus, we have from (3.2.58), (3.2.56), (3.2.57):

$$[K]\{u\} + [C]\{\dot{u}\} + [M]\{\ddot{u}\} = \{0\} \tag{3.2.59}$$

From this it can be shown that -for the generic eigenmode- a solution of the form

$$\{u\} = \{U\}_0 \, e^{-\lambda t} \, e^{i\omega t} , \tag{3.2.60}$$

with $\{U\}_0$ = real modal shape of the undamped system ($[C] = [0] \rightarrow \{U\}_0$), i.e. free oscillations <u>all in phase</u> and consisting of exponentially damped harmonic functions of time, is obtained <u>only if</u>:

$$[C] = 2\alpha_1 \frac{[K]}{\omega_0} + 2\alpha_2\omega_0[M] , \tag{3.2.61}$$

where α_1 and α_2 = numerical coefficients, $\omega_0 / 2\pi$ = frequency of the corresponding undamped mode; i.e., $\{U\}_0$ and ω_0 satisfy the equation:

$$[K]\{U\}_0 - \omega_0{}^2[M]\{U\}_0 = \{0\} \tag{3.2.62}$$

Indeed, by posing

$$\{u\} = \{U\}_0 \, e^{\sigma t} , \tag{3.2.63}$$

with: $\sigma = i\,\omega - \lambda$ (3.2.64)

and keeping into account (3.2.62), (3.2.64) it is easy to verifiy that (3.2.59) is satisfied with:

$$\omega=\omega_0\sqrt{1-(\alpha_1+\alpha_2)^2}\quad,\quad \lambda=(\alpha_1+\alpha_2)\omega_0$$

$$\nu=\frac{\lambda}{\omega}=\frac{\alpha_1+\alpha_2}{\sqrt{1-(\alpha_1+\alpha_2)^2}}\text{ ,(constant for all modes)} \tag{3.2.65}$$

It is implicitly assumed that $\alpha_1 + \alpha_2 < 1$ (subcritical damping).
In all other cases, i.e. when:

$$[C] \neq 2\alpha_1\frac{[K]}{\omega_0}+2\alpha_2\omega_0[M]\ , \tag{3.2.66}$$

the generic modal shape becomes complex: $\{U\}$ is formed by a set of complex amplitudes, i.e. the damped harmonic motions of the different d.o.f. of the modal shape will have different phases [49]. Besides, new modes can appear which do not correspond to any of the undamped solutions.
In the case of free vibrations of an arch dam modeled by F.E.M. at empty reservoir and with rigid foundation boundary, the only energy dissipation that can be represented in the model is the above treated structural damping in each volume element of the concrete and of the rock. Radiation damping through the rock deep boundary is, indeed, prevented by the "rigid boundary" constraints.
In physical reality, radiation energy losses will be incurred by emission of outwardly elastic waves in the indefinite rock mass. However, physically consistent 3-D radiation conditions are not easy to model [50], so generally this particular mechanism of energy dissipation is not taken into account, at least in an explicit way (from an approximate, global point of view it can somehow be included in an overall damping factor). It must be said, though, that this kind of radiation losses (implying energy transmission through the dam-foundation contact) is less efficient for arch dams than for other dam types, given the relatively small surface of contact.

[49] Strictly speaking, such is the case even for the isotropic viscous damping defined by (3.2.52)....(3.2.54), unless the Poisson ratio of the dam and foundation materials be taken equal to zero.

[50] It would be necessary to impose a matrix linear relationship between dynamic rock boundary deformations and boundary velocities (instead of zero displacements, zero velocities), which is cumbersome to express if "real", 3-D physical-geometric conditions are to be respected.

It is now necessary to see how the treatment developed sub- a) for an undamped system ($[C] = [0]$) can be extended to the case of finite damping [51].
In general (this is evident under the special condition (3.2.61), see above) one can still apply a modified orthogonality condition for the (complex) eigenshapes, i.e. $\{\psi^*\}^T_j [M] \{\psi\}_l = 0$ with $\omega_j = \omega_l$; if the damping factor ν is small (say of the order of 1 to 5% of the critical, which is the normal range of values found in arch dams) one can still treat the responses of the individual modes as those of the indipendent, linear, damped single d.o.f. oscillators (in general, with different damping factor for each mode). The most practical way to perform, e.g., a ground-shaking analisys is to use the Housner response spectra, which are already parametrized for different degrees of damping: if the modal damping factors are known (or can be assumed), it is then easy to obtain the response of each mode to the assigned ground-shaking. The individual modal responses are then combined to estimate the overall response by the SRSS technique already introduced for the undamped case, see (3.2.48), if the response spectrum is in terms of maximum velocities, or (3.2.50), if the response spectrum is in terms of maximum accelerations $A(\omega)$.

c) Influence of impounded water in the full (or partially full) reservoir: the problem of fluid/structure interaction

The presence of water in the reservoir changes considerably (at least in principle) the problem of dynamic behaviour. Physically-wise, the phenomena that arise are those typical of dynamically coupled continua, and for a "complete" modelling of the "real" situation one would have to represent three interacting extended systems (or continua), see fig. 74:

- the dam proper (including its "near-field"foundation rock);
- the water masses in the reservoir, interacting with the dam motion through displacements and pressures on the upstream dam face;
- the "far-field" rock mass, including all the "wet" impoundment bottom and the relevant substrata, which interact with the water masses in motion.

---

[51] If $[C]$ is the generic damping matrix, as a first approximation we still postulate $\{u\} \cong \{U\}_0 e^{\sigma t}$ with $\sigma = i\omega - \lambda$ and $\omega = \omega_0\sqrt{(1-\nu_0^2)}$, $\lambda = \omega_0\nu_0$ where $\omega_0/2\pi$ is the zero-damping frequency of mode $\{U\}_0$, i.e. $[K]\{U\}_0 = \omega_0^2[M]\{U\}_0$ and $\nu_0 \cong \dfrac{\{U\}_0^T[C]\{U\}_0}{2\omega_0\{U\}_0^T[M]\{U\}_0}$.

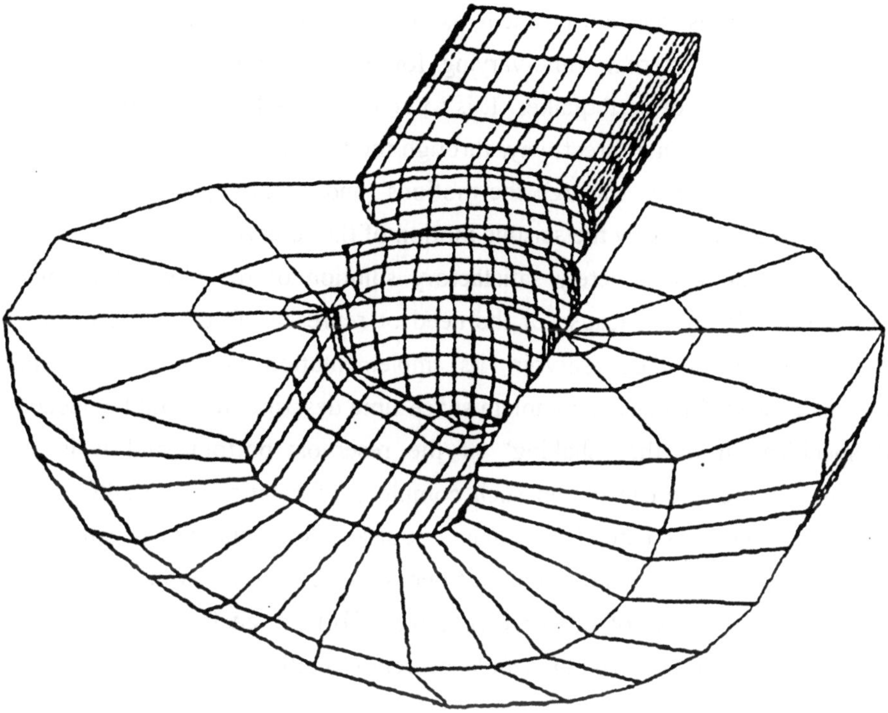

Dam-Reservoir-Foundation System of
Karun I Arch Dam.

Fig. 74
Dynamic behaviour involves interaction of three continua (coupled problem)

As in the preceding treatment for empty reservoir, the problem is best dealt with in stages. First the free-vibration modelling yields the eigenfrequencies and eigenshapes of the coupled modes. Then the response to the external excitation is investigated. In this second part the most relevant problem, as in the case of empty reservoir, is usually not so much that of external, time-varying forces, as the situation of "ground - shaking", representative of earthquakes. It is to be considered that the ground shaking excites the dam not only in a direct way, through the time variations of the position of its basis, but also indirectly, through the motions (and attending dynamic pressures) induced in the water masses by the seismic motion of the reservoir bottom.

It is fair to say that a completely satisfactory solution of the physical problem is theoretically possible. But, because of the need ot represent all of the geometry and physical parameters of the reservoir, its implementation would be very heavy computationally-wise. Besides, a complete space-time description would be needed for detailed specification of the "shaking" of the reservoir bottom; and it is surely unrealistic to assume one can obtain such complete information, especially for large reservoirs that sometimes extend over many square kilometers. So, because of the simplifications and assumptions that must of necessity be introduced in the input data, the topic is still the subject of intensive research efforts, even for the purely linear-elastic case which, strictly speaking, would be representative of reality only for low-intensity shaking.

Also in this case, it should be mentioned that "good" results [52], see fig.75, can be obtained not only by F.E.M., but also with the "classical" methods that formed the object of the first sections of the present course (e.g. with modified versions of the Ritter method, or with the T.L.M.). However, for the sake of brevity and for unity of treatment we will discusse the fluid/structure interaction problem only in the framework of the F.E.M. formulation.

As a final qualitative remark of these introductory notes, it is good to point out that much of the following material is applicable in general to almost all types of dams (at least concrete dams); however, for arch dams, by reason of their relatively small mass, the alterations in dynamic behaviour introduced by the presence of water in the reservoir can be somewhat more pronounced than for other structural types.

The presence of water, and the coupling between the dam, the water and the reservoir bottom will be seen to affect all three of the modal "signatures" of the dam dynamic behaviour, namely:

[52] i.e. results in good agreement with data obtained from analysis of in-situ dynamic tests.

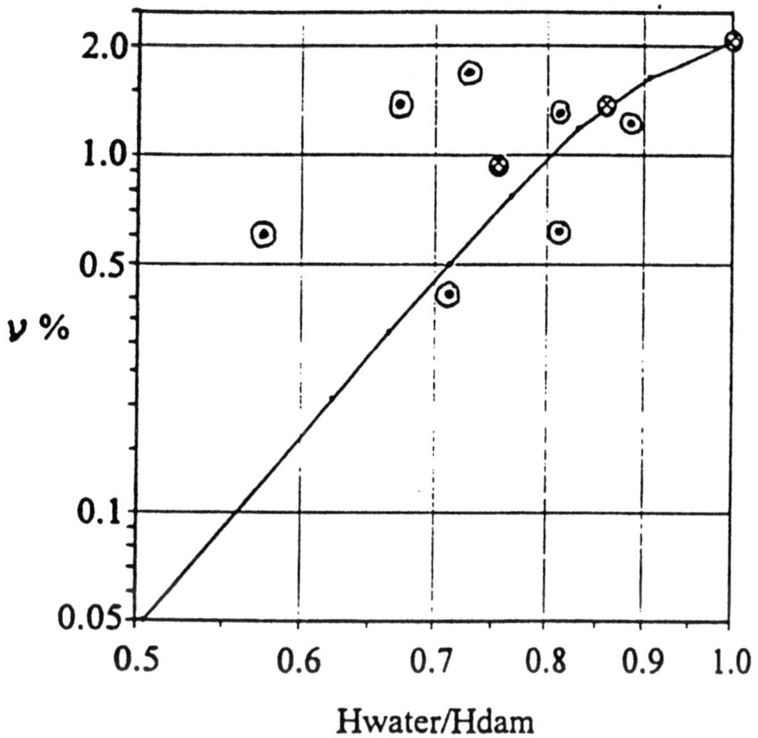

—— Numerical results of the computation (3.2.c_3) of ν, according to a Ritter model, as a function of water elevation in the reservoir, for the TALVACCHIA dam, first symmetric mode. (Ec= 3.6e+04 MPa, Er=1.2e+04 MPa)

⊗ = approximate experimental evaluation of reservoir-induced damping variation with Hw/Hd

⊙ = results reported by Masayuki Kashiwayanagi & Tatsuo Ohmachi: "Observed Effects of Reservoir Water on Dynamic Characteristics of an Arch Dam", Intnl Symp. on Arch Dams, Nanjing, 17-20 oct., 1992 (elaboration)

Fig. 75
Dam-reservoir-foundat. interaction: bottom-radiation damping, Ritter model

-the eigenfrequencies (modes more or less corresponding to those at "empty" reservoir will have their frequencies decreased by the participation of water masses in the motion; new modes will appear);

-the eigenshapes (as a consequence of the different space distribution of the participating masses);

-the modal damping factors (mainly as an effect of the transmission of energy from the dam to the water masses, and from these to the bottom: the bottom strata, dynamically excited through the water/rock contact surface, will indeed emit acoustic waves propagating downwards. This implies a steady loss of energy by radiation).

The only formal difference -with respect to the above treatment of dam vibrations at empty reservoir- in the F.E. vibration equations of the structure, where we neglect for the moment the structural damping, is the presence of a nodal-force term $\{F\}_p$ coming from the hydrodynamic pressures acting on the upstream face of the dam:

$$[K]\{u\}+[M]\{\ddot{u}\}=\{F\}\,, \tag{3.2.67}$$

where: $$\{F\}=\{F\}_e+\{F\}_p\,, \tag{3.2.68}$$

$\{F\}_e$ being the time-varying nodal forces corresponding to "live" loads, if any, and $\{F\}_p$ the nodal forces coming from surface integration of the hydrodynamic pressures acting on the wet surface of the dam upstream facing:

$$\{F\}_p = \iint_{S_u(\Omega)} [N]^T\{p\}dS\;; \quad \text{here } \{p\}_m = p_m \begin{Bmatrix} \cos \hat{nx} \\ \cos \hat{ny} \\ \cos \hat{nz} \end{Bmatrix} \text{ at node } m \tag{3.2.69}$$

$\vec{n}$ being the (internal) local normal to the surface S_u, see fig.76; and $p = (x, y, z)$, nodal pressure, being a function of the fluid motion. This means that additional equations are needed to make the problem a mathematically "closed" one.

In the water mass we can assume, in the linear approximation for small motions and under the assumption of inviscid fluid, the Helmholtz equation for p (deriving from the Euler equations of motion together with the conservation of mass equation and the equation of state for water):

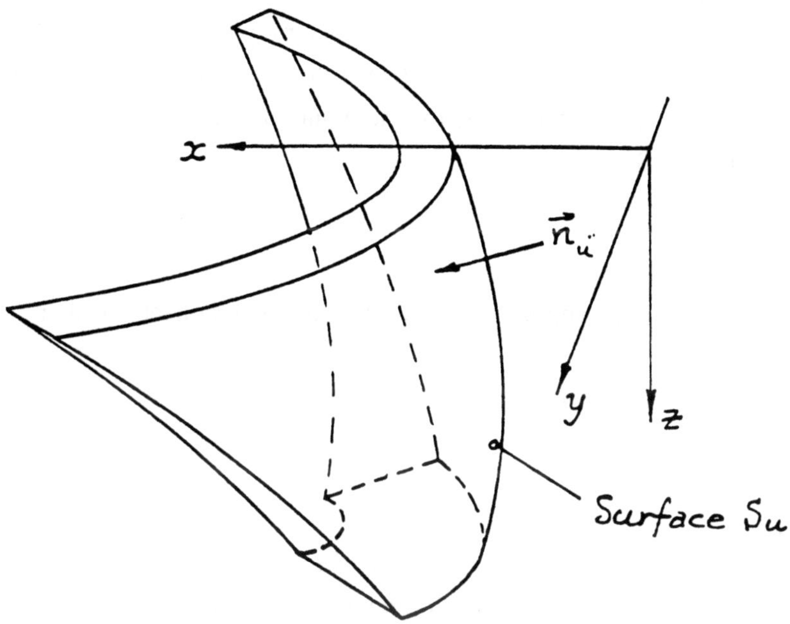

Fig. 76
Upstream dam force S_u and its normal $\vec{n}_u$

$$\nabla^2 p = \frac{\partial^2 p}{\partial x^2} + \frac{\partial^2 p}{\partial y^2} + \frac{\partial^2 p}{\partial z^2} = \frac{1}{C_a^2}\frac{\partial^2 p}{\partial t^2}, \qquad (3.2.70)^{53}$$

where $C_a = \sqrt{\frac{\varepsilon}{\rho_a}} \cong 1440\,\mathrm{ms}^{-1}$ = celerity of sound waves in water (ε = compressibility of water $\cong$ 2000 MPa, ρ_a = specific mass of water $\cong$ 1000 $\mathrm{Kgm^{-3}}$).

To (3.2.70) we have now to add suitable B.C. for the water motion.

Always under the assumption of <u>small motions</u> (small displacements, small velocities) and <u>inviscid fluid</u> one can write after Euler, for the points of the fluid domain Ω_w in contact with the dam face S_u :

$$\frac{\partial p}{\partial n} = -\rho_a \ddot{u}_n \qquad (3.2.71)$$

($\vec{n}$ is the internal normal as above, $u_n = \vec{u} \times \vec{n}$ the component of dam face displacement $\vec{u}$ along $\vec{n}$)

In a like manner at bottom wet surface S_u :

$$\frac{\partial p}{\partial n_b} = -\rho_a \ddot{u}_b \qquad (3.2.72)$$

($\vec{n}_b$ is the internal normal to S_b , $u_b = \vec{u}_R \times \vec{n}_b$ the component of bottom surface displacement $\vec{u}_R$ along $\vec{n}_b$); besides, $p = 0$ on the free surface S_0.

Since $\vec{u}_R$ has to represent the dynamic compliance of rock bottom under the hydrodynamic pressures, yet another condition on S_b is required in order to formally "close" the problem. Barring more complicated choices, that could be:

i) either $u_b = 0$ (<u>rigid bottom</u>), (3.2.73)

or:

ii) an "<u>acoustic radiation condition</u>" (see above) for a compliant bottom; under several simplifying assumptions[54] one could write e.g. a Sommerfeld radiation equation like:

---

[53] From $\frac{\partial p}{\partial x} \cong -\rho_a \frac{\partial Vx}{\partial t}$, $\frac{\partial p}{\partial y} \cong -\rho_a \frac{\partial Vy}{\partial t}$, $\frac{\partial p}{\partial z} \cong -\rho_a \frac{\partial Vz}{\partial t}$, $C_a^2 = \frac{\varepsilon}{\rho_a}$ and $\frac{\partial Vx}{\partial x} + \frac{\partial Vy}{\partial y} + \frac{\partial Vz}{\partial z} = -\frac{1}{\varepsilon}\frac{\partial p}{\partial t}$ (ε = bulk compressibility of water) (3.2.70) can be easily derived.

[54] S_b not too different from a horizontal plane; elastic waves (propagating downwards in the rock strata) approximated locally by quasi-plane P waves.

$$\rho_r c_r \dot{u}_b = p_b \,, \tag{3.2.74}$$

where specific mass of the rock = ρ_r , c_r = celerity of P-waves in the rock $\cong \sqrt{\dfrac{E_r}{\rho_r}}$;

E_r = Young modulus of the rock (see fig. 77).

As we already saw in many different contexts, the above coupled problem can be treated at different levels of approximation. It is instructive to review briefly some of the earlier, drastically simplified formulations before attempting to cope with the "complete" problem.

c_1) The Westergaard (added masses) approach[55]

This consists in putting $\frac{1}{c_a} = 0$ in (3.2.70) (incompressible fluid) and $\frac{1}{c_r} = 0$ (rigid rock) in (3.2.74). In that case we have from (3.2.70):

in the fluid: $$\nabla^2 p = \frac{\partial^2 p}{\partial x^2} + \frac{\partial^2 p}{\partial y^2} + \frac{\partial^2 p}{\partial z^2} = 0 \quad \text{(Laplace equation)}, \tag{3.2.75}$$

and

on S_b $$\dot{u}_b = 0 \,,\; \ddot{u}_b = 0 \,,\; \frac{\partial p}{\partial n_b} = 0 \quad \text{from (3.2.74), (3.2.72);} \tag{3.2.76}$$

the conditions (3.2.71) on S_u remaining unchanged, as the condition $p = 0$ on S_0.

Various techniques can be used to eliminate the variables $p(x,y,z,t)$ between (3.2.67) and (3.2.75). All are more or less related to the determination of singular solutions of (3.2.75). Either boundary elements (entailing discretization of S_u only) can be used, especially if the reservoir shape can be approximated by a simple geometry (e.g. in the case of a prismatic reservoir), or the reservoir impounded volume Ω_w can be discretized by a F.E. mesh[56].

We will take, in the following, this last option for unity of treatment.

Let us, then, assume that the fluid-filled volume Ω_w has been discretized with a F.E. mesh. Let the single-row matrix of shape functions for the scalar quantity p be $\langle N \rangle_w$

[55] The presentation given herein is, in reality, a modified version of the original Westergaard formulation. However, the basic assumptions being the same, we retain this classical namesake.

[56] Or a F.E. mesh can be adopted near the dam face and a B.E. mesh fitted on the outermost nodes of the F.E. one, see figure 77bis, to represent the rest of the reservoir.

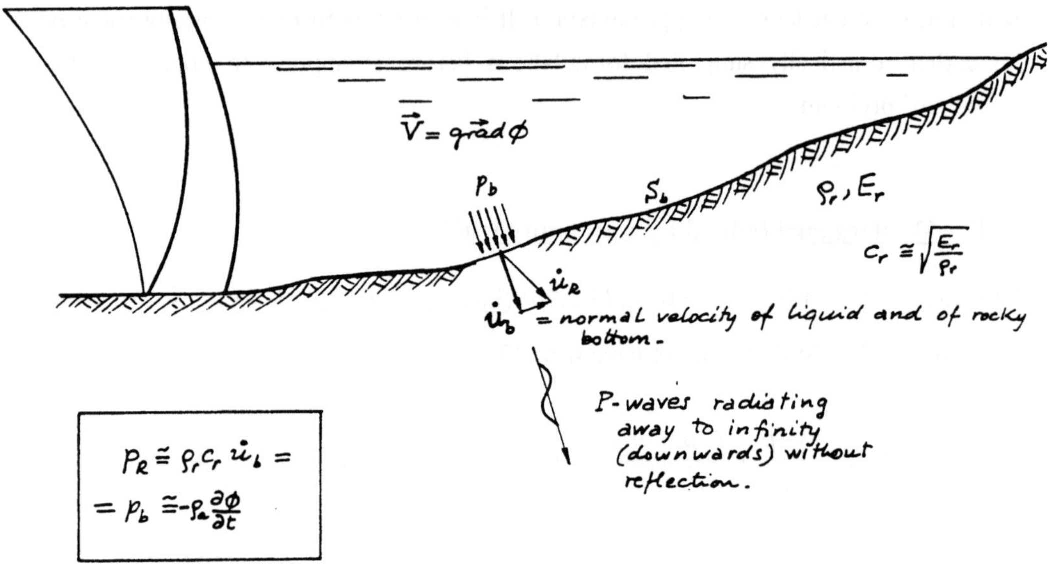

Fig. 77
Dam-reservoir-foundat. interaction: bottom radiation condition

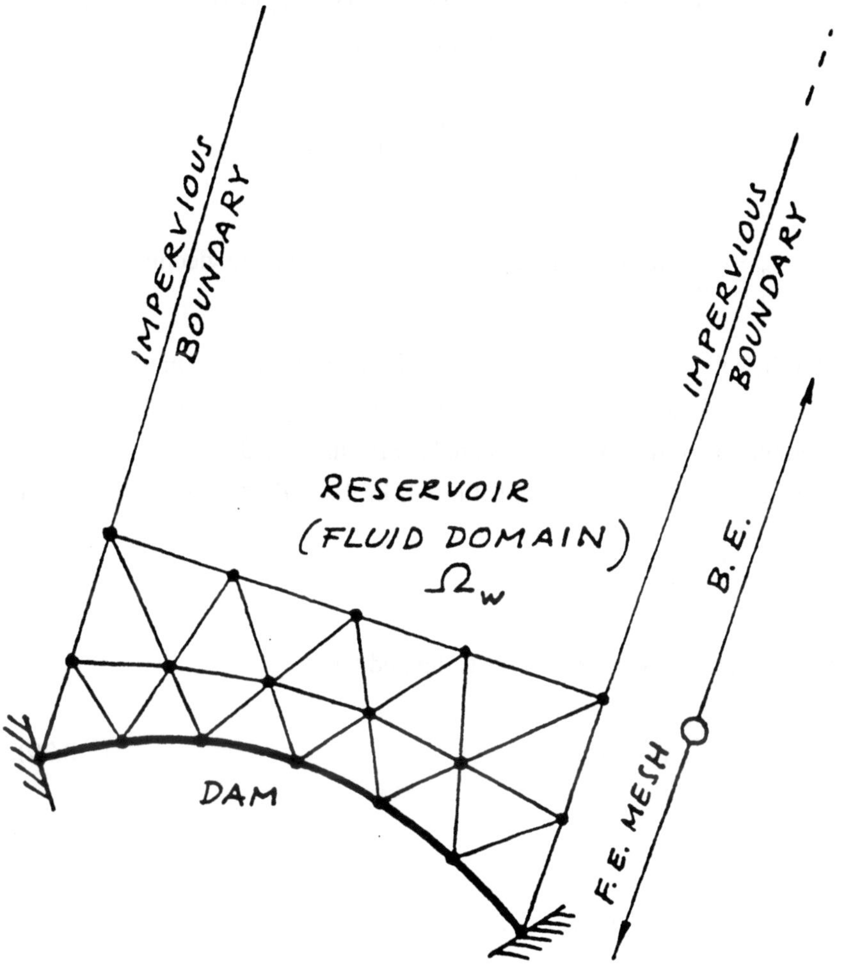

Fig. 77 bis
Matching of a F.E. grid with B.E. for flow field (plan view)

(defined formally over the whole domain). The assumption of inviscid, incompressible fluid and of small motions make it expedient to represent the flow field by an irrotational flow model. This means that we postulate the existence in Ω_w of a velocity potential $\phi(x,y,z,t)$: in that case the fluid velocity components are given by:

$$
\begin{gathered}
{}_x = \frac{\partial\phi}{\partial x}\ ,\ V_y = \frac{\partial\phi}{\partial y}\ ,\ V_z = \frac{\partial\phi}{\partial z} \quad \text{within } \Omega_w \ ; \\
{}_n = \frac{\partial\phi}{\partial n} \ \text{ on } S_u \ ; \ p \cong -\rho_a \frac{\partial\phi}{\partial t} \ \text{ anywhere in } \Omega_w \text{ neglecting } V^2
\end{gathered}
\tag{3.2.77}
$$

The surface S_u will contain faces and nodes of the reservoir F.E. mesh. We assume that on S_u the nodes of the fluid mesh coincide with the nodes of the dam mesh, and that the shape functions of the two meshes are "conformal" on S_u .

The Laplace equation (3.2.75) is substituted, in the F.E. discretization, by the matrix equation:

$$[T]\{\phi\} = \{0\} \tag{3.2.78}$$

where the "transmissivity matrix" $[T]$ is defined by:

$$[T] = \iiint_{\Omega_w} [\Delta]^T [\Delta] dV \ , \tag{3.2.79}$$

with

$$[\Delta] = \begin{Bmatrix} \frac{\partial}{\partial x} \\ \frac{\partial}{\partial y} \\ \frac{\partial}{\partial z} \end{Bmatrix} \langle N(x,y,z) \rangle_w \ , \tag{3.2.79bis}$$

and $\{\phi\}$ = vector of the nodal values of the velocity potential $\phi = \{\phi(t)\}$.

Under the simplifying assumptions made so far, we have to solve (3.2.78), with B.C. (3.2.71) on S_u and (3.2.76) on S_b , which in terms of the potential ϕ become:

on S_u : $$-\frac{\partial}{\partial n}(\rho_a \frac{\partial\phi}{\partial t}) = -\rho_a \ddot{u}_n \quad (p = -\rho_a \frac{\partial\phi}{\partial t} \text{ neglecting the square of velocities});$$

on S_b : $$\frac{\partial\phi}{\partial n_b} = 0 \quad \text{(impermeable boundary);} \tag{3.2.80}$$

besides, $\phi=0$ on S_0 [57].

From (3.2.80) one gets easily the transformed B.C. on S_u :

$$\text{on } S_u: \qquad \frac{\partial \phi}{\partial n} = \dot{u}_n , \qquad (3.2.80\text{bis})$$

which shows that this single condition satisfies also the continuity of normal velocities at S_u : fluid normal velocity $= \frac{\partial \phi}{\partial n}$; dam facing normal velocity $= \dot{u}_n$.

Now in order to get rid of (3.2.78) when solving (3.2.67) for the dam motion, we proceed first to build up as many spatial "elementary solutions" $\phi_j\ (x,y,z,t)$ as there are nodes (j_u) on the interface S_u ($j_1 \le j_u \le j_{Mu}$),
with B.C. defined as follows:

on S_0: $\quad \phi_j = 0$;

on S_u : $\quad \dfrac{\partial \phi_j}{\partial n} = N_{ju}$, shape function of node j_u :

$$\frac{\partial \phi_j}{\partial n} = \text{ (unit normal velocity) at node } j_u ,$$

$$\frac{\partial \phi_j}{\partial n} = 0 \text{ at all other nodes of } S_u ;$$

on S_b : $\quad \dfrac{\partial \phi_j}{\partial n_b} = 0$ on the wet bottom surface S_b. (3.2.81)

Once the $\phi_j\ (x,y,z,t)$ have been determined by solving (3.2.78) with B.C. (3.2.81), it is immediate to see that the solution of (3.2.78) with B.C. (3.2.80) will be given by:

$$\phi = \phi(x,y,z,t) \cong \sum_{j_u} \dot{u}_n (j_u) \phi_j (x,y,z) \qquad (3.2.82)$$

As a consequence, the interface pressures on S_u will be:

$$\text{on } S_u: \qquad p = -\rho_a \frac{\partial \phi}{\partial t} \cong -\rho_a \sum_{j_u} \ddot{u}_n (j_u) \phi_j (x,y,z) \qquad (3.2.83)$$

In particular, at node l on S_u it will be:

---

[57] From $p = 0$ on the free surface S_0, and $p = -\frac{\partial \phi}{\partial t}$, ϕ=const. on S_0; since the choice of the constant is arbitrary, we put $\phi = 0$ on S_0.

$$p_l = -\rho_a \sum_{j_u} \ddot{u}_n(j_u)\phi_j(l) \ , \tag{3.2.84}$$

and at all nodes on S_u :

$$\{p\} = -\rho_a \left[\phi_{jl}\right]^T \{\ddot{u}\}_n \ , \tag{3.2.85}$$

where $[\phi_{jl}]$ is a square matrix containing the values of the "elementary solutions" ϕ_j at all nodes l of S_u (order $M_u \times M_u$ if M_u s the number of nodes lying on S_u).
Now the more direct way would seem to compute $\{F\}_p$ in (3.2.68) from (3.2.85), thus completing the formal elimination of p. Instead, it is convenient to consider the physical system "dam + impounded water masses" as a single composite dynamic system. The motion of this system can be seen (e.g. by the above results) to be univocally determined (at least in our F.E. discretization model) once the nodal displacements $\{u\}$ and velocities $\{\dot{u}\}$ of the dam are specified. Then the Lagrangian equations of motion for the system will be:

$$\frac{\partial}{\partial t}\left\{\frac{\partial \mathcal{E}_c}{\partial \dot{u}}\right\} = -\left\{\frac{\partial \mathcal{E}_p}{\partial u}\right\} \ , \ \text{with} \ \{\dot{u}\} = \left\{\frac{\partial u}{\partial t}\right\} \tag{3.2.86}$$

and: $\mathcal{E}_c = \mathcal{E}_{cd} + \mathcal{E}_{cw}$
kinetic energy stored in the dam ($\mathcal{E}_{cd}$) + kinetic energy stored in the water ($\mathcal{E}_{cw}$);

$\mathcal{E}_p = \mathcal{E}_{pd} + \mathcal{E}_{pw}$
potential energy of the mechanical system ($\mathcal{E}_{pd}$)[58] + potential energy of the fluid system ($\mathcal{E}_{pw}$).

The potential energy of the fluid, due to the incompressibility assumption, reduces to the gravity field potential energy. It can be shown that the <u>first-order</u> gravitational variations of $\mathcal{E}_{pw}$ during an incrementally small motion (i.e. $\Delta\mathcal{E}_{pw}{}^{(1)}$) are exactly offset by equal and opposite first-order variations $\Delta\mathcal{E}_{pd}{}^{(1)} = -\Delta\mathcal{E}_{pw}{}^{(1)}$ of the elastic energy stored in the mechanical system, considering that the motion takes place around an equilibrium state between the elastic reactions of the mechanical system and the hydrostatic forces caused by the impounded water in its rest state. So, all we have to consider for the dynamic equilibrium are the <u>second-order</u> variations:

[58] The potential energy of the pressure forces acting on the interface S_u need not be considered, since in the composite system these are internal forces performing zero net work.

-variations in the kinetic energies: $\Delta\mathcal{E}_{cd}$, $\Delta\mathcal{E}_{cw}$;

-second-order variations in the elastic deformation energy stored in the mechanical system: $\Delta\mathcal{E}_{pd}^{(2)}$

Let us evaluate $\Delta\mathcal{E}_{cw}$. By definition:

$$\Delta\mathcal{E}_{cw} = \frac{\rho_a}{2}\iiint_{\Omega_w} V_w^2 d\Omega_w \text{, with } \bar{V}_w = \text{gr}\bar{\text{a}}\text{d } \phi \tag{3.2.87}$$

it comes successively, by subdividing Ω_w into "stream tubes" elements $ds \times dA$ (see fig. 78):

$$\Delta\mathcal{E}_{cw} = \frac{\rho_a}{2}\iiint_{\Omega_W} \text{gr}\bar{\text{a}}\text{d}\phi \times \text{gr}\bar{\text{a}}\text{d}\phi \, d\Omega_w$$

$$= \frac{\rho_a}{2}\iiint_{\Omega_w}\left|\frac{\partial f}{\partial s}\right| \times \left|\frac{\partial f}{\partial s}\right| \times ds dA = \tag{3.2.88}$$

(by partial integration)

$$= \frac{\rho_a}{2}\left[\iint_{S_u}\frac{\partial\phi}{\partial s}\phi dA - \iint_{S_0}\frac{\partial\phi}{\partial s}\phi dA - \iiint_{\Omega_W}\phi\frac{\partial}{\partial s}\left(\frac{\partial\phi}{\partial s}dA\right)ds\right],$$

since all the streamlines originate on the free surface S_0 (equipotential $\phi = 0$) and terminate on S_u (S_b is a stream surface because on it $\frac{\partial\phi}{\partial n_b} = 0$);

now $\iint_{S_0}\frac{\partial\phi}{\partial s}\phi dA = 0$, because $\phi = 0$ on S_0 ;

and $\iiint_{\Omega_W}\phi\frac{\partial}{\partial s}\left(\frac{\partial\phi}{\partial s}dA\right)ds = 0$,

because $\frac{\partial\phi}{\partial s}dA$ = constant along a streamline s as a consequence of the incompressibility assumption; at last one obtains:

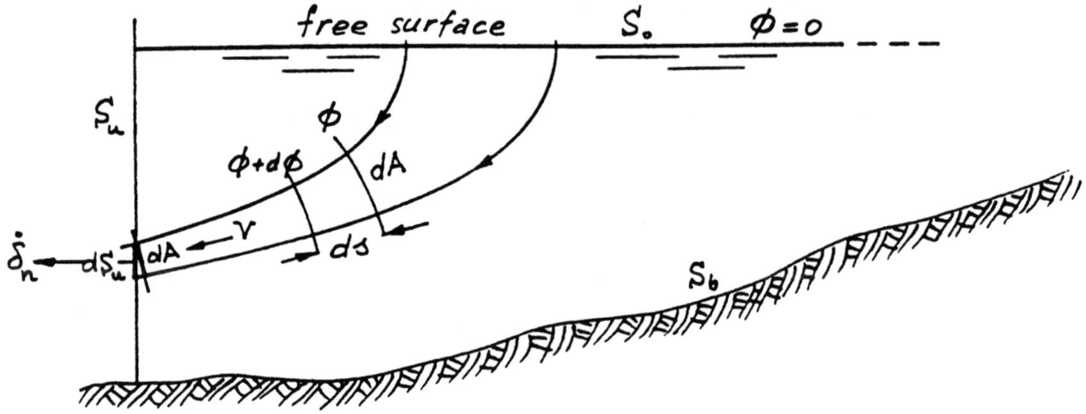

Fig. 78
Irrotational flow field in the reservoir: stream tube

$$\Delta\mathcal{E}_{cw} = \frac{\rho_a}{2}\iint_{S_u} \frac{\partial\phi}{\partial s}\phi dA = \frac{\rho_a}{2}\iint_{S_u} \dot{u}_n \phi \, dS_u \,, \tag{3.2.89}$$

since (see fig.79) $\frac{\partial\phi}{\partial s} dA = \dot{u}_n dS_u$ on S_u. .

Considering that on S_u :

$$\dot{u}_n(x,y,z,t) = \langle N(x,y,z,t)\rangle_w \{\dot{u}_n\}_{S_u},$$

$$\phi(x,y,z,t) = \langle N(x,y,z)\rangle_w \left[\phi_{ij}\right]^T \{\dot{u}_n\}_{S_u}, \tag{3.2.90}$$

see (3.2.82), and furthermore

$$\{\dot{u}_n\}_{S_u} = \left[\cos n\hat{\ }i\right]^T \{\dot{u}\}_{S_u}$$

where

$$\left[\cos n\hat{\ }i\right] = \begin{bmatrix} \cos n_{j_1}\hat{\ }x & 0 & \cdots & 0 \\ \cos n_{j_1}\hat{\ }y & 0 & & 0 \\ \cos n_{j_1}\hat{\ }z & 0 & & 0 \\ 0 & \cos n_{j_2}\hat{\ }x & & \vdots \\ 0 & \cos n_{j_2}\hat{\ }y & & \vdots \\ \vdots & \cos n_{j_2}\hat{\ }z & & \vdots \\ \vdots & \vdots & & \cos n_{j_{Mu}}\hat{\ }x \\ \vdots & \vdots & & \cos n_{j_{Mu}}\hat{\ }y \\ 0 & 0 & \cdots & \cos n_{j_{Mu}}\hat{\ }z \end{bmatrix} \text{(order } 3M_u \times M_u\text{)} \tag{3.2.91}$$

and that one could equally well write:

$$\Delta\mathcal{E}_{cw} = \frac{\rho_a}{2}\iint_{S_u} \{\dot{u}\}_{S_u}^T \left[\cos n\hat{\ }i\right]\{N\}_w \langle N\rangle_w \left[\phi_{ij}\right]^T \left[\cos n\hat{\ }i\right]^T \{\dot{u}\}_{S_u} dS_u \,, \qquad \text{or:}$$

$$\Delta\mathcal{E}_{cw} = \frac{\rho_a}{2}\iint_{S_u} \{\dot{u}\}_{S_u}^T \left[\cos n\hat{\ }i\right]\left[\phi_{ij}\right]\{N\}_w \langle N\rangle_w \left[\cos n\hat{\ }i\right]^T \{\dot{u}\}_{S_u} dS_u \,, \tag{3.2.92}$$

posing:

$$\left[\mu_a^*\right] = \frac{\rho_a}{2}\iint_{S_u} \left[\cos n\hat{\ }i\right]\{N\}_w \langle N\rangle_w \left[\phi_{ij}\right]^T \left[\cos n\hat{\ }i\right]^T dS_u$$

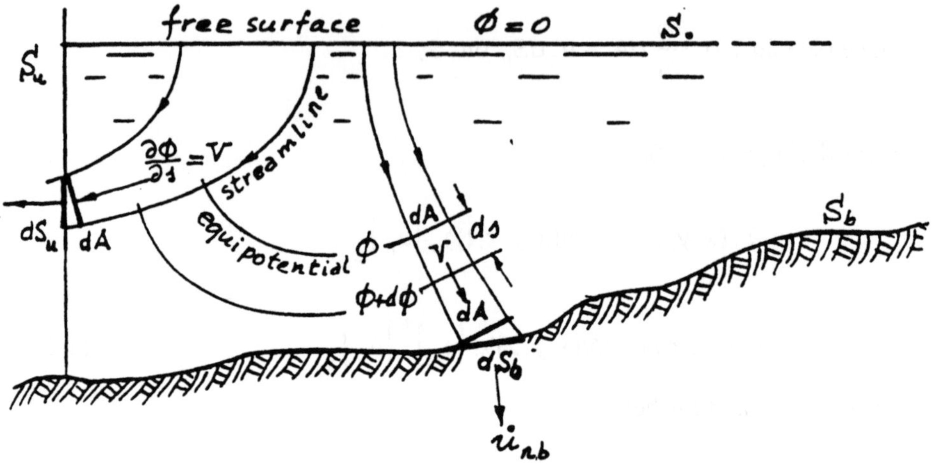

Fig. 79
Irrotational flow field in the reservoir: B.C. on S_u, S_b
N.B. For rigid bottom ($\dot{u}_{nb}$=0) all stream lines terminate on S_u

it appears convenient to take:

$$\Delta\mathcal{E}_{cw} = \frac{1}{2}\{\dot{u}\}^T_{S_u}\left[\bar{\mu}_a\right]\{\dot{u}\}_{S_u}, \text{ with } \left[\bar{\mu}_a\right] = \frac{1}{2}\left(\left[\mu_a^*\right] + \left[\mu_a^*\right]^T\right) \quad (3.2.93)$$

(a full $3M_u \times 3M_u$ matrix)

This can be formally expressed in terms of the full vector $\{\dot{u}\}$ for the complete mesh of the mechanical system thanks to the incidence matrix $[\,I\,]_{Su}$:

$$\{\dot{u}\}_{S_u} = [\mathrm{I}]_{S_u}\{\dot{u}\}, \left([\,I\,]_{Su} \text{ of order } 3M_u \times 3M_n\right) \quad (3.2.94)$$

it comes at last:

$$\Delta\mathcal{E}_{cw} = \frac{1}{2}\{\dot{u}\}^T\left[\mu_a\right]\{\dot{u}\}, \text{ with } \left[\mu_a\right] = [\mathrm{I}]^T_{S_u}\left[\bar{\mu}_a\right][\mathrm{I}]_{S_u}; \quad (3.2.95)$$

$[\mu_a]$ ("added mass" matrix) is now of the same order as $[M]$.
As defined above, $[\mu_a]$ is a square, symmetric, positive-definite matrix.
The kinetic energy of the dam is given by:

$$\Delta\mathcal{E}_{cd} = \frac{1}{2}\{\dot{u}\}^T[M]\{\dot{u}\}, \quad (3.2.96)$$

and the second-order variation of the elastic deformation energy stored in the mechanical system is given by:

$$\Delta\mathcal{E}_{pd} = \frac{1}{2}\{u\}^T[K]\{u\}, \quad (3.2.97)$$

where $\{u\}$ is the dynamic, i.e. time-varying, component of the total displacement field (the static component $\{u_0\}$ is not considered here for the above-explained reasons).
The motion equations (Lagrange eqs.) for the composite system read now, in terms of the mechanical system variables only:

$$[K]\{u\} + \left([M] + \left[\mu_a\right]\right)\{\ddot{u}\} = \{F\}_e \quad (3.2.98)$$

It appears that under the Westergaard assumptions (incompressible fluid, reservoir with rigid bottom) the presence of an impounded fluid is felt by the dam through an "added mass" effect. Of course $[\mu_a]$ will change when the water level in the reservoir varies, whereas $[M]$ remains always the same.
In the case $\{F\}_e = \{0\}$ and in presence of a known uniform ground shaking $u_g\ (t)$ in a given direction $\vec{g}$ near the dam, by posing as in (3.2.33)

$$\{u(t)\} = u_g(t)\{1\} + \{u_r(t)\}, \tag{3.2.99}$$

the eq. of motion (3.2.98) takes on the form:

$$[K]\{u_r\} + ([M] + [\mu_a])\{\ddot{u}_r\} = -([M] + [\mu_a])\ddot{u}_g\{1\}, \tag{3.2.100}$$

assuming that the reservoir bottom is not shaking appreciably along the local normals to its surface; otherwise excitation terms originating from the bottom shaking would have to be added on the R.H.S. of eq. (3.2.100). For the effect of reservoir bottom shaking, see further on ($\mathbf{c}_2$).
The determination of eigenfrequencies and eigenshapes in presence of water in the reservoir will be carried out exactly as in the case of empty reservoir, to the exception of substituting $[M]$ with $[M] + [\mu_a]$ (if one accepts the Westergaard approximation).
It is to be noted that, within the limitations caused by the initial assumptions (incompressible fluid, rigid bottom) and the approximations introduced by the F.E. discretization, the added mass matrix herein defined is a consistent one, i.e. it respects at all nodes of interface S_u the compatibility of fluid and structural motions as well as the equilibrium of interface pressures. In the earlier versions of the Westergaard approach, the matrix $[\mu_a]$ was computed assuming $\{u_n\}$ = constant over the whole interface S_u ; this corresponds to "lumping" the elements of each row of $[\mu_a]$ onto the principal diagonal. The matrix then becomes a diagonal, non-consistent, lumped-mass one. The assumption $\{u_n\} \cong$ cost. is more acceptable for the rigid gravity dams than for the flexible arch dams.
Note that under the above assumptions $[\mu_a]$ is independent from the frequency of the free vibrations or of the external excitation.
In general, the application of the "added mass" approximation to the problem of fluid/structure interaction gives quite satisfactory results for arch dams, at least until

the frequency of the mode (or the frequency of excitation) is low in comparison with the first frequency of the spontaneous reservoir oscillation tied to water compressibility ($f_1 \cong \frac{c_a}{4H}$), see fig.80; beyond that limit, compressibility effects become non-negligible, see further on ($\mathbf{c_4}$).

NOTE - It is worth mentioning that analytic solutions exist for $\nabla^2 \phi = 0$ with B.C. $\phi = 0$ on S_0, $\frac{\partial \phi}{\partial n_b} = 0$ on S_b and an arbitrary distribution $\dot{u}(z)$ along the vertical interface S_u of a prismatic, flat-bottom, infinitely long reservoir (see fig. 81), provided that the interface S_u is flat (unsuitable for arch dams without cumbersome tricks) and moves uniformly along each horizontal line.

E.g. $\phi_{S_u}(z) = -\frac{1}{\pi} \int_{\zeta=0}^{H} \dot{u}(\zeta) \ln \frac{\left| \mathrm{tg}^2 \frac{\pi z}{4H} - \mathrm{tg}^2 \frac{\pi \zeta}{4H} \right|}{1 - \mathrm{tg}^2 \frac{\pi z}{4H}\, \mathrm{tg}^2 \frac{\pi \zeta}{4H}} d\zeta$; $p(z) = \frac{\rho_a}{\pi} \int_{\zeta=0}^{H} \ddot{u}(\zeta) \ln \frac{\left| \mathrm{tg}^2 \frac{\pi z}{4H} - \mathrm{tg}^2 \frac{\pi \zeta}{4H} \right|}{1 - \mathrm{tg}^2 \frac{\pi z}{4H}\, \mathrm{tg}^2 \frac{\pi \zeta}{4H}}$

3-D analytical solutions can also be found for quarter-spherical reservoirs and a flat upstream face, separately for symmetric and antisymmetric motions[59].

c_2) Effects of far-field autonomous bottom shaking in the WESTERGAARD approach

Let the prescribed, autonomous motion of the bottom surface, S_b, be described by:

$$u_b(x,y,z,t) = \text{a known function of space-time on } S_b\,; \qquad (3.2.101)$$

from this, knowing the geometry of S_b, it is easy to derive the component of bottom surface velocity normal to S_b:

$$\dot{u}_{nb} = \frac{\partial u_{nb}}{\partial t} \qquad (3.2.102)$$

In a way analogous to what already seen for S_u, we take the nodes on S_b of the F.E. mesh covering Ω_w and choose as our basic variables $\{\dot{u}_{nb}\}$ at those nodes. The discretized form of (3.2.102) will be:

[59] Such analytical solutions, more than interesting in themselves, can be useful to "calibrate" and "validate" the numerical methods (F.E.M. or other) that are used for more realistic reservoir geometries.

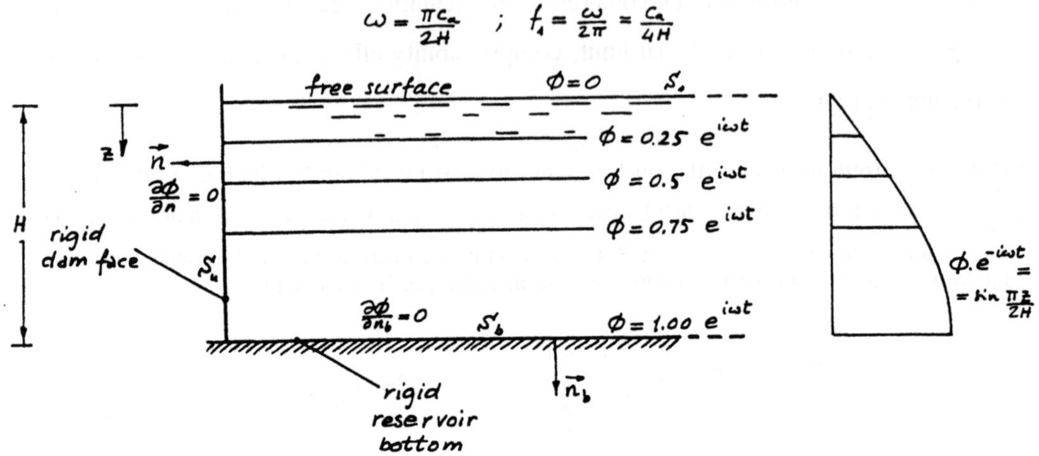

Fig. 80
Compressibility effects: first frequency of spontaneous reservoir modes

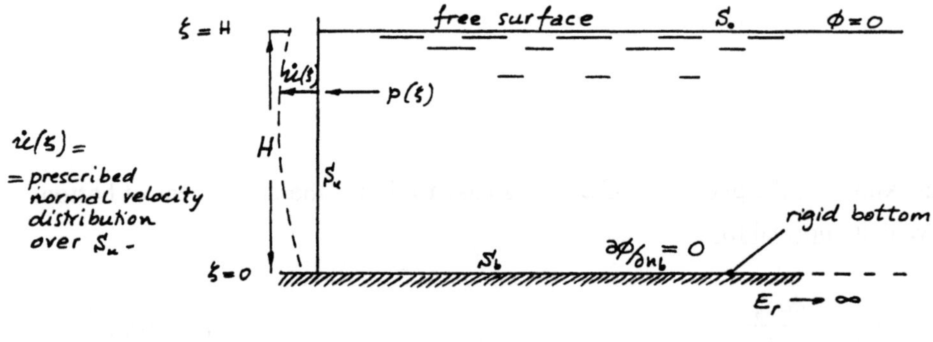

Fig. 81
Irrotational flow of an incompressible liquid in a prismatic reservoir.
Analytic solution for arbitrary motion of Su (dam face) is known.
$p(\zeta)$ is a <u>functional</u> of $\dot{u}_n(\zeta)$

$$\dot{u}_{nb}(x,y,z,t) \cong \langle N_b \rangle \{\dot{u}_{nb}\}, \qquad (3.2.103)$$

where $\langle N_b\ (x,y,z,t)\rangle$ = shape functions of the Ω_w mesh on S_b and $\{\dot{u}_{nb}\}=\{\dot{u}_{nb}(t)\}$= known nodal velocities at S_b nodes.

Now let us solve (3.2.78) for a generic "unit solution" ϕ_{bj} related to the generic mode j on S_b with the following B.C.:

on S_0: $\phi_{bj} = 0$; on $S_u : \dfrac{\partial \phi_{bj}}{\partial n_u} = 0$; on $S_b : \dfrac{\partial \phi_{bj}}{\partial n_b} = N_{jb}$, with:

N_{jb} = shape function N_b for node j_b ; i.e. $N_{jb} = 1$ on j_b and $N_{jb} \equiv 0$ on all other nodes on S_b.

By solving (3.2.78) successively for ϕ_{bj} with $j_b = j_{b\,1} \ldots. j_{b\,B}$ (all bottom nodes) we generate as many elementary solutions ϕ_{bj} as there are mesh nodes on S_b. Let us collect all the values of ϕ_{bj} on nodes I_u of S_b : one can thus build up a matrix:

$$\left[\phi_{bj}(I)\right] \text{ (order } M_b \times M_u \text{)} \qquad (3.2.104)$$

(the rows being made up by the values of ϕ_{bj} at all nodes I_u for a fixed j_b).

Then on S_u we have, for the component part due to bottom shaking, the potential:

$$\text{on } S_u : \quad \phi_b(x,y,z,t) = \langle N \rangle_w \left[\phi_{bj}(I)\right]^T \{\dot{u}_{nb}\},\ N_w = N_w(x,y,z); \qquad (3.2.105)$$

and, consequently, the pressure due to bottom shaking:

$$\text{on } S_u : \quad p_{bu}(x,y,z,t) = -\rho_a \frac{\partial \phi_b}{\partial t} = -\rho_a \langle N \rangle_w \left[\phi_{bj}(I)\right]^T \{\ddot{u}_{nb}\} \qquad (3.2.106)$$

From this pressure distribution on S_u it is easy to compute the <u>nodal</u> external (autonomous, or "live") <u>forces</u> acting on the dam upstream face as a consequence of "bottom shaking":

$$\{F_b\} = \iint_{S_u} [N]^T \{\cos n_u \hat{\ } i\} p_{bu}(x,y,z,t) dS_u = -[BE]\{\ddot{u}_{nb}\}, \qquad (3.2.107)$$

where:

$$\{\cos n_u \hat{\ } i\} = \begin{Bmatrix} \cos n_u \hat{\ } x \\ \cos n_u \hat{\ } y \\ \cos n_u \hat{\ } z \end{Bmatrix}, \qquad (3.2.108)$$

$[N] = [N(x,y,z)]$ = shape functions of the dam mesh (3 rows and 3 × tot. number of d.o.f. columns)[60],

and
$$[BE] = \rho_a \left(\iint_{S_u} [N]^T \{\cos n\hat{}i\} \langle N \rangle_w dS_u \right) \left[\phi_{bj}(l) \right]^T \qquad (3.2.108b)$$

Then the Lagrangian equations of motion for the dam take on the form (see 3.2.99):

$$[K]\{u_r\} + ([M] + [\mu_a])\{\ddot{u}_r\} = -([M] + [\mu_a])\{\ddot{u}_g\} - [BE]\{\ddot{u}_b\} \qquad (3.2.109)$$

In (3.2.109) the first term on the R.H.S. is the excitation originated by the near-field shaking:

$\{\ddot{u}_g\} = \ddot{u}_g\{1\}$, see (3.2.100), and the second term is the excitation due to bottom shaking.

NOTE - As in the case of the "added masses", one can find analytical solutions if one is prepared to accept drastic simplifications in the geometry. E.g., in the case of a prismatic, infinitely long valley of rectangular cross-section, flat dam face, flat reservoir bottom, one finds the following integral expression for the fluid motion potential $\phi(y)$ (y being the elevation of the generic point on S_u above the bottom):

$$\phi(y) = -\frac{1}{\pi} \int_0^\infty \dot{f}_n(x) \ln \frac{\mathrm{Ch}\frac{\pi x}{2H} - \cos\frac{\pi y}{2H}}{\mathrm{Ch}\frac{\pi x}{2H} + \cos\frac{\pi y}{2H}} dx,$$

where x is the abscissa from the dam face into the reservoir, and $f_n(x,t)$ is the prescribed vertical component of motion along the bottom (see fig.82).
One finds consequently for the pressure on the dam face:

$$p(y) = \frac{\rho_a}{\pi} \int_0^\infty \ddot{f}_n(x) \ln \frac{\mathrm{Ch}\frac{\pi x}{2H} - \cos\frac{\pi y}{2H}}{\mathrm{Ch}\frac{\pi x}{2H} + \cos\frac{\pi y}{2H}} dx.$$

Note that $f_n(x,t)$ is the vertical motion of the bottom plane taken as positive in a downward direction. The theoretical results above can be useful, as already said, to "validate" numerical codes based on discretization methods; moreover, note that the above $p(y)$ depends on the vertical components $\ddot{f}(x,t)$ of ground acceleration. This shows that vertical components of earthquake can generate seismic pressures on the dam.

[60] In (3.2.107), it is intended that $\cos n_u\hat{}x$, $\cos n_u\hat{}y$, $\cos n_u\hat{}z$, are functions of the place and that each shape function $N(x,y,z)$ is formally defined over all the domain (dam + near foundation); the 3 rows of $[N]$ containing the $N(x,y,z)$ in staggered formation to allow matricial product with the column of director cosines of $\vec{n}_u$ to be effected.

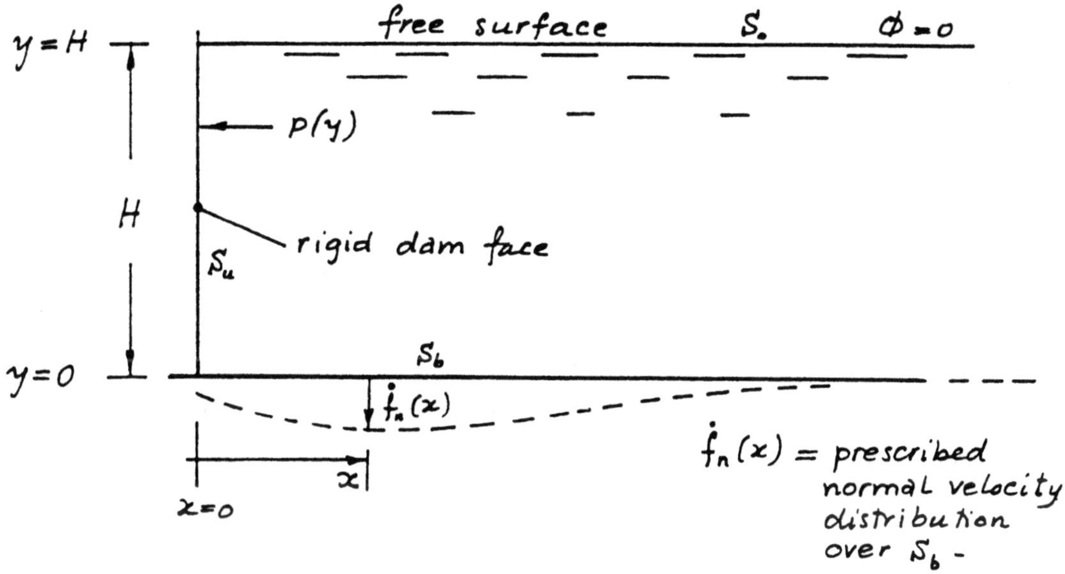

Fig. 82
Irrotational flow of an incompressible liquid in a prismatic reservoir.
Analytic solution for arbitrary motion of S_b (reservoir bottom) is known.
p(y) is a <u>functional</u> of $\dot{f}_n(x)$

c$_3$) **Damping associated with bottom compliance (bottom radiation effect)**

If the reservoir bottom is not rigid, besides autonomous bottom shaking (see **c$_2$** above) one is led to consider the effect of bottom compliance on the damping of dam vibrations, as explained earlier on in qualitative terms (see beginning of **c**) .
To try and formulate the problem in a simple way, let us consider the F.E. modelling of damped free vibrations in the presence of water in the reservoir and of a compliant bottom, retaining however the assumptions of incompressible fluid and irrotational motion in the reservoir.
Under these assumptions, the modal dam motion will consist of damped harmonic functions of time, generally with different phases from point to point; in complex notation we can write:

$$\{\delta\} = \{\Delta\} e^{\sigma t} \tag{3.2.110}$$

where $\{\Delta\}$ = complex-valued vector of the nodal amplitudes of motion, and

$$\sigma = i\omega - \lambda = i\omega(1+i\,\nu)\ , \tag{3.2.111}$$

On the wet dam surface one can write:

$$\delta(x,y,z,t) = [N(x,y,z)]\{\Delta\} e^{\sigma t}, \tag{3.2.112}$$

$[N(x,y,z)]$ being the shape functions of the wet surface nodes.
Let us look for an irrotational fluid motion, i.e. for a potential ϕ , matching the normal component of the water/concrete interface motion at the nodes of the F.E. mesh:

$$\{\phi(t)\} = [\phi_{jl}]^T [\cos n\hat{}j]^T \{\dot{\delta}\} = [\phi_{jl}]^T [\cos n\hat{}j]^T \{\Delta\}\sigma e^{\sigma t}\ , \tag{3.2.113}$$

where evidently the column $\langle\phi_{jl}\rangle^T$ is the vector of the nodal values of ϕ on S_u when the normal motion of the interface is given by the boundary condition:

$$\dot{\delta}_n(x,y,z)_{S_u} = [N]_{S_u} \begin{Bmatrix} 0 \\ 0 \\ \cdots \\ 1 \\ \cdots \\ 0 \\ 0 \end{Bmatrix} \leftarrow j^{th} \text{ row} = 1; \text{ all other rows} = 0 \qquad (3.2.114)$$

(the vector of the nodal normal components is everywhere null except for the jth node of the wet interface, see fig.83; $[N]_{Su}$ is the reduced set of shape functions for the interface nodes).

The "unit solutions" $\{\phi_j\ (x,y,z)\}$ are obtained by solving repeatedly the Laplace equation:

$$\nabla^2 \phi = 0 \qquad (3.2.115)$$

within Ω_w , with B.C. as follows:

on the free surface S_0: $\quad \phi = 0;$

as for (3.2.114) on S_u : $\quad \dfrac{\partial \phi}{\partial n_u} = \dot{\delta}_n(x,y,z);$

on S_b : $\quad p_b(x,y,z,t) = \rho_r c_r \dfrac{\partial \phi}{\partial n_b} = -\rho_a \dfrac{\partial \phi}{\partial t} = -\rho_a \sigma\, \phi\ , \qquad (3.2.116)$

with $c_r \cong \sqrt{\dfrac{E_r}{\rho_r}}$, see (3.2.74).

Since σ is complex, see (3.2.111), each unit solution $\phi_j\,(x,y,z)$ will take on complex values. The common factor $e^{\sigma t}$ can be dropped from the equations.

It is evident, then, that for a generic motion (3.2.112) of the wet face the potential $\phi=\phi(x,y,z,t)$ inside the fluid domain Ω_w will be given by the following linear combination of "unit solutions" ϕ_j :

$$\phi(x,y,z,t) = \{\phi_j(x,y,z)\}^T [\cos n\hat{}j]^T \{\Delta\}_{S_u}\, \sigma\, e^{\sigma t}, \qquad (3.2.117)$$

where $[\cos n\hat{}j]^T$ is the $M_u \times 3\ M_u$ matrix of director cosines of the wet surface at the M_u wet nodes of S_u ($\vec{n}$ directed from the water into the dam face), see (3.2.91). Indeed, (3.2.117) defines a fluid motion matching the wet face normal displacements as well as the Sommerfeld radiation condition (last eq. of 3.2.116) on the reservoir bottom.

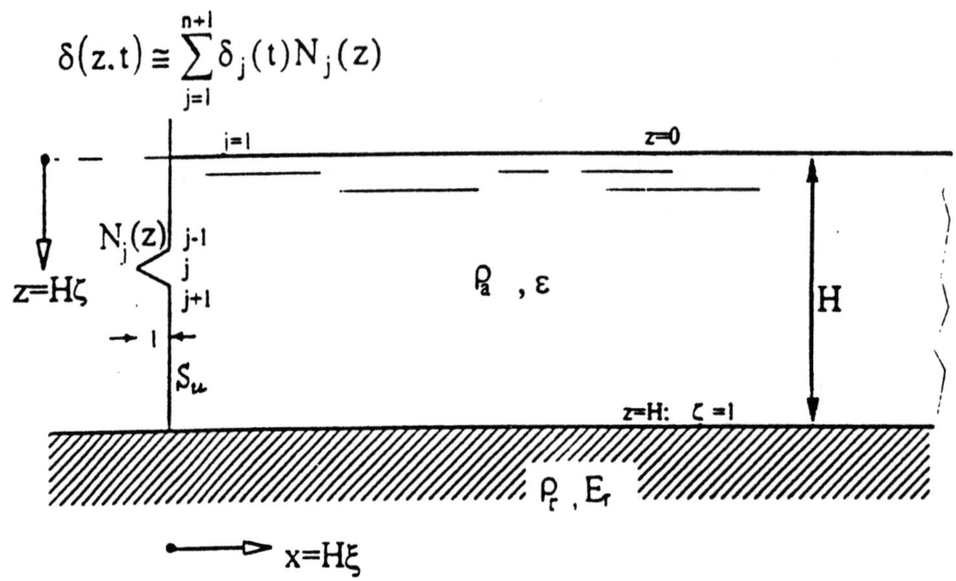

Reference scheme for the fluid motion

Fig. 83
Fluid-structure interaction: B.C. on S_u

Now let us separate, for each complex unit solution ϕ_j , the real (p_j) parts and imaginary ($-q_j$) parts; one obtains the following B.C.:

on S_b with $\phi_j = p_j - i\, q_j$; from the last of (3.2.116) and from (3.2.111):

$$\frac{\partial p_j}{\partial n_b} - i\frac{\partial q_j}{\partial n_b} = -\frac{\rho_a}{\rho_r c_r}(i\omega-\lambda)(p_j - q_j) = -\frac{i\omega-\lambda}{c}(p_j - iq_j), \qquad (3.2.118)$$

putting $c = \frac{\rho_r c_r}{\rho_a}$ for brevity.

Separating the real from imaginary parts in (3.2.118) gives:

$$\frac{\partial p_j}{\partial n_b} = -\frac{\omega}{c}q_j + \frac{\lambda}{c}p_j$$

$$\frac{\partial q_j}{\partial n_b} = \frac{\omega}{c}p_j + \frac{\lambda}{c}q_j, \text{ coupled B.C. on } S_b; \qquad (3.2.119)$$

on S_u one will have to take:

$$\frac{\partial p_j}{\partial n_u} = [N]_{S_u}\begin{Bmatrix}0\\0\\\cdots\\1\\\cdots\\0\\0\end{Bmatrix} \leftarrow j^{th} \text{ row},$$

$$\frac{\partial q_j}{\partial n_u} = 0 \quad \text{B.C. on } S_u; \qquad (3.2.120)$$

finally $p_j = 0$, $q_j = 0$ are the B.C.on S_0; moreover, $\begin{cases}\nabla^2 p_j = 0\\ \nabla^2 q_j = 0\end{cases}$ in Ω_w (3.2.121)

Within Ω_w the Laplace equation for $\phi_j = p_j - i\, q_j$ will give indeed:

$\nabla^2 p_j = 0$ and $\nabla^2 q_j = 0$ separately, i.e.: (see 3.2.78)

$$[T]_{\Omega_w}\{p_j\} = \{0\}\ ,\ [T]_{\Omega_w}\{q_j\} = \{0\} \qquad (3.2.122)$$

It is evident from the above that for $\frac{1}{c} = 0$, or $\frac{1}{c_r} = 0$, see (3.2.118), one reverts to the "rigid bottom" model and in that case $q_j = 0$, everywhere: σ and $\{\Delta\}$ are real. Now the pressures on the dam/fluid interface nodes will be given by:

$$\{p_u\} = -\rho_a \left\{\frac{\partial \phi}{\partial t}\right\}_{S_u} = -\rho_a \left[\phi_{jl}\right]^T \left[\cos n\hat{\ } j\right]^T \{\ddot{\delta}\} = -\rho_a \left[\phi_{jl}\right]^T \left[\cos n\hat{\ } j\right]^T \{\Delta\}\sigma^2 e^{\sigma t}, \tag{3.2.123}$$

where $\phi_{jl} = \phi_j(x_l, y_l, z_l)$ So the nodal forces $\{F^*\}_{Su}$ applied at the nodes of the dam wet face will be:

$$\{F^*\}_{S_u} = \iint_{S_u} [\cos n\hat{\ } j][N]^T_{S_u} [N]_{S_u} \{p_u\} dS_u$$

$$= -\rho_a \sigma^2 \left(\iint_{S_u} [\cos n\hat{\ } j][N]^T_{S_u} [N]_{S_u} \left[\phi_{jl}\right]^T \left[\cos n\hat{\ } j\right]^T dS_u \right) \{\Delta\}_{S_u} e^{\sigma t}$$

from which successively

$$\{F^*\}_{S_u} = -\sigma^2 \left(\left[\overline{\mu}_a\right] - i\left[\overline{\nu}_a\right]\right)\{\Delta\} e^{\sigma t}, \tag{3.2.124}$$

with:

$$\left[\mu_a{}^*\right] = \rho_a \iint_{S_u} [\cos n\hat{\ } j][N]^T_{S_u} [N]_{S_u} \left[p_{jl}\right]^T \left[\cos n\hat{\ } j\right]^T dS_u \; ; \quad [\,\overline{\mu}_a] = \tfrac{1}{2}([\mu^*{}_a] + [\mu^*{}_a]^T)$$

$$\left[\nu_a{}^*\right] = \rho_a \iint_{S_u} [\cos n\hat{\ } j][N]^T_{S_u} [N]_{S_u} \left[q_{jl}\right]^T \left[\cos n\hat{\ } j\right]^T dS_u \; ; \quad [\,\overline{\nu}_a] = \tfrac{1}{2}([\nu^*{}_a] + [\nu^*{}_a]^T)$$

Note that $[\,\overline{\mu}_a]$, although formally identical with (3.2.93), is now different because the boundary conditions for q_j are different from (3.2.80) on S_b .
In particular, because now the B.C. (3.2.119) contain ω, the matrices $[\,\overline{\mu}_a]$, $[\,\overline{\nu}_a]$ become <u>frequency dependent</u>.
By the use of the incidence matrix $[\,I\,]_{Su}$ of the nodes on S_u with respect to the general ordering of all dam nodes, see (3.2.94), one gets as usual for the whole F.E. mesh of the dam:

$$\{F\} = -\sigma^2\left([\mu_a] - i[\nu_a]\right)\{\Delta\}e^{\sigma t}, \tag{3.2.125}$$

where:

$$[\mu_a] = [I]^T_{S_u}\,[\bar{\mu}_a]\,[I]_{S_u}\ ,$$

$$[\nu_a] = [I]^T_{S_u}\,[\bar{\nu}_a]\,[I]_{S_u}\ .$$

Introducing the above results into the matrix equations of motion of the dam:

$$[K]\ \{\delta\} + [M]\ \{\ddot{\delta}\} = \{F\} \tag{3.2.126}$$

(δ is used in place of u to remind the reader that one is dealing here with complex values), we get, keeping into account (3.2.110) and dropping the common factor $e^{\sigma t}$:

$$[K]\ \{\Delta\} + \sigma^2[M]\ \{\Delta\} = -\sigma^2\left([\mu_a] - i[\nu_a]\right)\ \{\Delta\}, \tag{3.2.127}$$

or:

$$[K]\ \{\Delta\} + \sigma^2\left([M] + [\mu_a]\right)\ \{\Delta\} - i\sigma^2[\nu_a]\ \{\Delta\} = \{0\}, \tag{3.2.128}$$ [61]

where $[\mu_a]$, $[\nu_a]$ are functions of the unknown $\sigma = i\omega - \lambda$ through B.C. (3.2.119). Eq. (3.2.128) poses an eigenvalue problem in the complex field, which can be solved by successive approximations.

E.g. as a first approximation one may take $\omega \cong \omega_0$, ω_0 being an eigenfrequency for the undamped system, $\dfrac{\partial p_j}{\partial n_b} = 0$ as for rigid bottom, $\dfrac{\partial q_j}{\partial n_b} = \dfrac{\omega_0}{c} p_j$ on S_b ; the approximate value of λ is then :

$$\lambda_0 \cong \frac{\omega_0}{2}\frac{\{U_0\}^T[\nu_a]\{U_0\}}{\{U_0\}^T\left([M] + [\mu_a]\right)\{U_0\}}, \tag{3.2.129}$$

---

[61] It can be shown that it would be more natural to write, in place of (3.2.126), $[K]\{\delta\} + [M]\{\ddot{\delta}\} = -[\mu_a]\{\ddot{\delta}\} + [\nu'_a]\{\dddot{\delta}\}$ with $[\nu'_a] = \dfrac{i[\nu_a]}{\sigma}$, or, after (3.2.110), $[K]\{\Delta\}+\sigma^2([M]+[\mu_a])\{\Delta\}-\sigma^3[\nu'_a]\{\Delta\}=0$ in place of (3.2.128).

$\{U_0\}$ being the real-valued eigenvector corresponding to ω_0 in the undamped system, i.e.:

$$[K]\{U_0\} = \omega_0^2\left([M] + [\mu_a]\right)\{U_0\}\ ; \qquad (3.2.130)^{62}$$

Eq. (3.2.129) is derived from energetic considerations, see fig.84. In fact, the negative work done by hydrodynamic forces $\{F\}_p$ during one cycle of vibration, as a consequence of the presence of components of hydrodynamic pressures that are in-quadrature with respect to the displacements, and in opposition of phase with the velocities, causes a decrease of the total (kinetic + deformation) energy stored in the mechanical system: from approximate evaluation of these two terms (3.2.129) is derived.

With $\sigma = i\omega_1 - \lambda_0$ one can then find a complex first-approximation eigenvector $\{\Delta\}$ from (3.2.128), taking $\omega_1 \cong \omega_0\sqrt{1-\left(\frac{\lambda_0}{\omega_0}\right)^2}$ and re-computing $[\mu_a]$, $[\nu_a]$, with B.C. (3.2.119) into which $\omega = \omega_1$, $\lambda = \lambda_0$ are introduced[63]. A second-approximation solution is thus found; the process can be repeated until satisfactory convergence of results is obtained.

Another cause of damping associated with the presence of impounded water is the occurence of <u>dissipative layers</u> in the bottom of the reservoir (e.g. muddy layers with high viscosity), see fig.85. Also this case can be modelled by adopting ad-hoc simplifying assumptions; it will not, however, be treated in the framework of the present course for sake of brevity.

---

[62] If we include in (3.2.126) also the <u>structural damping</u>, through matrix $[C]$:

$[K]\{\delta\} + [C]\{\dot\delta\} + [M]\{\ddot\delta\} = -([\mu_a] - i[\nu_a])\{\ddot\delta\}$ or, for a damped mode:

$[K]\{\Delta\} + \sigma[C]\{\Delta\} + \sigma^2([M]+[\mu_a])\{\Delta\} - i\sigma^2[\nu_a]\{\Delta\} = \{0\}$, the first-approximation value for λ is:

$$\lambda_0 \cong \frac{\{U_0\}^T\left([C]+\omega_0[\nu_a]\right)\{U_0\}}{2\{U_0\}^T\left([M]+[\mu_a]\right)\{U_0\}} .$$

[63] As soon as the real eigenvector $\{U_0\}$ is replaced by a complex one $\{\Delta\}$, in place of (3.2.129) the following equation should be used:

$$\lambda \cong \frac{\omega}{2}\frac{\{\Delta^*\}^T[\nu_a]\{\Delta\}}{\{\Delta^*\}^T\left([M]+[\mu_a]\right)\{\Delta\}} ,$$

where Δ^* is the complex-conjugate value of Δ : i.e. if $\Delta = A+iB$, $\Delta^* = A - iB$.

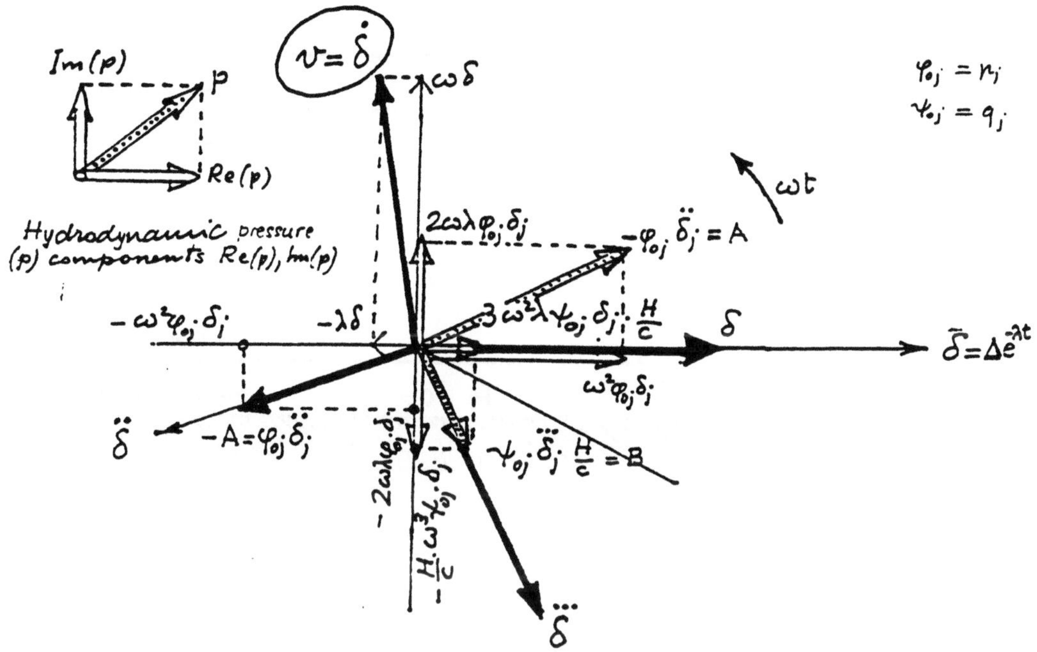

Fig. 84
Dam-reservoir-foundat. interaction: rotating-vector representation of pressures

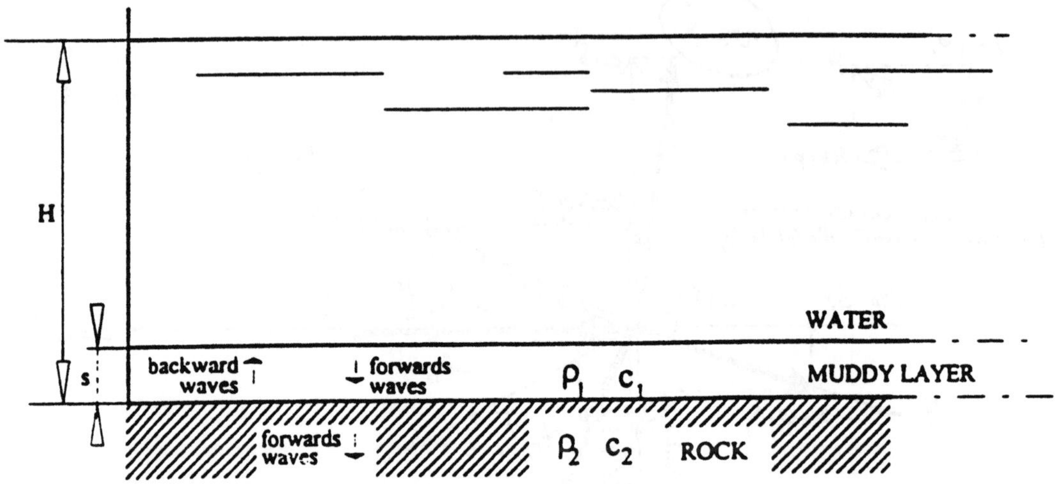

Fig. 85
Scheme for damping effect of dissipative bottom Layers and bottom radiation

c_4) Compressibility effects

If the incompressibility assumptions for water is relinquished, i.e. if $1/C_a \neq 0$,the only noteworthy change in the above procedure for the computation of $[\mu_a]$ and $[\nu_a]$ consists in replacing the indefinite Laplace partial differential equation and its discretized counterpart, see (3.2.122), with the two coupled Poisson equations:

$$\nabla^2 \phi_j = \frac{\sigma^2}{C_a^2}\phi_j \tag{3.2.131}$$

i.e.

$$-\nabla^2 p_j = \frac{\omega^2 - \lambda^2}{C_a^2} p_j + \frac{2\lambda\omega}{C_a^2} q_j \; ;$$

$$-\nabla^2 q_j = \frac{\omega^2 - \lambda^2}{C_a^2} q_j - \frac{2\lambda\omega}{C_a^2} p_j$$

i.e., in the F.E. discretization,

$$[T]_{\Omega_w}\{p_j\} = \frac{\omega^2 - \lambda^2}{C_a^2}[m_1]\{p_j\} + \frac{2\lambda\omega}{C_a^2}[m_1]\{q_j\};$$

$$[T]_{\Omega_w}\{q_j\} = \frac{\omega^2 - \lambda^2}{C_a^2}[m_1]\{q_j\} - \frac{2\lambda\omega}{C_a^2}[m_1]\{p_j\}, \quad (3.2.132)[64]$$

where $[m_1]$ is defined as follows:

$$[m_1] = \iiint_{\Omega_w} \langle N\rangle_w \langle N\rangle_w^T \, d\Omega_w \; ;$$

the B.C. remain unchanged, see (3.2.119)........(3.2.121).

It is seen from the above that both the bottom-radiation damping and the finite value of C_a deriving from the fluid compressibility make the added-mass matrix frequency-dependent; also the damping, through matrix $[\nu_a]$, will be frequency-dependent, so that the damping factors of individual modes will be different.

[64] Eqs. (3.2.131) are to be used in place of (3.2.122) for building up the "unit" singular solutions $\{\phi_j(x,y,z)\}$ and matrix $[\phi_{jl}]=[p_{jl}] - i[q_{jl}]$, see (c_3).

It is also apparent that the compressibility effects will become especially important near the reservoir eigenfrequencies, i.e. the values of ω (or of $\sigma = i\omega - \lambda$ in the case of compliant bottom) which are eigenvalues of (3.2.131), see fig.86. Under these conditions the Westergaard classical formulation, with a frequency-independent added-mass matrix $[\mu_a]$, shows its inherent shortcomings; damping from bottom radiation is also greatly increased near these particular frequencies. These effects can be quite relevant in forecasting the dam behaviour under seismic excitation with full, or partially full, reservoir. However, as already said, the difficulty of gaining sufficiently "complete" information (in order that the input of the computations be "realistic" enough) explains why the fluid/structure interaction problem is still the subject of intense academic and engineering research (see e.g. CHOPRA et al.), especially with the aim of finding sufficently simplified, yet reliable mathematical formulations.

c_5) Response to near-field ground shaking and to bottom shaking by modal analysis

Remembering all of the above results [in particular eqs. (3.2.108), (3.2.109)] one can write in general, for simultaneous near-field ground shaking and bottom shaking, $u_g(t)$ and $u_b(x,y,z,t)$ on S_b :

$$[K]\{u_r\}+[C]\{\dot{u}_r\}+\left([M]+[\mu_a]\right)\{\ddot{u}_r\}=-\left([M]+[\mu_a]\right)\ddot{u}_g\{1\}-[BE]\{\ddot{u}_b\}, \quad (3.2.133)^{65}$$

where $\ddot{u}_g(t)$ and $\{\ddot{u}_b\}=\{\ddot{u}_b(t)\}_{S_b}$ are supposed to be known quantities.

Along the same lines already illustrated for undamped systems, one can deal separately with each individual mode of vibration of the structure as if it were a simple, one d.o.f. linear oscillator with a defined frequency $\omega_m/2\pi$, a damping factor (relative to critical

[65] It is supposed, for simplicity sake, that the damping associated with fluid/structure interaction (see headings c_3 and c_4) has been somehow incorporated into the 2nd term on the L.H.S. of eq. (3.2.133). This may be difficult in practice, since matrices $[\mu_a]$ and $[\nu_a]$ above can be defined without ambiguity only if the motion is a sinusoidal one (of frequency $\omega/2\pi$ known). However, since we will use modal analisys to solve the problem, for each mode $[\mu_a]$ and $[\nu_a]$ will be defined and the formal incorporation of all types of modal damping into $[C]$ becomes possible.

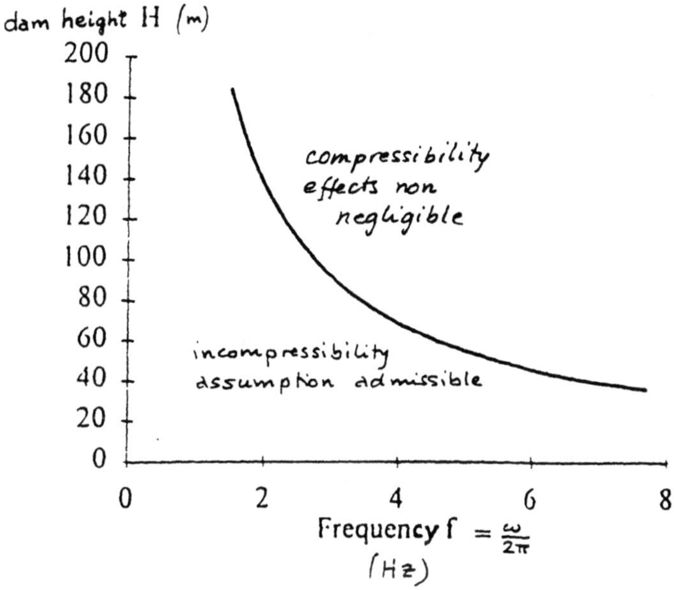

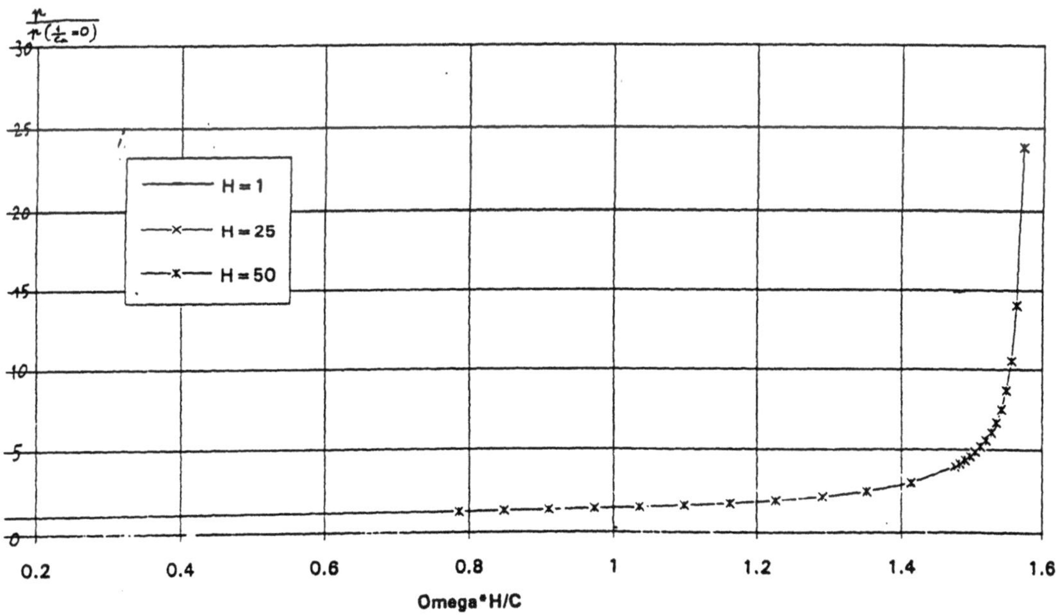

Fig. 86
Compressibility effects on fluid/structure interaction

damping) ν_m and a (complex) eigenshape $\{\psi\}_m$ (which is supposed to have been normalized according to (3.2.19) suitably modified: $\{\psi^*\}^T{}_m$ in place of $\{\psi\}^T$ and $([M] + [\mu_a])$ in place of $[M]$).
Then the <u>response of each damped mode</u> $\{\psi\}_m$ to the <u>near-field ground shaking</u> $u_g(t)$ is best evaluated by (3.2.47), (3.2.48), i.e. by the Housner response spectra technique (of course the Housner spectrum of $u_g(t)$ for the modal damping factor has to be known).
As for the <u>response to bottom shaking</u>, one can take the following way.
Let us take the Fourier transforms of vectors $\{u_r(t)\}$ and $[BE]\{\ddot{u}_b(t)\}$ (see 3.2.107, 3.2.108) and decompose them onto the base of the normalized (complex) modes $\{\psi\}_m$ [note that both are $3\,M_u \times 1$ - order vectors, i.e. of the same order as $\{\psi\}_m$]; we denote the Fourier transform of $u_r(t)$ with the symbol $\mathcal{U}(\omega)$:

$$\{\mathcal{U}_r(s)\} = \int_0^\infty e^{-st}\{u_r(t)\}dt$$

where $s = i\omega$ [and likewise for $\{\mathcal{U}_b(\omega)\}$ from $\{u_b(t)\}$].
So it is:

$$\{\mathcal{U}_r(\omega)\} = X_1(\omega)\{\psi\}_1 + X_2(\omega)\{\psi\}_2 + \ldots + X_m(\omega)\{\psi\}_m + \ldots \quad (3.2.134)$$

the complex quantities $X_j(\omega)$being unknown; now if:

$$\mathcal{L}_{Bj}(\omega) = \{\psi^*\}_j^T[BE](-\omega^2)\{U_b(\omega)\},$$

from the Fourier transform of the dynamical equation we get for mode :

$$\chi_j X_j + 2i\omega\lambda_j\mu_j X_j - \omega^2\mu_j X_j = -\mathcal{L}_{Bj}(\omega)$$

[instead of $\omega^2\mu_j\, U_g(\omega)$as in the case of near-field ground shaking], with χ_j = modal stiffness, $\lambda_j = \nu_j\,\omega_j$ = modal damping, μ_j = modal mass of mode $\{\psi\}_j$ having frequency $\omega_j/2\pi$, $\omega_j^2 = \frac{\chi_j}{\mu_j}(1-\nu_j^2)$, one gets the Fourier transform of the modal response and, by anti-transforming, the time-domain response:

$$X_j(\omega) = -\frac{\mathcal{L}_{Bj}(\omega)}{\mu_j\left(\frac{\omega_j^2}{1-\nu_j^2} + 2i\omega\lambda_j - \omega^2\right)}$$

and
$$x_j(t) = \frac{1}{2\pi i}\int_{-\infty}^{\infty} e^{i\omega t} X_j(\omega)\, d\omega = \frac{1}{\pi}\int_0^{\infty} A_j(\omega)\cos\omega t\, d\omega + \frac{1}{\pi}\int_0^{\infty} B_j(\omega)\sin\omega t\, d\omega, \tag{3.2.135}$$

where:

$$A_j = \mathrm{Re}\left[\frac{\omega^2\{\psi^*\}^T [BE]\{\mathcal{U}_b(\omega)\}}{\mu_j\left(\frac{\omega_j^2}{1-\nu_j^2} + 2i\omega\lambda_j - \omega^2\right)}\right],$$

$$B_j = \mathrm{Im}\left[\frac{\omega^2\{\psi^*\}^T [BE]\{\mathcal{U}_b(\omega)\}}{\mu_j\left(\frac{\omega_j^2}{1-\nu_j^2} + 2i\omega\lambda_j - \omega^2\right)}\right]$$

(Remember that a normalized mode $\{\psi\}_j$, together with its modal mass μ_j and stiffness χ_j, is defined for a given direction $\vec{g}$, see (3.2.19); but (3.2.135) needs not be written three times, (one for each of the three coordinate axes), because, it will be recalled, the response is independent from $\vec{g}$ (as far as the modal parameters do not vanish for $\vec{g}$) . We will develop the treatment, therefore, as if only one (arbitrary) direction were involved), as far as bottom shaking is concerned.

Eq. (3.2.135) is equivalent to (3.2.40), with a suitable excitation $-\beta_j(t)$ in place of $-\mu_j \ddot{u}_g(t)$.

We can consider $\frac{\beta_j(t)}{\mu_j}$ as an "equivalent near-field ground shaking" $\ddot{u}_{gbj}$ for mode $\{\psi\}_j$ [note that the $\ddot{u}_{gbj}$ are now different for each mode $\{\psi\}_j$, contrary to the case of an actual near-field ground shaking $\ddot{u}_g(t)$].

Eventually one has to build up the response of the single oscillator with damping λ_j and frequency $\omega_j/2\pi$ to the equivalent support shaking

$$\ddot{u}_{gbj}(t) = \frac{\beta_j(t)}{\mu_j} \tag{3.2.136}$$

Classical simple-oscillator theory then gives for the modal amplitude $x_{jb}(t)$:

$$x_{jb}(t) = -\int_{\tau=t_0}^{\tau=t} \frac{\beta_j(\tau)}{\omega_j \mu_j} e^{-\lambda_j(t-\tau)} \sin\omega_j(t-\tau)\, d\tau, \qquad (3.2.137)$$

where t_0 is the time of beginning of bottom shaking; $t \geq t_0$; and for the modal velocity, $\dot{x}_{jb}(t)$:

$$\dot{x}_{jb}(t) = -\frac{1}{\mu_j}\left[\int_{\tau=t_0}^{\tau=t} \beta_j(\tau) e^{-\lambda_j(t-\tau)} \cos\omega_j(t-\tau)d\tau - \frac{\lambda_j}{\omega_j}\int_{\tau=t_0}^{\tau=t} \beta_j(\tau) e^{-\lambda_j(t-\tau)} \sin\omega_j(t-\tau)d\tau\right]$$

(3.2.138)

The latter could be added to the completely analogous formula for modal velocity response (along the coordinate axis being considered, see above) $\dot{x}_{jg}(t)$ to near-field ground shaking defined by accelerogram $\ddot{u}_g(t)$ (see equivalence 3.2.136); the total maximum modal response for velocity $\dot{x}_j(t)$ could then be evaluated as:

$$\max\,(\dot{x}_j) = V_{g+b}(\omega_j) = \max\left[\dot{x}_{jg}(t) + \dot{x}_{jb}(t)\right]; \qquad (3.2.139)$$

note that this is different from $\max[\dot{x}_{jg}(t)]$ + $\max[\dot{x}_{jb}(t)]$, since the two maxima cannot be assumed to take place at the same moment.

After this has been done for the suitable number of modes (and for all three coordinate directions, see above), (3.2.46)......(3.2.50) give the desired responses (SRSS technique).

Although all of the above considerations are formally consistent, in practice it would be difficult to carry out all the steps of the procedure as described. More than likely, indeed, not all of the necessary input data would be known; for instance, the space-time details of bottom shaking u_b (x,y,z,t) on S_b may be difficult to define; or $V_g(\omega)$may be known, but not the space-time details of $\ddot{u}_g(t)$ [66], thus making it difficult to calculate $\dot{x}_{jg}(t)$ to be added to $\dot{x}_{jb}(t)$ in (3.2.139).

However, from all of the foregoing discussion it is quite clear that the knowledge of eigenmodes (modal frequencies, modal shapes and modal damping factors) is an essential prerequisite to any attempt at forecasting (or hindcasting) the dynamic -in particular the seismic- performance of a dam. The F.E.M. allows to estimate the modal

[66] There are, however, mathematical routines to generate synthetic accelerograms $\ddot{u}_g(t)$ having a prescribed response spectrum $V_g(\omega)$.

characteristics of arch dams with good accuracy, see fig.87, to the exception of global modal damping, which in the present state of the art has to be evaluated by experience or by experimental means, i.e. through *in-situ* dynamic testing.

It is only fair to remark, in conclusion, that the F.E. models are the most powerful tools now available for numerical analysis of arch dams; yet they cannot be taken to be entirely representative of reality.

In particular, arch dams are built with construction joints between adjacent monoliths; these joints, although grouted at the end of erection, represent physical discontinuities whose real behaviour is very difficult to model in a satisfactory way. At empty reservoir, under strong seismic excitation (and especially in the cold season) joints may open, and close again, many times during the motion. This behaviour completely disattends the continuity assumptions of the F.E. modelling and makes for highly non-linear dynamics, with the distinct possibility of "chaotic" behaviour, which cannot be forecast effectively with ordinary (even non-linear) deterministic models.

At full (or partially full) reservoir, in the cold season the joints may be partially open near the downstream face, and again the geometry and stiffness of the arch dam may vary in a non-negligible way during a seismic motion, with attending non-linear effects. Moreover, the effects of a given seismic motion at the same dam (with the same impoundment level) will be different from season to season; see fig.88.

To a lesser degree, the existence of cracks and discontinuities in the foundation rock, which may tend to close under the arch thrust or to open due to pressurized water infiltration, may cause quite substantial departures from the current assumption of a continuous, homogeneous rock mass.

It is true that advanced, non-linear F.E. models can in principle cope with these situations. Adjacent monoliths could, in the limit, be individually represented; contact and impact conditions over the joint surfaces could be modelled to a degree, etc (see fig.89). However, such non-linear approaches are extremely heavy in terms of SW complication as well as of computer time. Moreover, it is known that such models are characterized by a very high sensitivity to even very small variations in initial assumptions. This undermines the reliance on a single non-linear simulation as representative of the spectrum of possibilities: many simulations, with different initial assumptions, would thus be needed to gain sensitivity as to the possible range of effect variations.

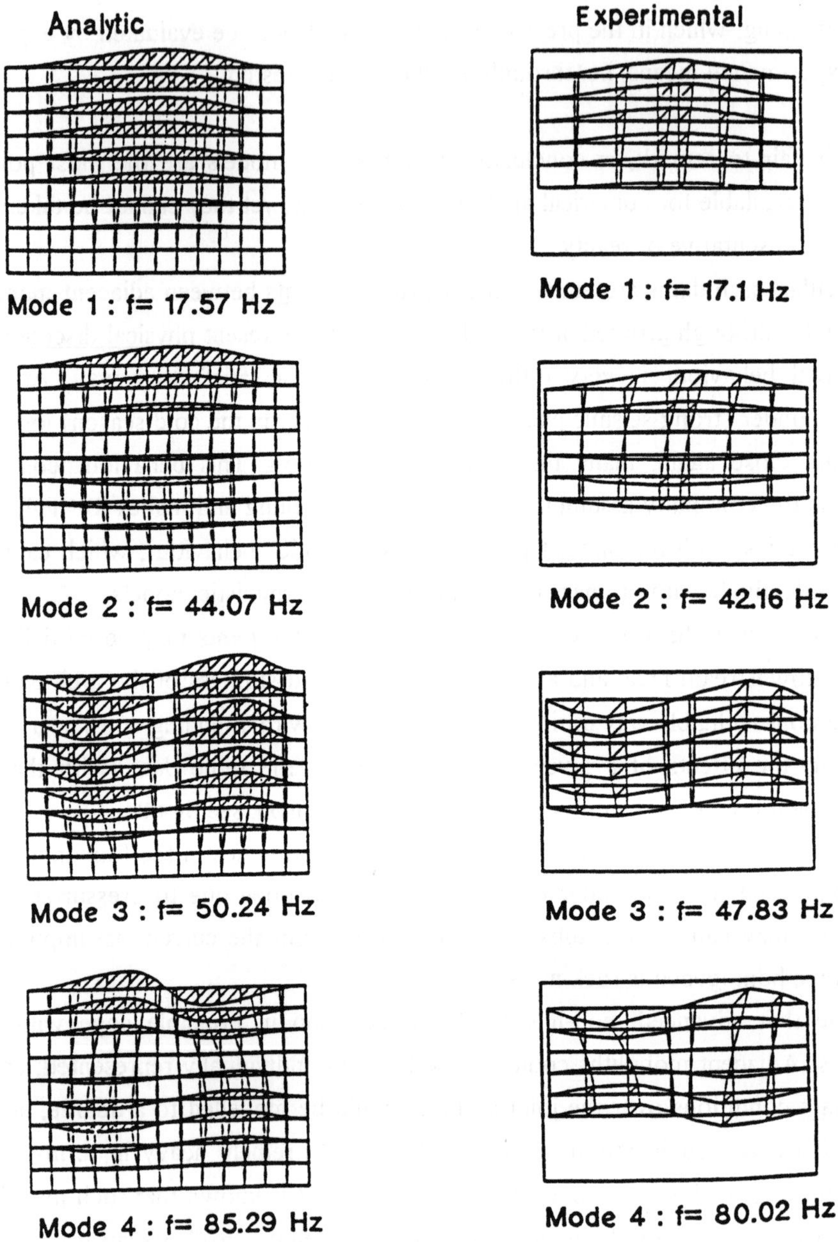

Fig. 87
Results of validation model tests for fluid/structure interaction

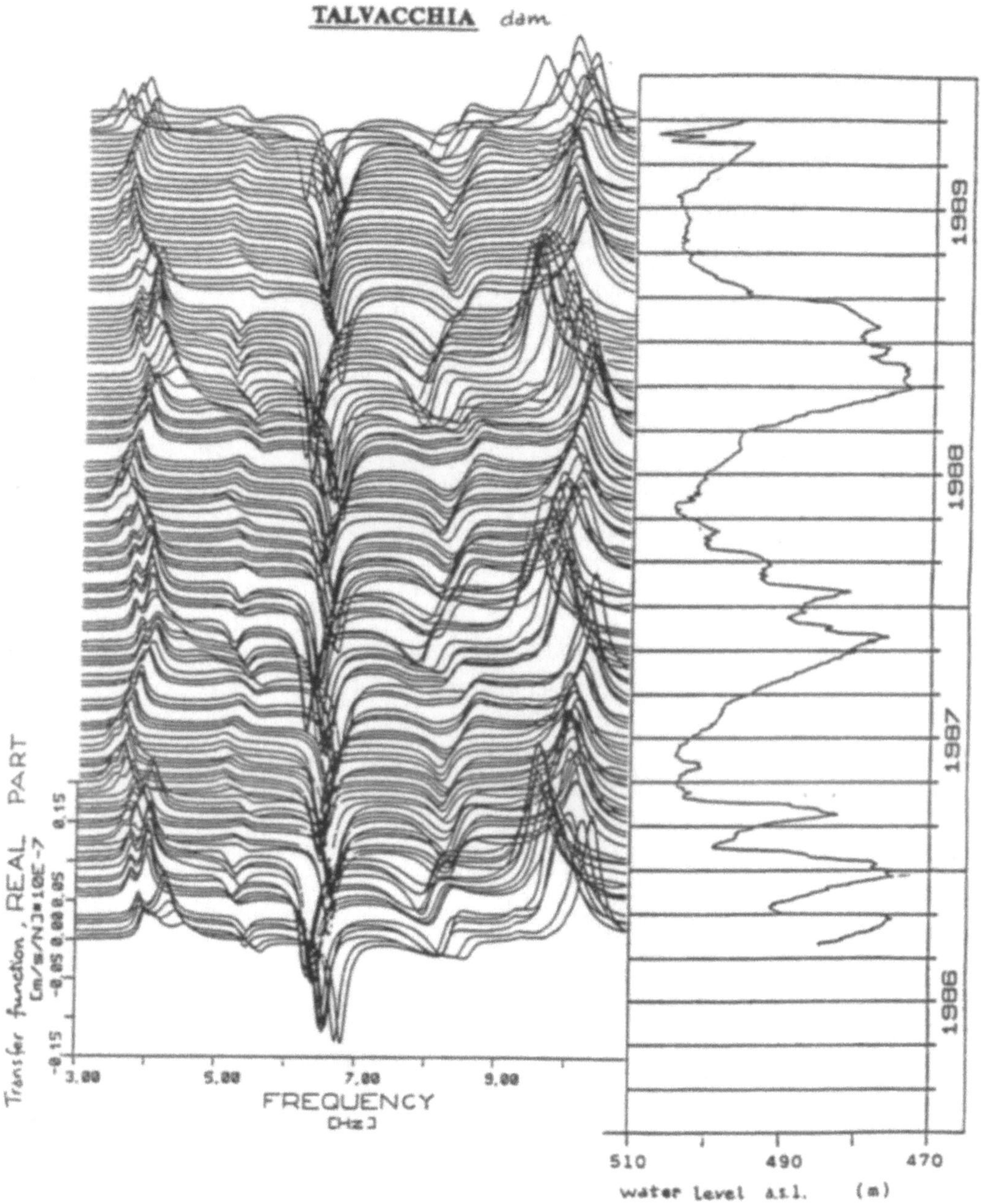

Fig. 88
The eingenfrequencies of an arch dam vary with the water level and with the season

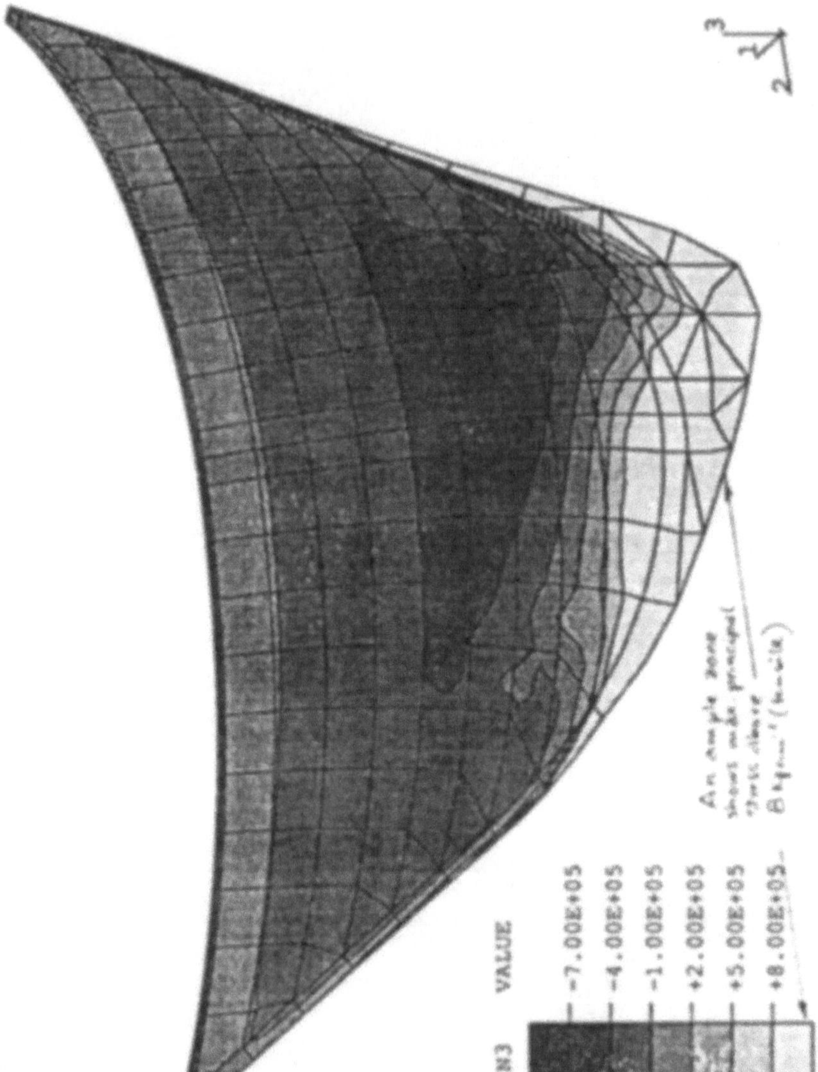

Fig. 89a
Elastic analysis (dead weight +hydrost. load).
Tensile stress > 8 Kgcm^2 at heel

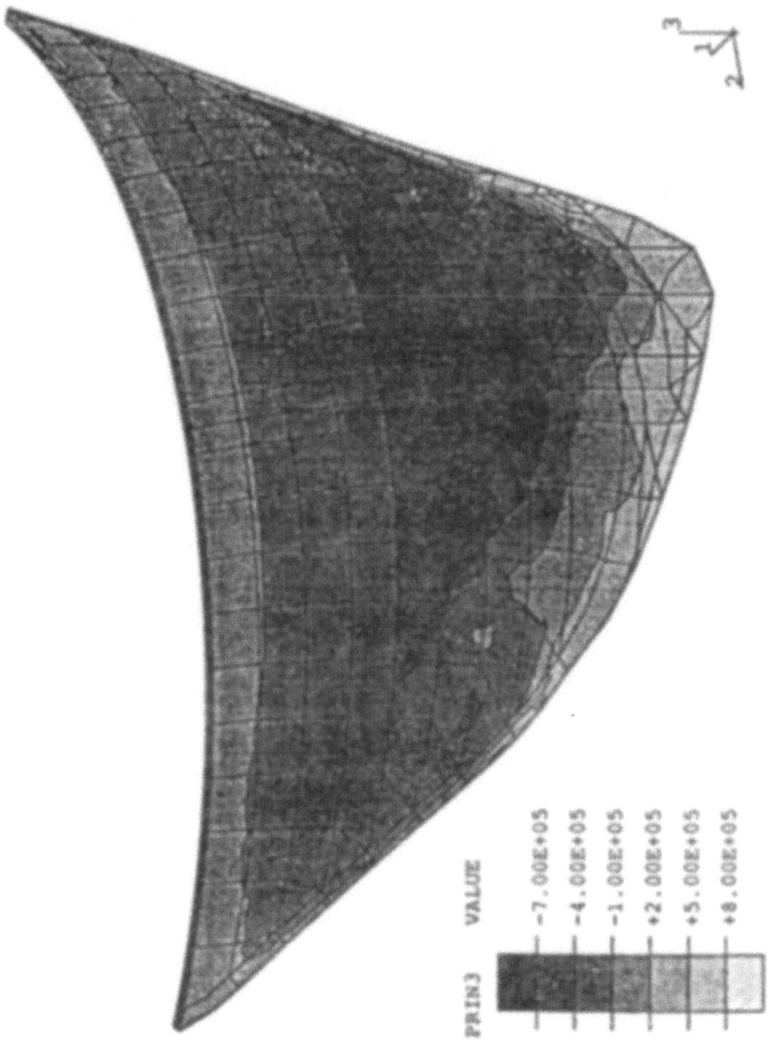

Fig. 89b
Elasto-plastic analysis (Drucker-Prager model with 8 $Kgcm^2$ cutoff).
Dead weight +hydrost. load.

The zones with $\sigma_3 > 8$ $Kgcm^2$ have all but disappeared

It is, thus, understandable that such "advanced" models are for the present more of a research tool than a practical engineering help. It is worth to retain firmly the concept that a dam is a structure changing continuously its mechanical (static and dynamic) properties with the level of impoundment, the thermal state, last but not least the many aging processes. All mathematical models of behaviour, and the attending response estimates, should be appraised against the background of this very complex reality, keeping always in mind all the drastic simplifications that are made in order to set up usable models, i.e. to get numerical results at reasonable costs in terms of input information, formalization effort, SW development and validation, computer time and costs, final result interpretation.

While the future will certainly witness a sustained pace of mathematical model development for arch dams, it is by no means certain that it will go on along the present lines of advanced research. Basic progresses in the field of solid mechanics and of complex non-linear dynamic behaviour are expected to avoid being drawn into a path of ever increasing awkwardness.

BIBLIOGRAFIC LIST

"Practice and Theory of Arch Dams"
Proceedings of the International Symposium on Arch Dams Nanjing, 17-20 Oct 1992

"Trial Load Method of Analyzing Dams"
Bureau of Reclamation, Bulletin 1, Denver 1938

F.Vogt
"Ueber die Berechnung der Fundamentdeformation"
Noske Videnskaps-Akademie, Oslo 1925

Boussinesq
"Application des potentiels à l'etude de l'équilibre et du mouvement des solides èlastiques"

Arch Dams. International Workshop on Arch Dams Coimbra 5-9 Apr. 1987 -Ed. A.Balkema,1990

"Arch Dams. A Review of British Research & Development"
Symposium, London, 20-21 March 1968 The Institution of Civil Engineerings - London, 1968

O.C. Zienkiewicz
"The Finite Element Method"
J.Wiley, Chichester & New York, 1977

M.Fanelli
"Rapid Preliminary Analysis of Arch Dams: The Ritter Method Revisited"
Dam Engeneering, Vol I, Issue 2 -Apr. 1990

M.Fanelli ,A.Fanelli, P.Bonetti
"Non-linear Analysis in the preliminary evaluation of arch dam design"
The International Journal of Hydropower & Dams, Vol I, Issue 6, Nov. 1994

M.Fanelli, A.Fanelli, C.Galimberti, P.Palumbo
"A simple evaluation of the damping of free vibrations of an arch dam caused by acoustic radiation from reservoir bottom"
Under review for publication on Hydropower & Dams

Westergaard
"Water pressures on dams during earthquakes"
Proc. 1931

APPENDIX

(The illustrations from a to i present a synthetic overview of some topics that it was not possible to develop at length within the limits of the main text)

(a)

SHELL MODELS:

THE SIMPLEST VERSION: PLANE FACETS ASSEMBLY.

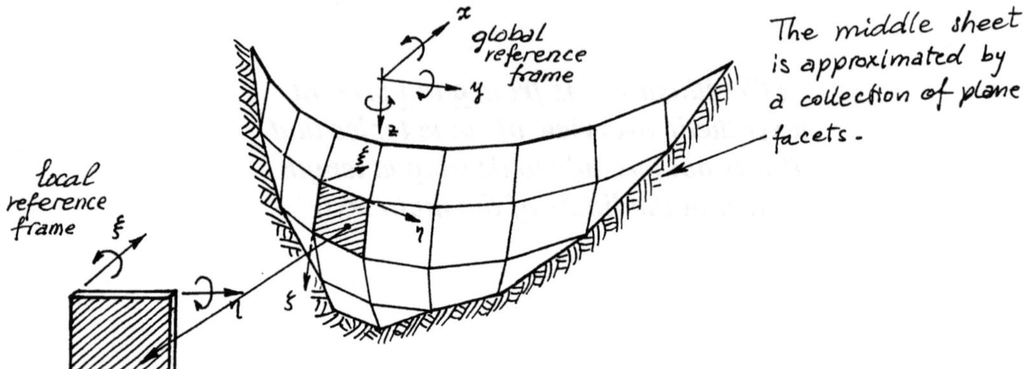

"Plane facet" generic element: elastic deformation energy:

$$W_{el} = \frac{1}{2}\{\delta_P\}^T_{el}\cdot[K_P]_{el}\cdot\{\delta_P\}_{el} + \frac{1}{2}\{\delta_B\}^T\cdot[K_B]_{el}\cdot\{\delta_B\}_{el} \quad \text{in local frame.}$$

($[K_P]_{el}$ → in-plane stiffness; $[K_B]_{el}$ → bending stiffness)

In-plane displacement vector $\{\delta_P\}_{el} = [CTP]_{el}\cdot\{\delta\}_g$,

Out-of-plane " " $\{\delta_B\}_{el} = [CTB]_{el}\cdot\{\delta\}_g$, } displacem. + rotations

where $\{\delta\}_g$ = vector of displacements <u>in global frame</u>,

$[CTP]_{el}$ = coördinate transformation/incidence matrix for generic element in-plane displacements,

$[CTB]_{el}$ = coördinate transformation/incidence matrix for generic element out-of-plane displacements.

If $[K]_g = \sum_{el}\left([CTP]^T_{el}[K_P]_{el}[CTP]_{el} + [CTB]_{el}\cdot[K_B]_{el}\cdot[CTB]_{el}\right)$,

then $[K]_g\{\delta\}_g = \{F\}_g$ expresses global equilibrium.

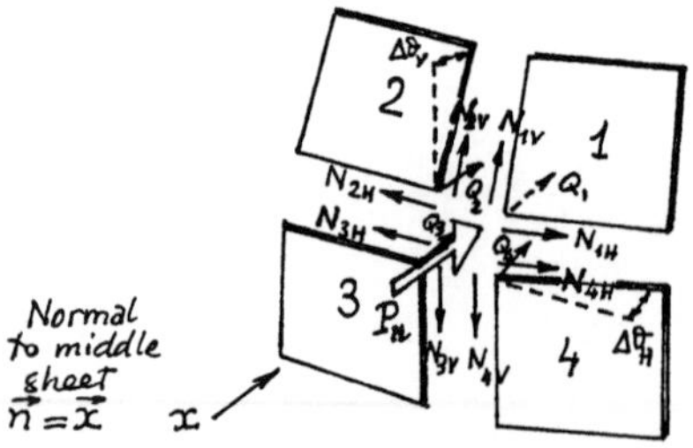

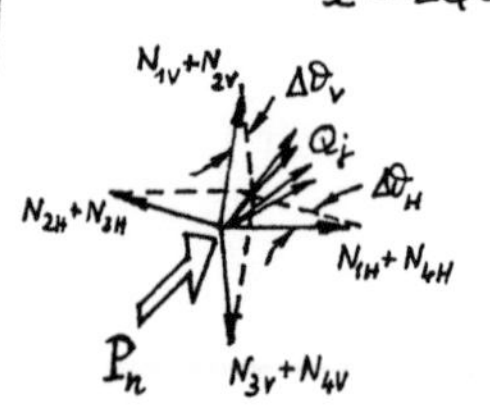

PHYSICAL MEANING: AT EACH NODE, x – EQUILIBRIUM CONDITION IS:

$$\frac{\sum^{1..4} N_{jH}}{2}\times\Delta\vartheta_H + \frac{\sum^{1..4} N_{jV}}{2}\times\Delta\vartheta_V + \sum^{1..4} Q_j \cos\widehat{n_{el}^{j} n} = P_n$$

(N from in-plane stiffnesses, Q from out-of-plane ").

THERMAL DIFFUSION ANALYSIS BY FINITE ELEMENTS

(b)

Orthotropic conductivity k_x, k_y, k_z:
The Heat conduction (FOURIER) equation is:

$$\frac{\partial}{\partial x}\left(k_x \frac{\partial \vartheta}{\partial x}\right) + \frac{\partial}{\partial y}\left(k_y \frac{\partial \vartheta}{\partial y}\right) + \frac{\partial}{\partial z}\left(k_z \frac{\partial \vartheta}{\partial z}\right) + q = c\,\dot{\vartheta},$$

where q = heat source intensity per unit volume ($q = q(x,y,z,t)$);

c = thermal capacity of the unit volume (supposed constant).

If $\dot{\vartheta} = 0$, the functional of the problem is:

$$F = \frac{1}{2}\left[k_x\left(\frac{\partial \vartheta}{\partial x}\right)^2 + k_y\left(\frac{\partial \vartheta}{\partial y}\right)^2 + k_z\left(\frac{\partial \vartheta}{\partial z}\right)^2\right] - q\vartheta$$

and its approximation by the F.E. technique is ($\{\vartheta\}$ = vector of the nodal temperature values):

$$[k_{x_i}] = \begin{bmatrix} k_x & 0 & 0 \\ 0 & k_y & 0 \\ 0 & 0 & k_z \end{bmatrix} \qquad F_{appr.} = \frac{1}{2}\{\vartheta\}^T\left[\frac{\partial N}{\partial x_i}\right]^T [k_{x_i}]_D \left[\frac{\partial N}{\partial x_i}\right]\{\vartheta\} - \{q\}^T[N]^T[N]\{\vartheta\}$$

from which the F.E. matrix equation:

$$\left\{\frac{\partial F_{appr.}}{\partial \vartheta}\right\} = [T]\{\vartheta\} - \{Q\},$$

with

$$[T]_k = \iiint_{\Omega_k} \left[\frac{\partial N}{\partial x_i}\right]^T \cdot [k_{x_i}]_D \cdot \left[\frac{\partial N}{\partial x_i}\right] dV,$$

$$\{Q\}_k = \iiint_{\Omega_k} [N]^T \cdot [N]\{q_i\}\, dV \quad \text{for element } \Omega_k.$$

($N(x,y,z)$ = shape functions; $\{q_i\}$ = nodal values of q).

By formally treating $-c\dot{\vartheta}$ as a heat source we get for the transient problem:

$$\left\{\frac{\partial F_{appr.}}{\partial \vartheta}\right\} = [T]\cdot\{\vartheta\} - \{Q\} + [C]\{\dot{\vartheta}\}, \text{ where}$$

$$[C]_k = \iiint_{\Omega_k} [N]^T \cdot [N] \cdot c\, dV.$$

$[T]\cdot\{\vartheta\} = -[C]\cdot\{\dot{\vartheta}\} + \{Q\}$ then expresses $\left\{\frac{\partial F_{appr.}}{\partial \vartheta}\right\} = \{0\}$.

INTEGRATION IN TIME OF A F.E. MODEL BY THE CRANK-NICHOLSON (TRAPEZE) RULE e.g. for the THERMAL ANALYSIS

Ⓒ

$[T]\cdot\{\vartheta\}=-[C]\cdot\{\dot{\vartheta}\}$ is the thermal diffusion F.E. eq. (1)

At time t_n : $[T]\cdot\{\vartheta\}_n=-[C]\cdot\{\dot{\vartheta}\}_n$

At time t_{n+1} : $[T]\cdot\{\vartheta\}_{n+1}=-[C]\cdot\{\dot{\vartheta}\}_{n+1}$ (*)

Knowing $\{\vartheta\}_n$, $\{\dot{\vartheta}\}_n$ the problem is to find

$\{\vartheta\}_{n+1}$, $\{\dot{\vartheta}\}_{n+1}$.

The F.D. "trapeze" approximation for integration gives :

$$\{\vartheta\}_{n+1}\cong\{\vartheta\}_n+\frac{\Delta t}{2}\left(\{\dot{\vartheta}\}_n+\{\dot{\vartheta}\}_{n+1}\right)$$

from which $\alpha\{\vartheta\}_n+\{\dot{\vartheta}\}_n=\alpha\{\vartheta\}_{n+1}-\{\dot{\vartheta}\}_{n+1}$, $\alpha=\frac{2}{\Delta t}$

($\Delta t=t_{n+1}-t_n$), and $\{\dot{\vartheta}\}_{n+1}=\alpha\left(\{\vartheta\}_{n+1}-\{\vartheta\}_n\right)-\{\dot{\vartheta}\}_n$ (**)

Substituting (**) in (*) gives:

$[T]\cdot\{\vartheta\}_{n+1}=-[C]\left(\alpha\{\vartheta\}_{n+1}-\alpha\{\vartheta\}_n-\{\dot{\vartheta}\}_n\right)$ and then:

$\left([T]+\alpha[C]\right)\{\vartheta\}_{n+1}=[C]\left(\{\dot{\vartheta}\}_n+\alpha\{\vartheta\}_n\right)$ - From this $\{\vartheta\}_{n+1}$ and, by use of (**), $\{\dot{\vartheta}\}_{n+1}$.

The process is repeated for the next step.

(1) B.C. are supposed to be incorporated into this matrix eq. for simplicity.

ITERATIVE PROCEDURE FOR EIGENVALUES/ /EIGENVECTORS DETERMINATION. (d)

a) <u>LOWEST EIGENVALUE</u> ω_ℓ AND ASSOCIATED EIGENVECTOR $\{\psi\}_\ell$.
By definition $[K].\{\psi\}_\ell = \omega_\ell^2.[M].\{\psi\}_\ell$ or $[A].\{\psi\}_\ell = \omega_\ell^{-2}.\{\psi\}_\ell$ with $[A]=[K]^{-1}.[M]$; $[K]$ = stiffness matrix with constraints taken account of.

It follows $[A].\{\psi\}_\ell = \omega_\ell^{-2}\{\psi\}_\ell$: $\omega_\ell, \{\psi\}_\ell$ are unknown.

a_1) CHOICE OF A TRIAL VECTOR $\{q_0\}_\ell$ as an approximation for $\{\psi\}_\ell$:
(ex.:

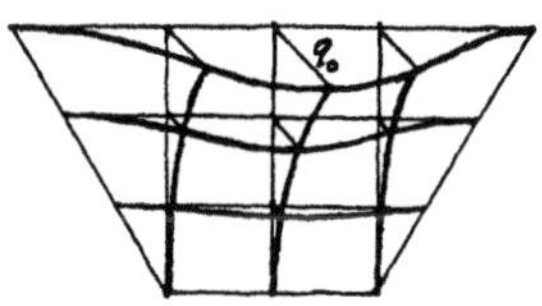

or

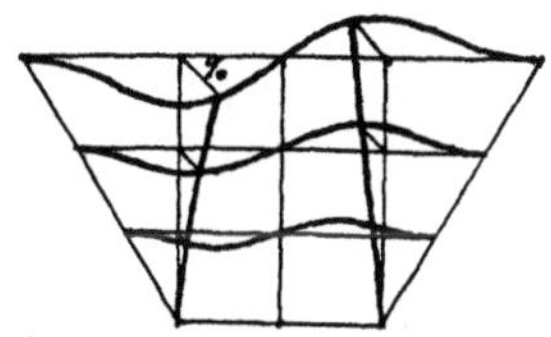

for symmetric lowest mode of a symmetric dam   for lowest antisymmetric mode of a symmetric dam).

a_2) ITERATIVE PROCEDURE : $[A].\{q_0\}_\ell = \{Q_1\}_\ell$ FIRST APPROXIMATION:

$$\omega_{\ell 1}^{-2} = \frac{\{q_0\}_\ell^T.\{Q_1\}_\ell}{\{q_0\}_\ell^T.\{q_0\}_\ell} \quad \text{AND} \quad \{q_1\}_\ell = \omega_{\ell 1}^2\{Q_1\}_\ell ;$$

$[A].\{q_1\}_\ell = \{Q_2\}_\ell$ SECOND APPROXIMATION:

$$\omega_{\ell 2}^{-2} = \frac{\{q_1\}_\ell^T.\{Q_2\}_\ell}{\{q_1\}_\ell^T.\{q_1\}_\ell} \quad \text{AND} \quad \{q_2\}_\ell = \omega_{\ell 2}^2\{Q_2\}_\ell ;$$

. . . .

$[A].\{q_{n-1}\}_\ell = \{Q_n\}_\ell$ n^{th} APPROXIMATION:

$$\omega_{\ell n}^{-2} = \frac{\{q_{n-1}\}_\ell^T.\{Q_n\}_\ell}{\{q_{n-1}\}_\ell^T.\{q_{n-1}\}_\ell} \quad \text{AND} \quad \{q_n\}_\ell = \omega_{\ell n}^2\{Q_n\}_\ell ;$$

$\omega_{\ell n} \to \omega_\ell$ AND $\{q_n\}_\ell \to \{\psi\}_\ell$ for $n \to \infty$.

b) <u>SUCCESSIVE EIGENVALUES</u> AND ASSOCIATED EIGENVECTORS : $\omega_m > \omega_\ell$, $\{\psi\}_m$.
As sub a) $[K]\{\psi\}_m = \omega_m^2[M]\{\psi\}_m$ or $[A].\{\psi\}_m = \omega_m^{-2}\{\psi\}_m$: $\omega_m, \{\psi\}_m$ unknown.

b_1) CHOICE OF A TRIAL VECTOR $\{p_0\}_m$ as an approximation for $\{\psi\}_m$: the trial vector must be "<u>cleaned</u>" of components non-orthogonal to $\{\psi$
(ex.:

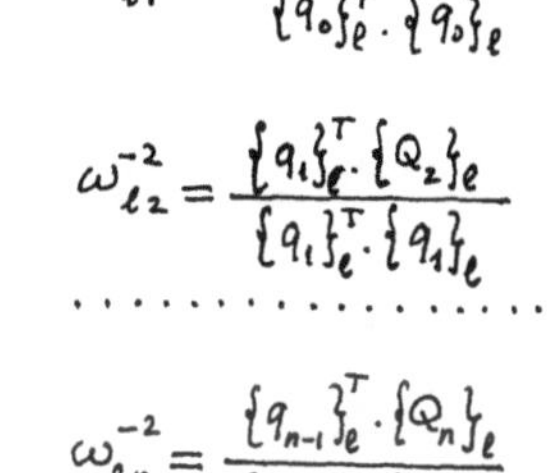

$\{p_0\}_m = \{p_0^*\}_m + b_0.\{\psi\}_\ell \to \{p_0\}_m^T.[K].\{\psi\}_\ell = b_0\{\psi\}_\ell^T.[K].\{$

so that $b_0 = \dfrac{\{p_0\}_m^T.[K].\{\psi\}_\ell}{\{\psi\}_\ell^T.[K].\{\psi\}_\ell}$ and $\{p_0^*\}_m = \{p_0\}_m - b_0\{\psi\}_\ell$;

THEN $[A].\{p_0^*\}_m = \{P_1\}_m \to$ FIRST APPROXIMATION :

for next-to-lowest symmetric mode of a symmetric dam).

$$\omega_{m1}^{-2} = \frac{\{p_0^*\}_m^T.\{P_1\}_m}{\{p_0^*\}_m^T.\{p_0^*\}_m} \quad \text{AND} \quad \{p_1\}_m = \omega_{m1}^2\{P_1\}_m .$$

b_2) ITERATIVE PROCEDURE AS SUB a_2) ABOVE, WITH "CLEANING":

$b_j = \dfrac{\{p_j\}_m^T.[K].\{\psi\}_\ell}{\{\psi\}_\ell^T.[K].\{\psi\}_\ell}$, $\{p_j^*\}_m = \{p_j\}_m - b_j\{\psi\}_\ell$ <u>at every step</u> (cleaning must be effected with respect to all modes lower in frequency than the one being determined) ; $\omega_{mj} \to \omega_m$, $\{p_j^*\}_m \to \{\psi\}_m$ for $j \to \infty$. Convergence is fast.

EIGENVALUES ω_k,
EIGENSHAPES $[\Psi]$ (e)

from $[K]\{\Psi_k\} = \omega_k^2 [M]\cdot\{\Psi_k\}$

$$[K]\cdot[\Psi] = [M]\cdot[\Psi]\cdot[\omega_k^2]_{diag.}$$; similarly:

$$[C]\cdot[\Psi] = [M]\cdot[\Psi]\cdot[2\omega_k \nu_k]_{diag.}$$

(effect the product:

$$\begin{bmatrix} \Psi_{11} & \Psi_{12} & \cdots & \Psi_{1N} \\ \Psi_{21} & \Psi_{22} & \cdots & \Psi_{2N} \\ \cdots & \cdots & \cdots & \cdots \\ \Psi_{N1} & \Psi_{N2} & \cdots & \Psi_{NN} \end{bmatrix} \cdot \begin{bmatrix} \omega_1^2 & 0 & \cdots & 0 \\ 0 & \omega_2^2 & \cdots & 0 \\ \cdots & \cdots & \cdots & \cdots \\ 0 & 0 & \cdots & \omega_N^2 \end{bmatrix} = \begin{bmatrix} \omega_1^2\Psi_{11} & \omega_2^2\Psi_{12} & \cdots & \omega_N^2\Psi_{1N} \\ \omega_1^2\Psi_{21} & \omega_2^2\Psi_{22} & \cdots & \omega_N^2\Psi_{2N} \\ \cdots & \cdots & \cdots & \cdots \\ \omega_1^2\Psi_{N1} & \omega_2^2\Psi_{N2} & \cdots & \omega_N^2\Psi_{NN} \end{bmatrix}$$

↑ I[st] eigenmode

↑ I[st] column of $[M]\cdot[\Psi]\cdot[\omega_k^2]_{diag}$

If the eigenvector normalization is made by requesting that $[\Psi]^T[M][\Psi] = [I]$, then it ensues:

$$[\Psi]^T[K][\Psi] = [\omega_k^2]_{diag.},$$

$$[\Psi]^T[C][\Psi] = [2\omega_k \nu_k]_{diag.}$$

Complex "transfer function" for force $f = 1e^{i\omega t}$ acting along s^{th} d.o.f. : at r^{th} d.o.f. the response (at regime) is (apart from factor $e^{i\omega t}$):

$$R_s(s) = \sum_m \frac{U_m(s)\cdot U_r(r)\cdot 1}{\omega_m^2 \{U\}_m^T [M]\{U\}_m \cdot \left[1 - \frac{\omega^2}{\omega_m^2} + 2i\nu_m \frac{\omega}{\omega_m}\right]},$$

$R_s(r)$ is called the "tranfer function" from d.o.f. s to d.o.f. r

LINEAR FRACTURE MECHANICS WITH F. E. (plane case) (f)

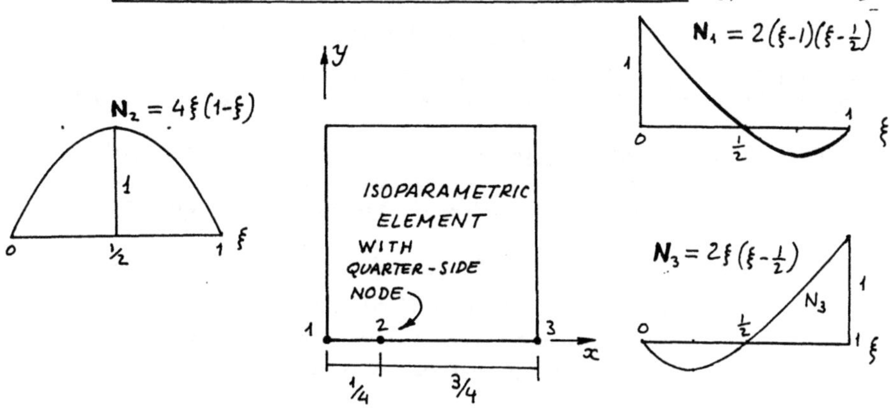

Transformation $\xi \to x$: $x = x_1 N_1 + x_2 N_2 + x_3 N_3 = 0 . N_1 + \frac{1}{4} . N_2 + 1 . N_3 =$

$$= \xi . (1-\xi) + 2 . \xi \left(\xi - \frac{1}{2}\right) = \xi\left(1 - \xi + 2\xi - 1\right) = \xi^2$$

$x = \xi^2$ ALONG SIDE $y = 0$;

$\xi = \sqrt{x}$ " " " ;

$$\frac{\partial N_i}{\partial x} = \frac{\partial N_i}{\partial \xi} \cdot \frac{\partial \xi}{\partial x} = \frac{\partial N_i}{\partial \xi} \cdot \frac{1}{2\sqrt{x}} \longrightarrow \infty \quad \text{for } x \to 0 .$$

Crack along $y = 0$ with apex in the origin:

$$u, v \sim \sqrt{z} = \sqrt{\rho}\, e^{\frac{i\vartheta}{2}}$$

$$\frac{\partial(u,v)}{\partial(x,y)} \sim \frac{\partial(\sqrt{z})}{\partial z} = \frac{1}{2\sqrt{z}} \xrightarrow[\rho \to 0]{} \infty$$

4 SINGULAR ELEMENTS WITH $\frac{1}{4}$-SIDE NODES 1, 2, 3, 4, 5 SURROUNDING CRACK TIP. (NOTE DOUBLED NODES ALONG CRACK PLANE)

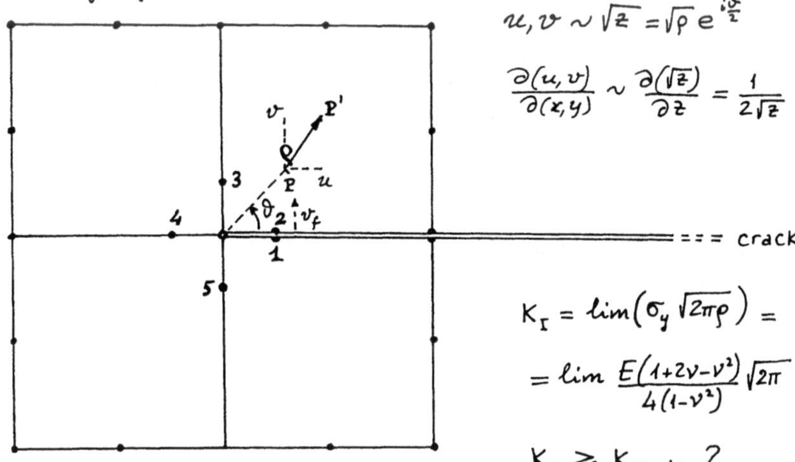

$$K_I = \lim\left(\sigma_y \sqrt{2\pi\rho}\right) =$$

$$= \lim \frac{E(1+2\nu-\nu^2)}{4(1-\nu^2)} \sqrt{2\pi}\, \frac{v_f}{\sqrt{\rho}}$$

$K_I \gtrless K_{I\,crit.}$?

dim. $[K] = f\ell^{-3/2}$

(order of magnitude 100 $Kgf\,cm^{-3/2}$ for concrete)

(9)

NON-LINEAR MODELS FOR ARCH DAMS

Geometric non-linearity (Large displacements) is usually not of practical interest for arch dams. The following is for material non-linearities

3-D "JOINTS": special pre-localized elements (of negligible thickness) able to transmit compression and shear but no tensile stress. Specific interest for arch dams.

NO-TENSION: elements with one of the principal stresses tensile get a modified stiffness that allows only compressive action along the other principal direction; non pre-localized. General purpose

ELASTO-PLASTIC MODELS

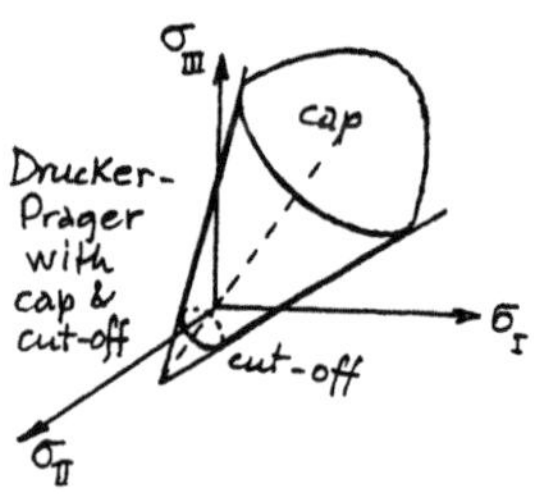

Definition of a limit surface and of a plastic deformation flow rule - non pre-localized.
example: Drucker-Prager models with tension cut-off and compression cap (from σ_t, σ_c)
- more advanced models (e.g. Bazant, ...) require many more parameters to be determined by tests.

General purpose model.

NON-LINEAR FRACTURE MECHANICS

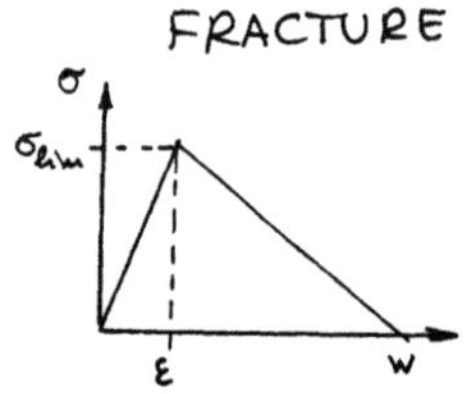

- Localized crack models (pre-localization of zones likely to develop cracks, e.g. at bottom upstream toe...)
e.g. cohesive or "fictitious crack tip" models (Hillerborg).

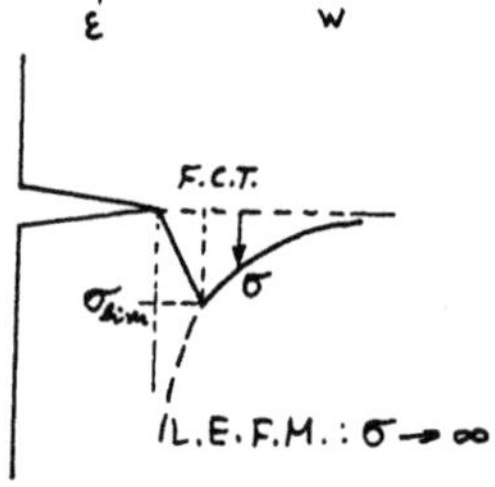

- "smeared crack" models - non-prelocalized
Elements with a tendency to crack get a modified stiffness. General purpose.

INCREMENTAL DAMAGE with softening and reduction of elastic modulus after damage.
Definition for each affected element of a damage index.
Non pre-localized. General purpose.

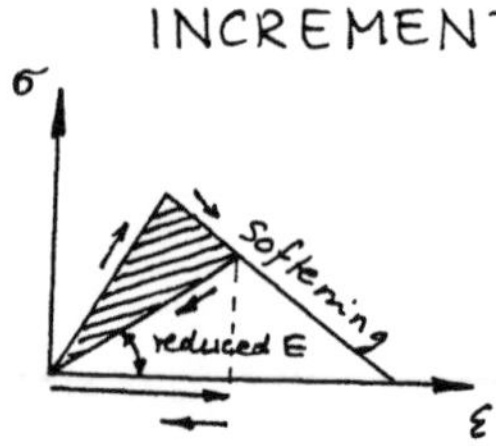

NON-LINEAR MODELS : THE NO-TENSION MODEL (plane case)

(h)

u, v = displacem. components along x, y.

ξ, η = principal stress axes (according to linear elastic computation) -

σ_I = elastic stress along ξ = tensile.

σ_{II} = elastic stress along η = compressiv

$\triangle_{1\,2}^{3}$ = = GENERIC ELEMENT WITH TENSILE STRESSES (thickness = 1 unit of length along normal to plane xy)

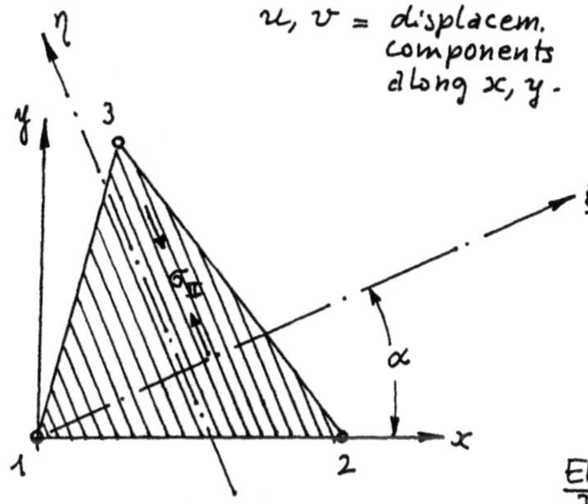

Elastic computations yield $\sigma_x, \sigma_y, \tau_{xy}$ - Then:

$$2\alpha = \operatorname{artg} \frac{2\tau_{xy}}{\sigma_x - \sigma_y} \; ; \quad \begin{cases} \sigma_I = \frac{\sigma_x+\sigma_y}{2} + \sqrt{\left(\frac{\sigma_x-\sigma_y}{2}\right)^2 + \tau_{xy}^2} \\ \sigma_{II} = \frac{\sigma_x+\sigma_y}{2} - \sqrt{\left(\frac{\sigma_x-\sigma_y}{2}\right)^2 + \tau_{xy}^2} \end{cases}$$

MOHR CIRCLE

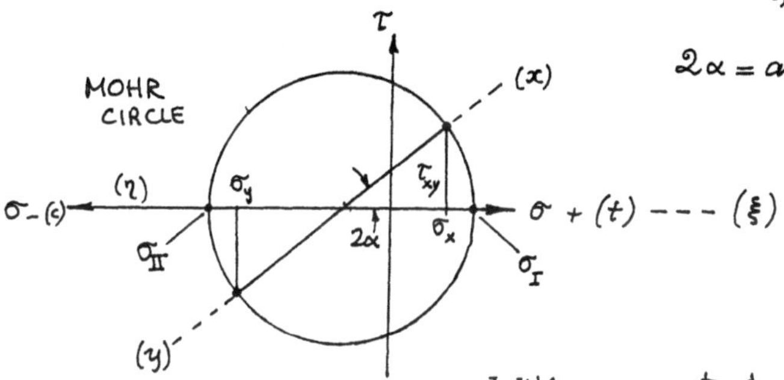

The material, along the direction ⊥ ξ, develops cracks due to the tensile stress. What is left is a set of rods oriented along η.

With respect to new axes ξ, η:

$$\begin{Bmatrix} \delta_\xi \\ \delta_\eta \end{Bmatrix} = \begin{bmatrix} \cos\alpha & \sin\alpha \\ -\sin\alpha & \cos\alpha \end{bmatrix} \begin{Bmatrix} u \\ v \end{Bmatrix} \; ; \quad \begin{Bmatrix} \varepsilon_\xi \\ \varepsilon_\eta \\ \gamma_{\xi\eta} \end{Bmatrix} = \begin{bmatrix} \frac{\partial}{\partial\xi} & 0 \\ 0 & \frac{\partial}{\partial\eta} \\ \frac{\partial}{\partial\eta} & \frac{\partial}{\partial\xi} \end{bmatrix} \cdot \begin{Bmatrix} \delta_\xi \\ \delta_\eta \end{Bmatrix} \text{ , where :}$$

$$\begin{Bmatrix} \delta_\xi \\ \delta_\eta \end{Bmatrix} = \begin{bmatrix} N_1 & 0 & N_2 & 0 & N_3 & 0 \\ 0 & N_1 & 0 & N_2 & 0 & N_3 \end{bmatrix} \begin{Bmatrix} \delta_{\xi 1} \\ \delta_{\eta 1} \\ \delta_{\xi 2} \\ \delta_{\eta 2} \\ \delta_{\xi 3} \\ \delta_{\eta 3} \end{Bmatrix} = [N]\{\delta\} \text{ , } \{\delta\} = \text{nodal displac. vector;}$$

$$\begin{Bmatrix} \varepsilon_\xi \\ \varepsilon_\eta \\ \gamma_{\xi\eta} \end{Bmatrix} = [B] \cdot \{\delta\} \quad , \qquad [B] = \begin{bmatrix} \frac{\partial N_1}{\partial\xi} & 0 & \frac{\partial N_2}{\partial\xi} & 0 & \frac{\partial N_3}{\partial\xi} & 0 \\ 0 & \frac{\partial N_1}{\partial\eta} & 0 & \frac{\partial N_2}{\partial\eta} & 0 & \frac{\partial N_3}{\partial\eta} \\ \frac{\partial N_1}{\partial\eta} & \frac{\partial N_1}{\partial\xi} & \frac{\partial N_2}{\partial\eta} & \frac{\partial N_2}{\partial\xi} & \frac{\partial N_3}{\partial\eta} & \frac{\partial N_3}{\partial\xi} \end{bmatrix} ;$$

$$[K]_{\xi\eta} = \int_{\triangle_{1\,2}^{3}} [B]^T [D] [B] \, d\xi \, d\eta \quad , \quad [D] = \begin{bmatrix} 0 & 0 & 0 \\ 0 & E & 0 \\ 0 & 0 & 0 \end{bmatrix} \text{, in the } \xi,\eta \text{ reference frame; indeed,}$$

$$\begin{Bmatrix} \sigma_\xi \\ \sigma_\eta \\ \tau_{\xi\eta} \end{Bmatrix} = [D] \begin{Bmatrix} \varepsilon_\xi \\ \varepsilon_\eta \\ \gamma_{\xi\eta} \end{Bmatrix} \text{ - In full, } [K]_{\xi\eta} = E \int_{\triangle_{1\,2}^{3}} \begin{bmatrix} 0 & 0 & 0 & 0 & 0 & 0 \\ 0 & \left(\frac{\partial N_1}{\partial\eta}\right)^2 & 0 & \frac{\partial N_1}{\partial\eta}\frac{\partial N_2}{\partial\eta} & 0 & \frac{\partial N_1}{\partial\eta}\frac{\partial N_3}{\partial\eta} \\ 0 & 0 & 0 & 0 & 0 & 0 \\ 0 & \frac{\partial N_2}{\partial\eta}\frac{\partial N_1}{\partial\eta} & 0 & \left(\frac{\partial N_2}{\partial\eta}\right)^2 & 0 & \frac{\partial N_2}{\partial\eta}\frac{\partial N_3}{\partial\eta} \\ 0 & 0 & 0 & 0 & 0 & 0 \\ 0 & \frac{\partial N_3}{\partial\eta}\frac{\partial N_1}{\partial\eta} & 0 & \frac{\partial N_3}{\partial\eta}\frac{\partial N_2}{\partial\eta} & 0 & \left(\frac{\partial N_3}{\partial\eta}\right)^2 \end{bmatrix} dS \text{ .}$$

From $[K]_{\xi\eta} \longrightarrow [K]_{xy}$ -

(i)

NON-LINEAR MODELS: ELASTO-PLASTIC MATERIAL BEHAVIOR (simplified presentation)

$\begin{Bmatrix}\sigma_1\\ \sigma_2\\ \sigma_3\end{Bmatrix}$ = principal stresses. (eg. in a constant-stress F. E.)

Cartesian space whose co-ordinates are proportional to the principal stresses $\sigma_1, \sigma_2, \sigma_3$:
The planes normal to the trisector axis ($\sigma_1+\sigma_2+\sigma_3$ = constant) are surfaces of constant isotropic stress $\frac{\sigma_1+\sigma_2+\sigma_3}{3}$.

Given $\{\sigma\} = \begin{Bmatrix}\sigma_1\\ \sigma_2\\ \sigma_3\end{Bmatrix}$,
the isotropic stress (first invariant) is:

$$I_1 = \frac{\sigma_1+\sigma_2+\sigma_3}{3} = \frac{\sigma_x+\sigma_y+\sigma_z}{3};$$

the second invariant (deviatoric stress) is:

$$I_2 = \sqrt{\sum_k \sigma_k^2 - 3I_1^2}\,.$$

The stress state must be represented by a point inside (or at most on) the limit surface $\varphi(\sigma_k)=0$.

The limit surface may be fixed (perfect plasticity) or vary with the amount of plastic deformation (hardening).

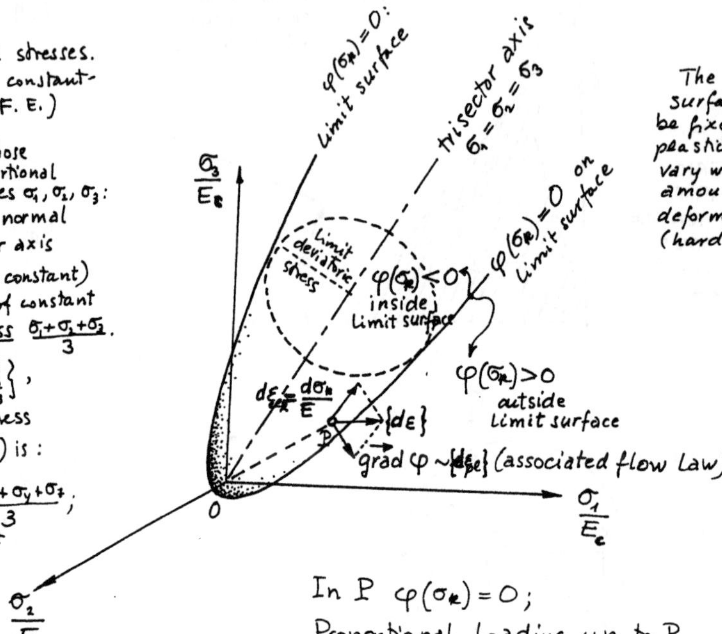

In P $\varphi(\sigma_k)=0$;
Proportional loading up to P (in first element to reach limit state); $\vec{OP}+\vec{d\varepsilon}=\vec{OP'}$ next step loading.
Beyond P the constitutive equations are:

$$\{d\varepsilon\} = \{d\varepsilon_{el}\} + \{d\varepsilon_{pl}\} = [B]\cdot\{du\},$$

$\{d\varepsilon_{pl}\}^T \times \{\vec{grad}\,\varphi\} \ge 0$ (with associated flow law $\{d\varepsilon_{pl}\} = \lambda\,\vec{grad}\,\varphi$)(*)

"elastoplastic stiffness matrix" $[K_{epl}]$:
$$[K_{epl}]\{du\} = \iiint_{vol} [B]^T\{d\sigma\}\,dV$$
(changes at every step)

$\longleftarrow$ $\{d\sigma\} = \frac{E_c}{1+\nu}[D]\cdot\{d\varepsilon_{el}\}$; $\{d\varepsilon_{el}\}^T \times \{\vec{grad}\,\varphi\} \le 0$;
i.e. $\{d\varepsilon_{el}\} \longrightarrow d\varphi \le 0$ (if $d\varphi<0$, $\{d\varepsilon_{pl}\}=\{0\}$).

The above equations give $\{d\varepsilon_{pl}\}$ from $\{d\varepsilon\}$, allowing to express the incremental non-linear relationship between $\{d\sigma\}$ and $\{du\}$ (which is $\{d\sigma\} = \frac{E}{1+\nu}[D]\cdot[B]\cdot\{du\}$ in the linear elastic case), whence the incremental stiffness matr
The resulting non-linear equilibrium equations are solved by iterative procedures (e.g. Crank-Nicholson + Newton-Raphson).

(*) The condition $\{d\varepsilon_{pl}\} = \lambda\,\vec{grad}\,\varphi$ (with $\lambda \ge 0$) can be viewed as a condition of minimum plastic deformation compatible with the total deformation vector $\{d\varepsilon\}$.

PART 3

DYNAMIC BEHAVIOUR OF ARCH DAMS

by

R.T. Severn

University of Bristol
United Kingdom

ESTIMATION OF EARTHQUAKE GROUND MOTION FOR ARCH DAM DESIGN

1 INTRODUCTION Probably the most important, and yet the most difficult aspect of designing arch dams for seismic forces is that of determining what such forces will be. As with all structures, determining seismic design parameters for a site requires the analysis of data from geologic, geophysical, seismological and geotechnical studies. For the site region, such studies are used to develop statistical and deterministic models to characterise the earthquake source, the wave propagation path and the local ground response. The required data is obtained from observational and instrumental data from earthquakes and aftershocks, regional seismicity networks, geologic mapping, geophysical and corresponding laboratory measurements and analytical models. Because of the uncertainties in such data, the design process normally adopts a conservative stance, examples of which include the following:

1 Using a rare event as the maximum possible earthquake because the location and magnitude of potential earthquakes are uncertain.

2 Using upper-bound values for the peak ground acceleration expected at the site because of uncertainties in attenuation laws.

3 Using upper bound values for the design response spectrum because of the variation in such calculated spectra for the same peak value of ground acceleration and/or intensity.

For the rational determination of appropriate ground motion to be used in designing an arch dam, there are seven principal characteristics which need to be addressed, and each will be considered in turn. The flow chart for the whole process is given in Fig. 1, taken from Ref. 1.

2 SEISMICITY This requires the compilation of data of all past earthquakes from which it is hoped that valid extrapolations can be made for future earthquake activity at a particular site. It has to be said that this is an uncommon occurrence. Recurrence laws have been formulated (Ref. 2) such as

$$\log N(M) = a-bM$$

where N(M) is the number of earthquakes occurring within a region in a given period having magnitude $\geq$ M. Constants a and b are obtained by least squares analysis, the former is a function of the size of the region considered, the level of activity and the period. For most parts of the World, $0.8 \leq b \leq 1.0$, and it appears to be related to the type of tectonic activity causing the earthquake. It is useful to note that a similar empirical relationship holds for earthquake intensity (I) because it is usual to have more data concerning I than M.

For the European area, a valuable source of data is the paper by Ambraseys in Ref 3. The same author has produced similar valuable data for Turkey and the Middle East.

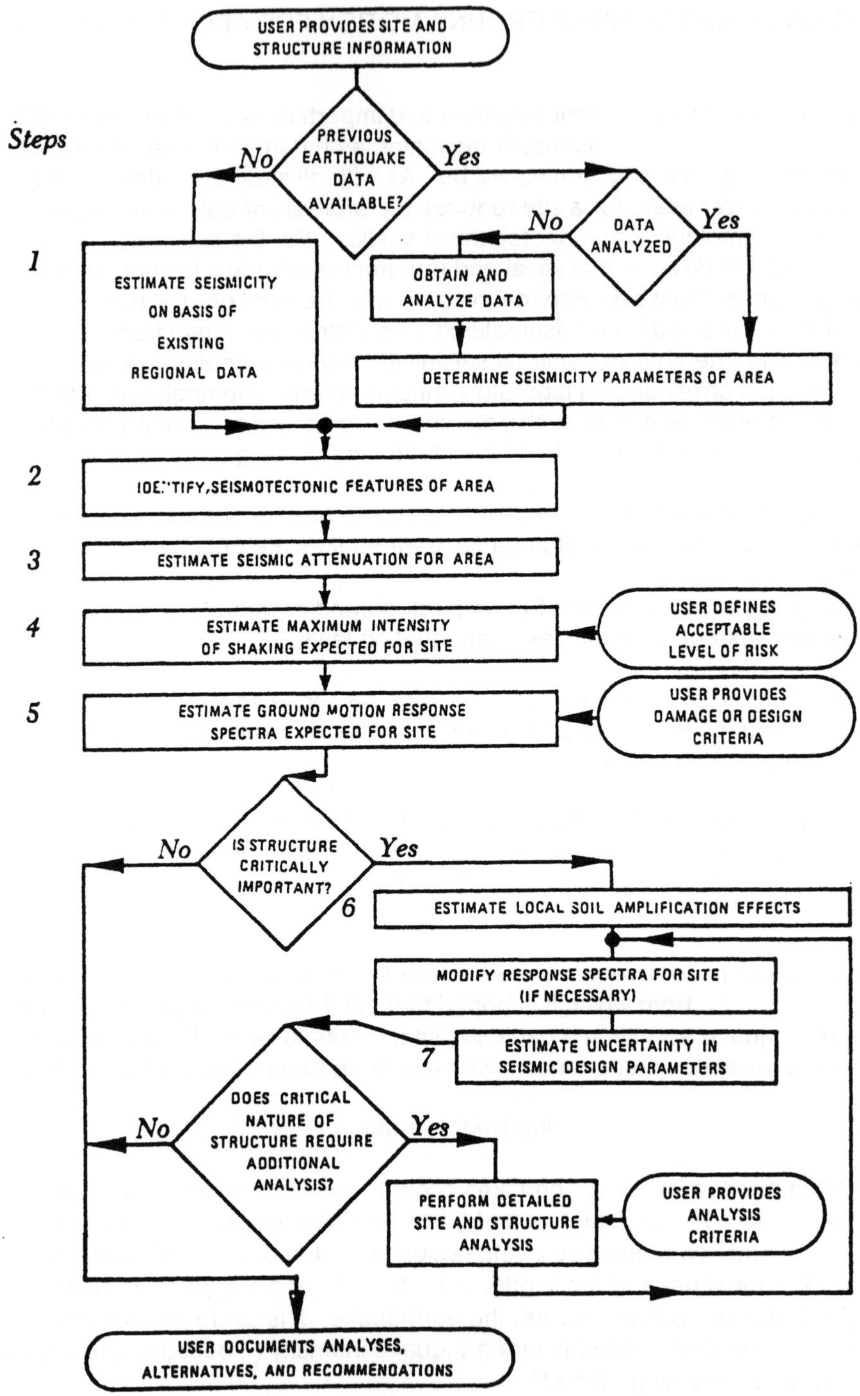

Fig.1 Steps in estimating ground motion

3 SEISMOTECTONIC FEATURES

In addition to historical seismicity it is necessary to map the location and characteristics of faults and other tectonic structures in the region. Active faults usually produce clear geological and seismological evidence; they are considered to be significant if they cross the dam site or are located near the site and are capable of generating large earthquakes. There is some tentative relationship between fault rupture length and earthquake magnitude: this suggests that for 32-40 km rupture length $5 \leq M \leq 7$; for 40-113 km $7 \leq M \leq 7.5$; for rupture lengths of 300 km or more, magnitude 8 earthquakes are produced.

4 SEISMIC ATTENUATION

Having identified the likely earthquake epicentre it is necessary to know how seismic waves attenuate from it to the construction site. The best basis for this is a detailed isoseismal map (Fig.2) from a past earthquake, which consists of contours of maximum Modified Mercalli intensity reported at various location (MMI has 12 scale points). Some care is needed in interpretation, however, because data is obtained from people, who mostly live in areas underlain by alluvium, and such sites often experience enhanced shaking compared to rock sites. The Mexico City event and this year's Kobe earthquake show clear evidence of this point.

5 EXPECTED GROUND SHAKING AT SITE

The earthquake process (Fig. 3) involves elasto-dynamic cracking at a rock interface, the energy from which is transmitted by a number of seismic waves (slide), which are either body waves (P, SH, SV) or Love and Rayleigh surface waves. The ground motion resulting at the site can be described either as a seismogram (time-history), as peak ground motion parameters (acceleration, velocity or displacement), or as spectra which show these parameters as functions of frequency.

(i) Seismograms. Several examples will be shown, in particular that for the S16E component of the 1971 San Fernando earthquake recorded at the Pacoima dam, for which the 1.2g maximum horizontal component is the largest yet recorded (Fig. 4). As indicated, velocity and displacement can be obtained by numerical integration (Fig.5) , but accumulated numerical and measurement error have to be removed to ensure zero resultant displacement. From seismograms a great deal of information can be deduced, including the various measures of magnitude; the well-known Richter magnitude is obtained from the difference in arrival times of the P and S waves as indicated in Fig. 6, whereas the body-wave (M_b) and (M_s) magnitudes are obtained from the 1 and 20 second period seismic waves. The three scales are empirically related. The seismic moment, $M_o = \mu \bar{u} A$ where μ is the rigidity of the source region, $\bar{u}$ is the average dislocation and A is the area of faulting, can also be obtained from the seismogram. M_o is sometimes regarded as a better means of characterising earthquake strength than any of the magnitude measures, particularly when M>8 because the rupture dimensions of the earthquake then exceeds the wavelength of the seismic waves used in the magnitude determination.

(ii) Peak Ground Acceleration This is commonly used in design because its multiplication by mass gives seismic force for high-frequency structural systems. [For medium and low frequency, ground

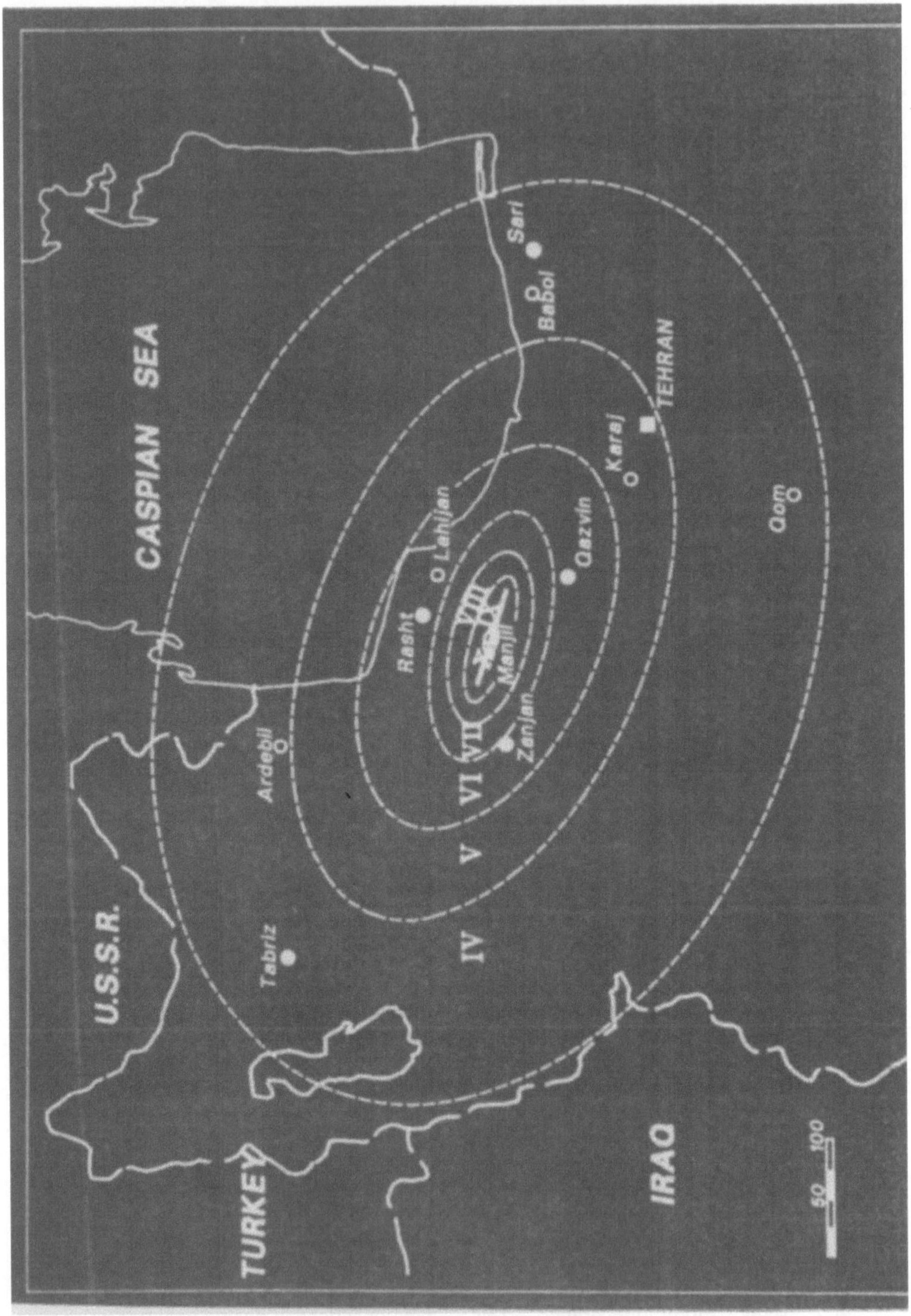

Fig.2 Isoseismal map, Manjil Iran 1990

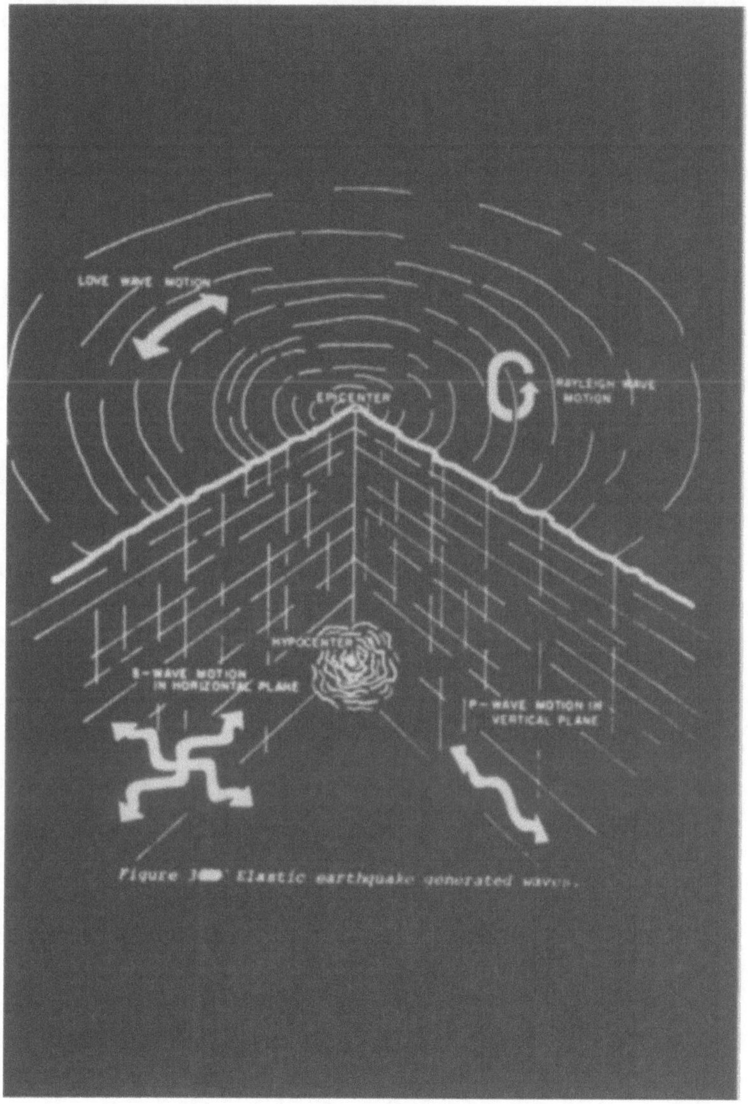

Fig.3 Earthquake waves

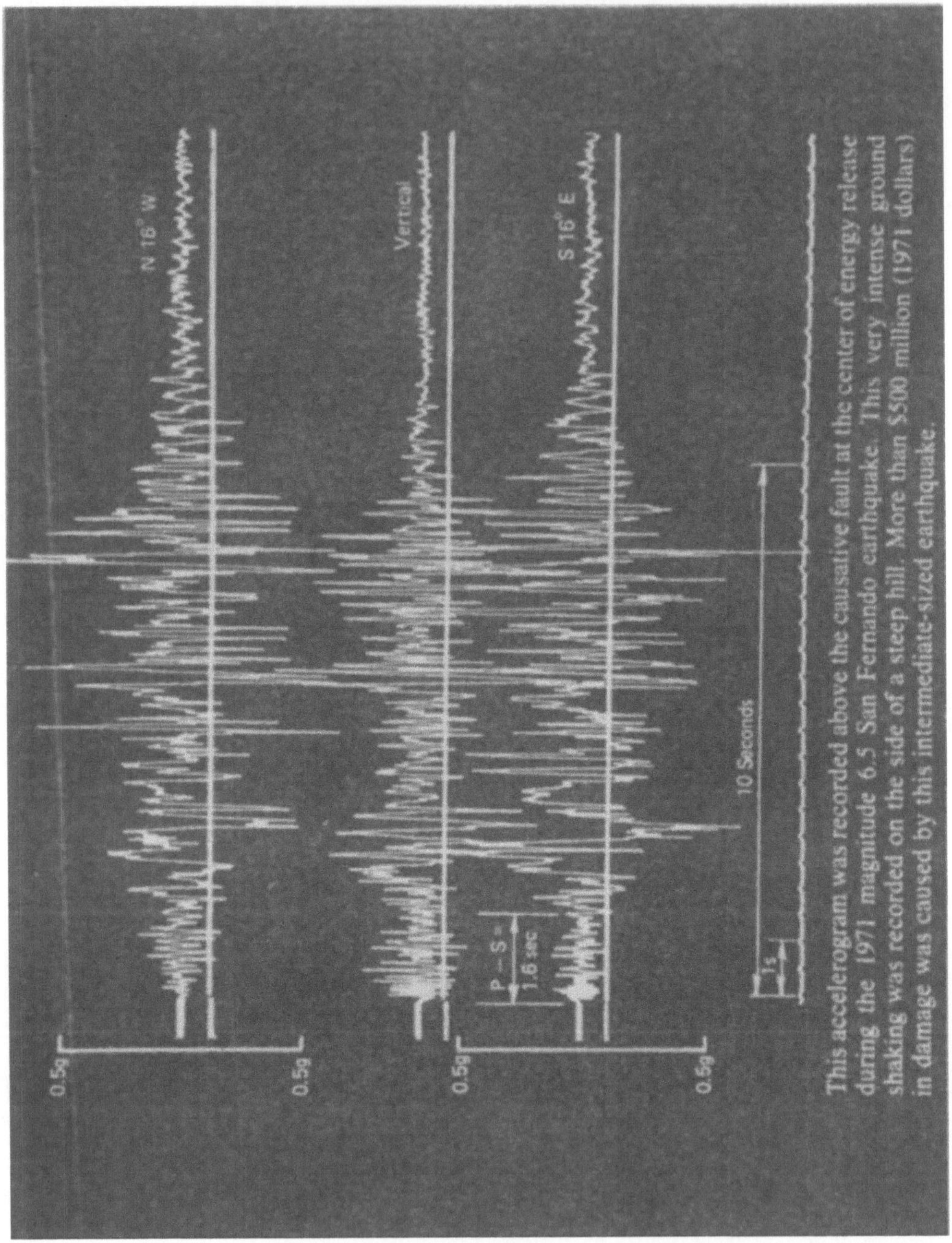

Fig.4 The 3-component time-history recorded at the Pacoima dam during the 1971 San Fernando earthquake

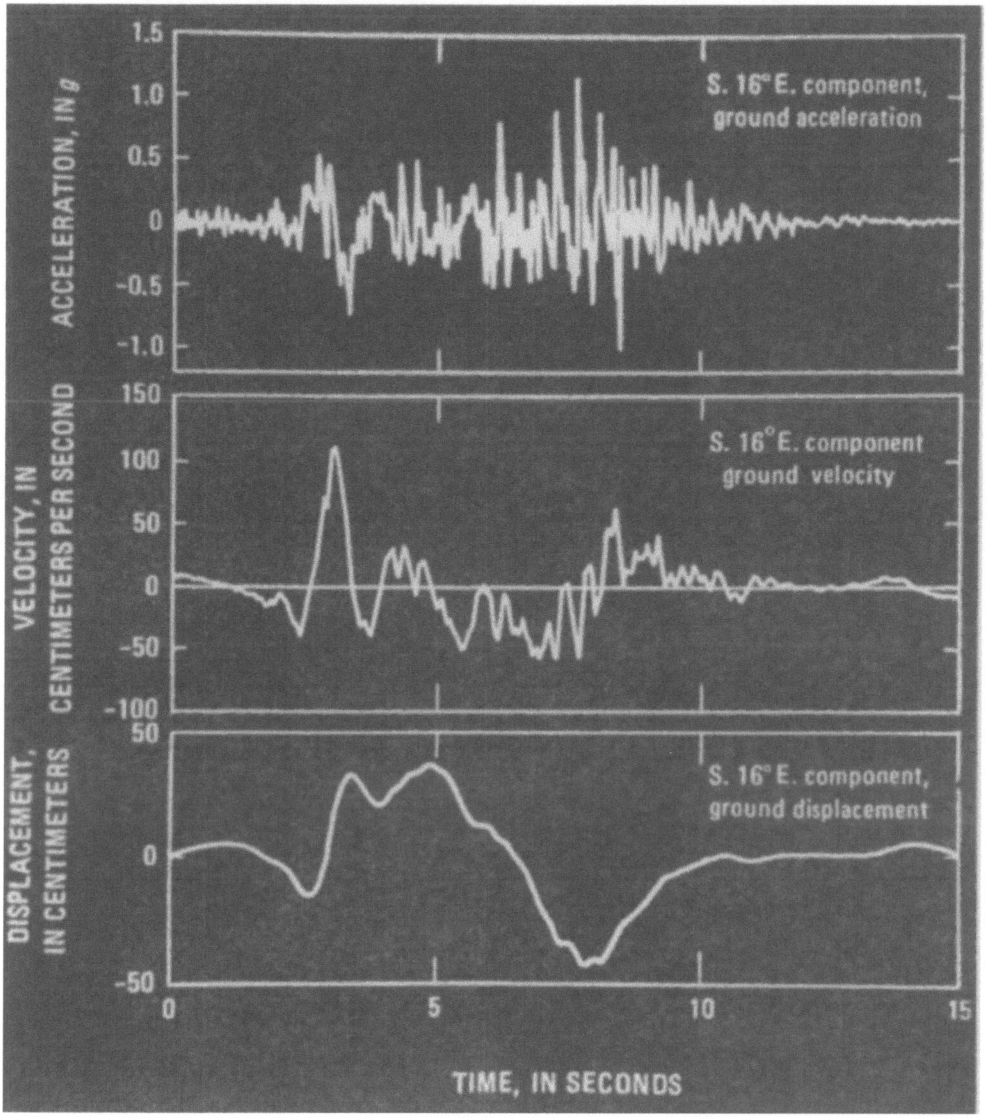

Fig.5 Integration of acceleration (Fig.4) to produce velocity and displacement

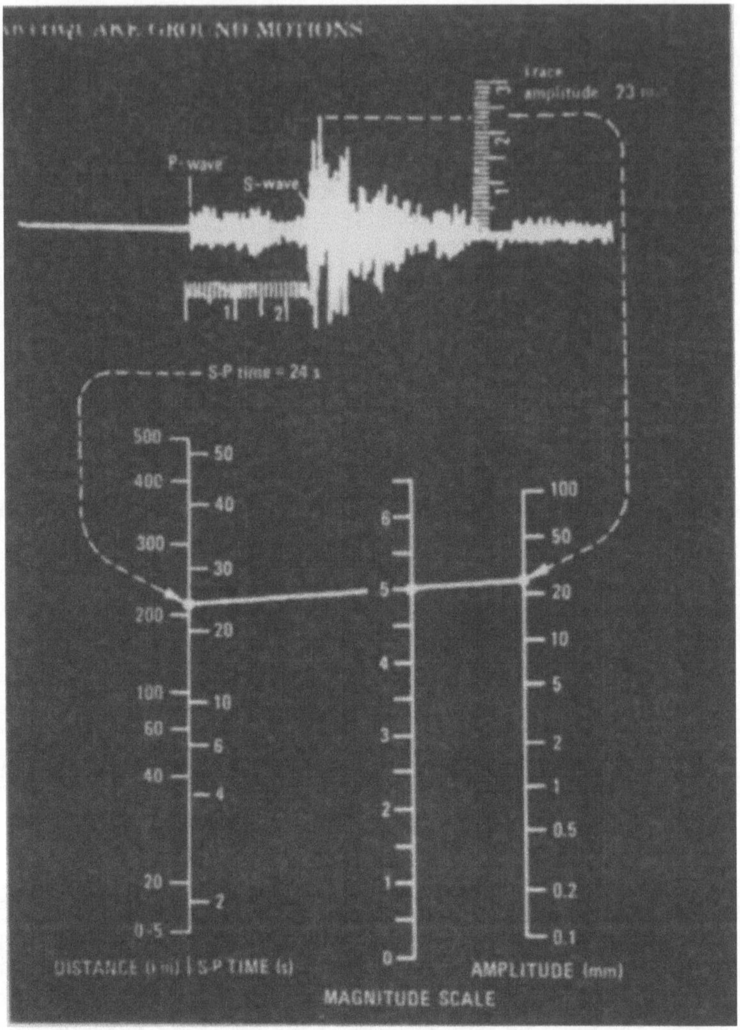

Fig.6 Earthquake Magnitude obtained from P and S-wave arrival times

velocity and displacement are more relevant.] For multi axis input it is usual to assume that peak accelerations in the two horizontal directions are equal, and that in the vertical direction is two-thirds the value. Design practice also assumes peak acceleration to be strongly dependent on magnitude, but this is doubtful, and evidence can be adduced both for, and against.

A major damage-causing parameter is duration of the earthquake. This is so for all ground conditions, but particularly so in liquifiable soils, which are often to be found in embankment dams. Associated with duration, as parameters causing structural damage, are amplitude and the dominant frequency at which the ground shakes. In view of its importance, a precise definition of duration has been sought. One such is the time during which the acceleration level is greater than or equal to some threshold such as 0.05g; another is based on 90% of the integral of the squared time-history,

From analysis of many earthquakes it has been concluded that average duration for soft alluvium sites (v_s < 600 m/s) is 5-6 seconds longer than for intermediate sites (V_s = 600-765 m/s) and 10-12 seconds longer than for rock sites (V_s > 765 m/s). Also, the average duration increases by 2 seconds for acceleration and 5 seconds for displacement for each unit increase in magnitude, and 1-1.5 seconds for every 10 km increase in epicentral distance.

6 RESPONSE SPECTRA

For linear behaviour computer software is available which will solve the equations of motion for any given time-history input. Numerical methods are required, which necessitate digitising this input at small intervals. The response output can be embarrassingly large, causing the designer some difficulty in utilising the information produced, particularly if a number of time-history inputs need to be considered. Quite often only the maximum values of response parameters (displacement, stress) are needed, and here the response spectrum approach becomes valuable. It effectively disengages any particular structure from any particular earthquake. The use of these spectra has been dealt with in other lectures of this course; here we are concerned only with aspects of their derivation and an analogue approach will be used for explanation although digital methods are now universally used.

In Fig. 7 taken from Ref.1, (a) is the time-history, velocity input which in analogue form is fed through a narrow band pass filter having 5% critical damping tuned to 0.33 Hz, as shown in (b), resulting in a maximum velocity response of about 22 cm/s; this is recorded as a single point in (d) as indicated by the dotted line. If the filter is retuned to 2 Hz, say, as in (c) a maximum response of about 36 cm/s is obtained which is also recorded in (d) as a single point. If the filter is re-tuned in small frequency increments, over a required range, each maximum response can be plotted as in (d), in total forming the irregular curve shown, which is referred to as the Spectral Velocity Curve for 5% damping for this particular earthquake. The damping can also be changed incrementally so that a family of such curves can be produced, an example of which is given in Fig.8. Because of the incremental method of derivation it is justifiable to smooth them, as indicated in Fig. 9. In structural terms, the filter represents a single degree of freedom system and the spectral velocity value, S_v, is the maximum velocity it would attain, having the specified damping, when acted upon by the given earthquake. Spectral

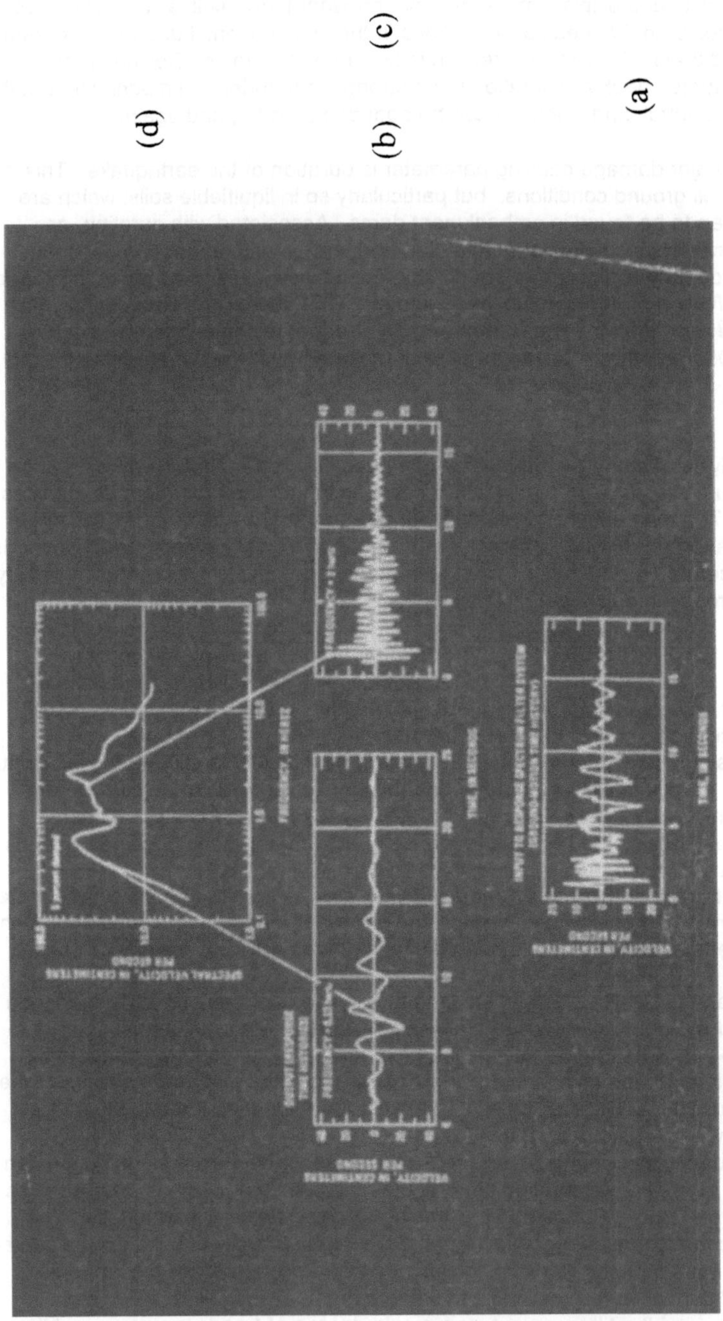

Fig.7 Analogue derivation of spectral velocity

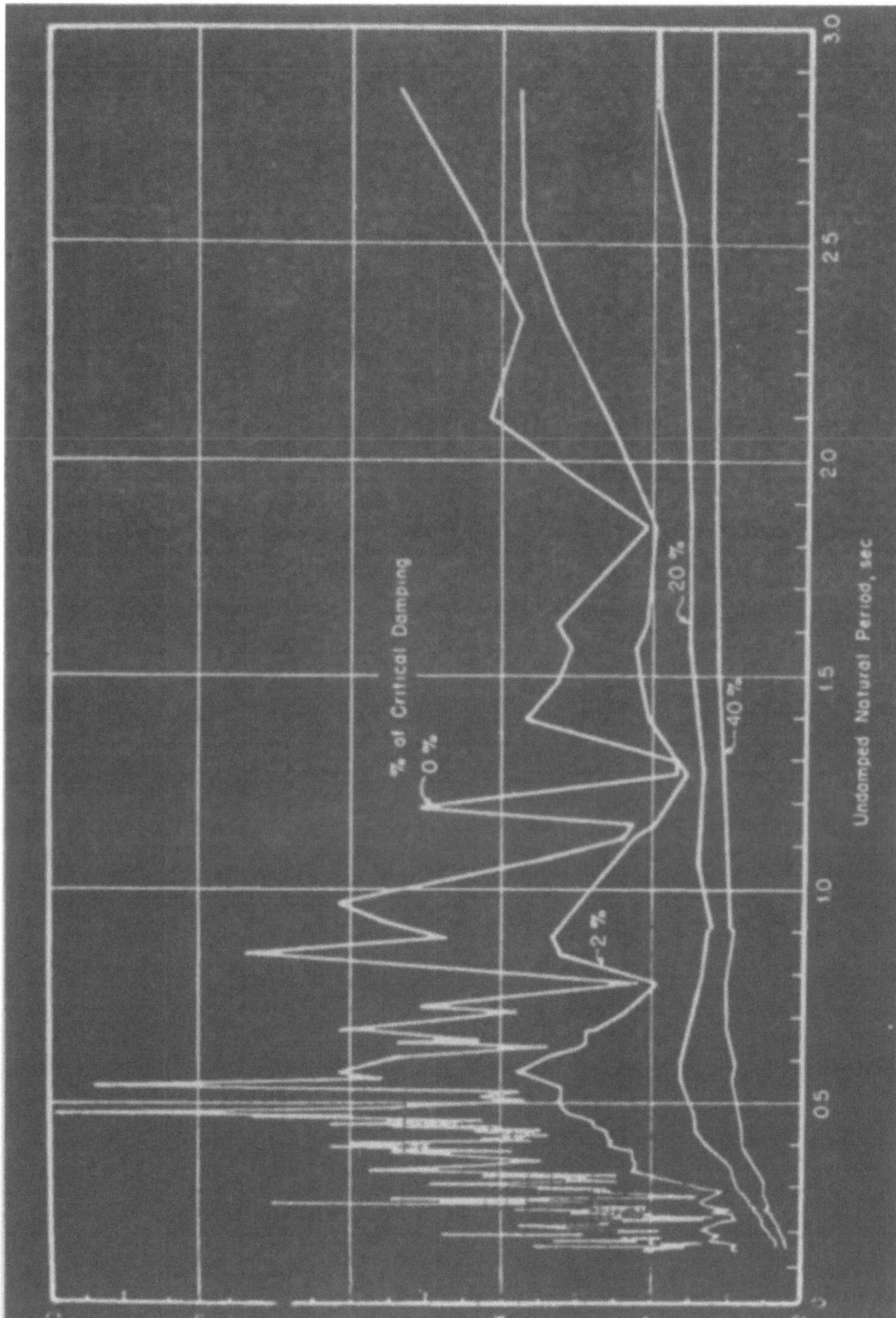

Fig.8 Spectral curves for different damping valves

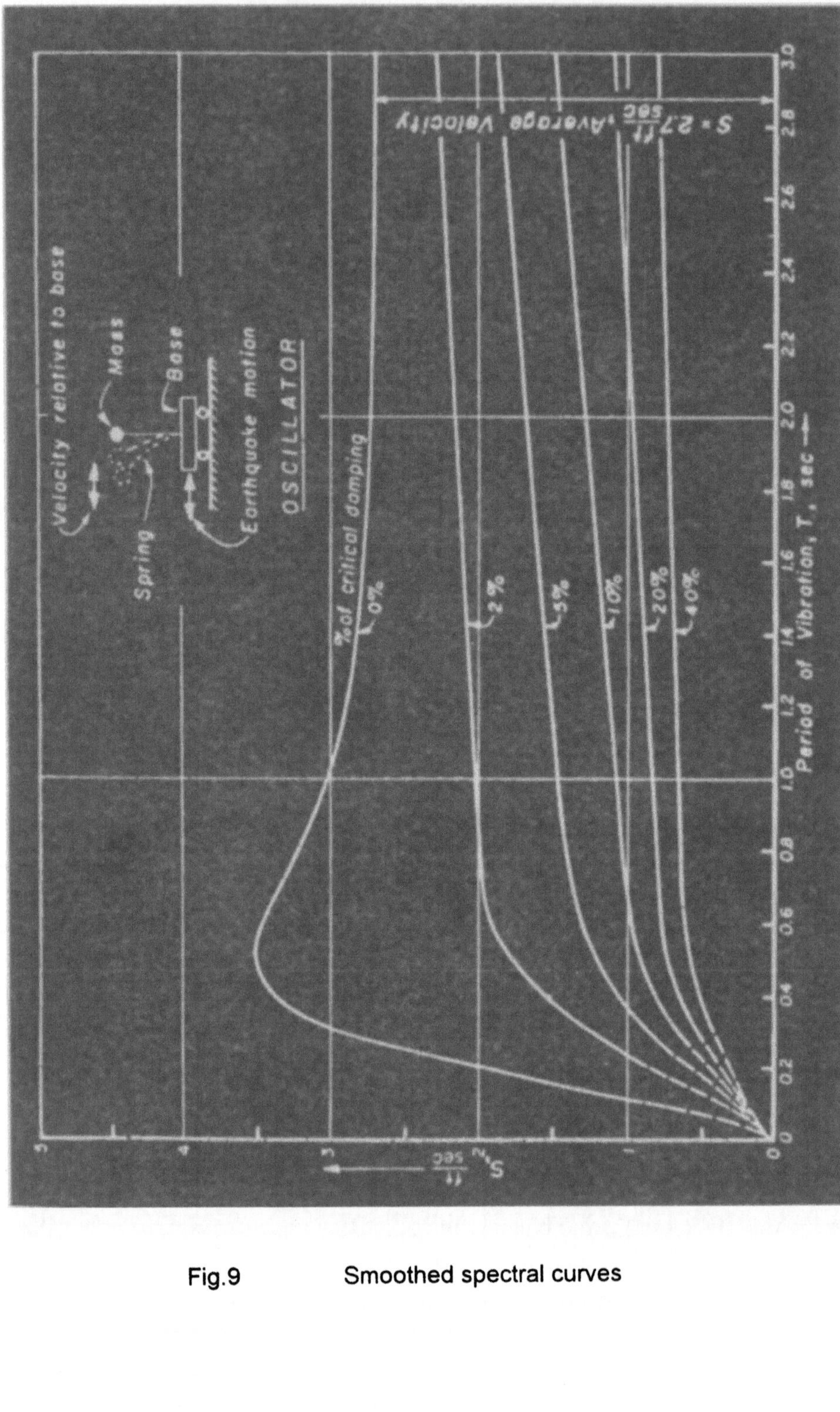

Fig.9 Smoothed spectral curves

acceleration, S_a, and spectral displacement, S_d, can also be found, and are related by

$$w^2 S_d = w S_v = S_a$$

so that the three spectra can be displayed on the tri-partite plot shown in Fig.10.

In view of the irregular nature of most spectral curves for a single earthquake it is usual for regions where sufficient earthquake records are available to average spectra, having first brought each earthquake record to a common base, say 1.0g maximum acceleration; shows such average curves. Taking the process a stage further, it is recognised that site conditions affect seismograms, and Fig.11 presents the results of analysis of 104 seismograms recorded on 4 types of site. The average acceleration spectra for soft to medium clay and sand sites are flat over a 0.3-1.0 second period range, whereas for rock they peak sharply at 0.25. Of course, the actual numbers here would change for a different set of seismograms, but the principles remain.

For very important structures such as dams or nuclear facilities, attempts would be made to produce site-specific spectra, but failing that, the generalised spectra which now appears in Eurocode 8 could be used. It has parameters which allow for different soil conditions, and also a 'behaviour factor' to allow for non-linear material behaviour.

It is noted that software exists which allows the process of obtaining response spectra from time-histories to be reversed. In fact, an infinity of such time histories can be produced from a given spectrum and some skill is required in choosing the most appropriate set for a particular purpose.

It is remembered that spectral curves give maximum values for a single mode of vibration. When many modes have to be considered, as for an arch dam, the most appropriate method of combining separate responses is still a matter for research. Some early studies (Ref.4) used the SRSS - or square-root of sum of squares of modal contributions method, but currently available software would use the CQC (Complete Quadratic Combination) approach (Ref.8), which covers the probability that modes of complex structures, like arch dams, are not clearly separated.

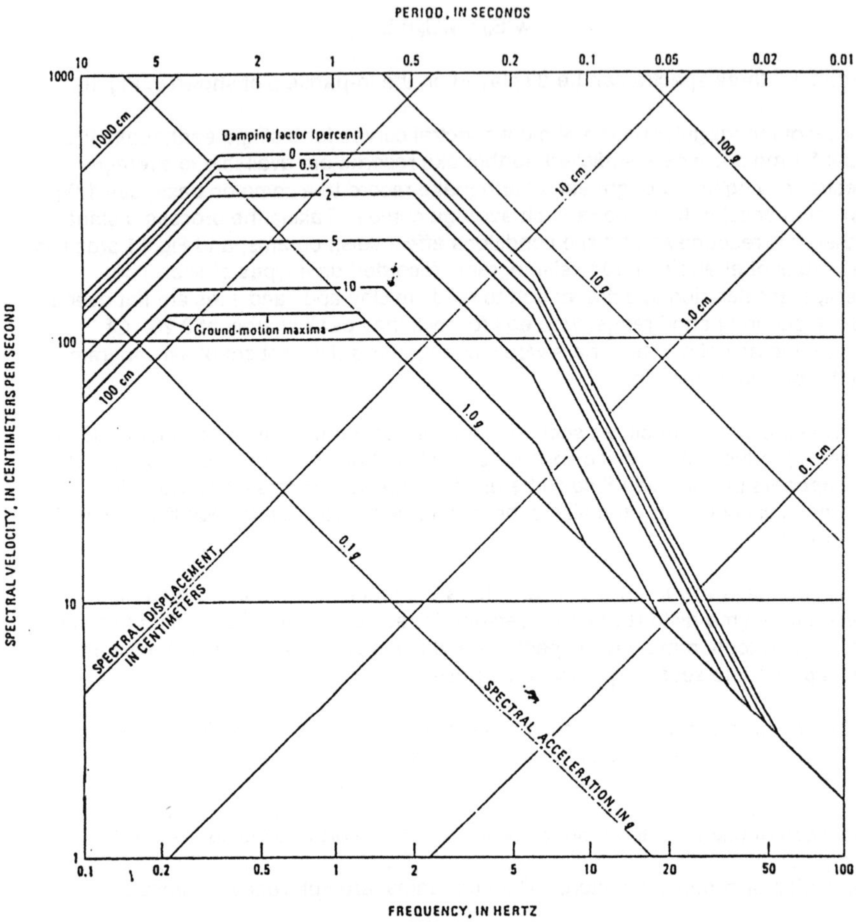

Fig.10 Tri-partite spectral curves; S_d, S_v and S_a

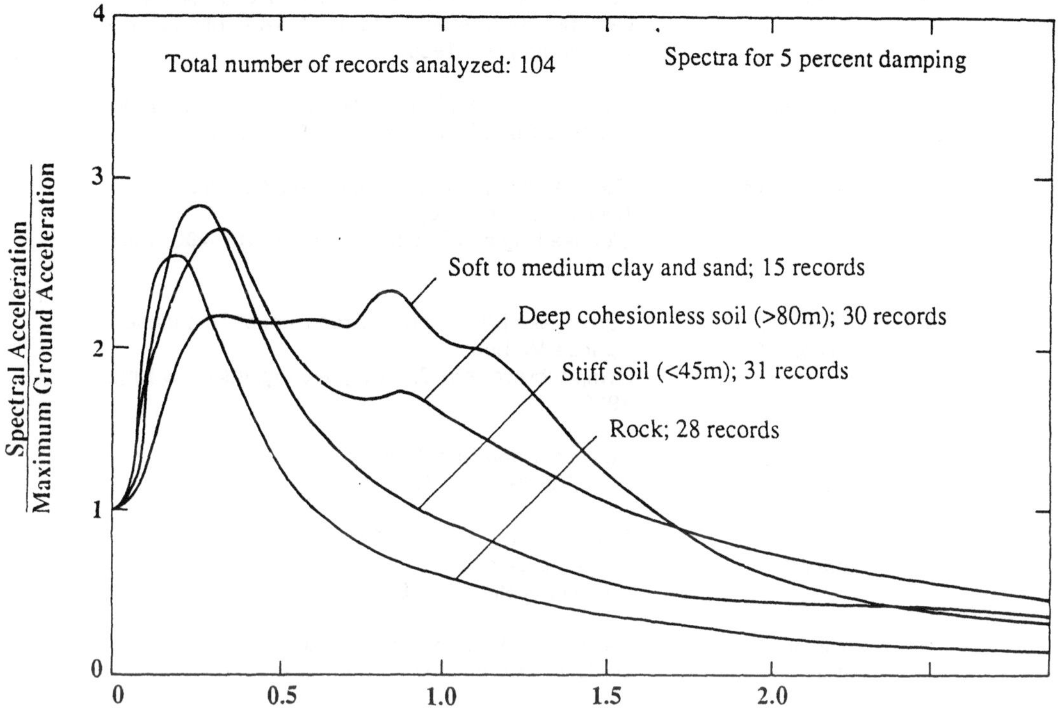

Fig. 11 Spectral accelerations for different soils

REFERENCES

(Includes background reading as well as those mentioned in the text)

1	Hays W W	Procedures for Estimating Earthquake Ground Motions Geol Svy, Prof Paper 1114, US Govt Printing Office, Washington D C 1980
2	Evernden J F	Study of Regional Seismicity and Associated Problems Bull Seis Sov of America V60, 393-446, 1970
3	Ambraseys N	The Prediction of Earthquake Peak Ground Acceleration in Europe E'quake Eng and Struct Dgn Vol 24, 267-490, 1995 April
4	Severn R T Taylor P R	Earthquake Effects on an Arch Dam by Response Spectra Methods Symp Vibrations in Civil Engineering, Instn Civ Engrs 1965
5	Lane R G T (editor)	Seismicity and Dam Design ICOLD Bulletin 46, 1983
6	Severn R T (editor)	Guidelines for Selecting Seismic Parameters for Large Dams ICOLD Bulletin 72, 1988
7	Morrison C et al	Earthquake Recordings on or Near Dams US COLD Committee on Earthquakes, Nov. 1977
8	Der Kiureghian A	Earthquake Eng. and Struct. Design Vol 22, p943, 1993

RESERVOIR INDUCED SEISMICITY (RIS)

BACKGROUND. As far as is known, no arch dam has been damaged by RIS, although a number have experienced moderate ($3<M<5.9$) earthquake caused by reservoir filling; a list is given in Fig 1. For concrete gravity dams, however, RIS caused a M=7.0 earthquake at Koyna (India) in 1967 which resulted in major cracks on the downstream face, and at Hsinfengkiang (China) dam in 1961 a M = 6.2 also caused major structural damage. Because many occurrences of RIS occur in assumed seismically quiet situations, as was Koyna, it is important to be aware of the possibility in arch dam design. Current attitudes to hazard analysis requires the recognition that damaging earthquakes can occur in intraplate areas and assurances are needed that logical analysis has been used to predict low probability of occurrence.

As a starting point, dam engineers have considered RIS to be a possibility when the impounded volume is greater than $10^9 m^3$ and/or the reservoir depth is greater than 100m. On this basis USCOLD 1977 (1) estimated that RIS affected 0.7% of dams on the World Register.

MECHANISM Most induced shocks have had focal depths less than 15Km, and it is generally agreed that whilst rapid loading due to reservoir filling may play a part in early foreshocks, major shocks involve a triggering role caused by water intrusion, and consequent reduction in effective stress on the fault plane. A case has also been made (2) that physico-chemical reaction between water and silicate rocks leads to fatigue and weakening.

RIS characteristics are observed to vary between cases, making it difficult to draw any general conclusions. For example, the onset of RIS sometimes occurs immediately after filling and sometime several years after. Also, no direct relationship has been observed between reservoir volume and depth and the maximum earthquake magnitude produced.

At Koyna, following impounding in 1962, approximately 35000 tremors were recorded in the immediate vicinity of the reservoir until 1979, the major event (M = 7.0) occurring in 1967. When analysed by the Gutenberg and Richter recurrence law, $\log N = a-bM$, annual b-value variations show consistent decrease over a 4-year period prior to the major event. Similar reductions have been observed in many other cases (2) of RIS.

The arch dam experiencing the most severe RIS is Kariba which probably has the World's largest artificial lake (240Km in length, 122m high dam). Situated in an area of no known seismicity, it experienced a M=5.7 shock in 1962, two years after initial impounding, followed by three more shocks of 6.1, 5.6 and 5.8. Since 1962 there has been a progressive decline in seismic activity in the lake area, though some RIS activity continues. Fig 2 is a summary of Kariba data.

A thorough study of RIS has been made at the 186m high Kurobe arch dam, in Japan (4), but although many events have been recorded, there is no clear correlation between water level and seismic activity.

EUROPEAN EXPERIENCE Yague (3) reports observations of RIS in 4 Spanish dam, two of which are gravity and two arch. At the Camarillas gravity dam more than 100 weak earthquakes were recorded in 1961, correlated with changes in reservoir level. Two other dams, the 150m high Canelles arch dam and the 101, high Santa Ana gravity dam are arranged in cascade on the same river, and it is argued that a rapid drawdown of $2hm^3$ daily over a period of 87 days from the upper reservoir, with consequent filling of the lower reservoir at almost the same rate, caused an earthquake of intensity (II-III) over a seven year period, almost half of which originated within 2.5Km of the recording stations. The tremors correlated with reservoir level, but with a time-lag of about 120 days.

In Fig 1 two large European arch dams are listed, Contra and Emosson, where the RIS experienced led to post construction seismic design studies which depended partly on experimental vibration measurements on each dam; these will be considered in another lecture.

The other European arch dam listed in Fig 1 is Vajont, for which, in view of the subsequent failure by rock-slide, it is of interest to note an M<3 event in 1963.

Dam	Height	Capacity	Magnitude	Year
Almendra, Spain	185	2.6	2	1972
Canelles, Spain	132	0.7	4.7	1962
Contra, Switz	190	0.1	3	1965
Emosson, Switz	170	0.2	3	1974
Grancarevo, Yugo	105	1.3	3	1970
Iddukki, India	160	1.5	3.5	1977
Kariba, Zambia	122	175	6.25	1963
Kerr, USA	54	1.5	4.9	1971
Kurobe, Japan	180	0.1	4.9	1961
Monteynard, France	125	0.3	4.9	1963
Schlegis, Austria	113	0.1	2	1973
Vajont, Italy	232	0.2	3	1963
Vouglans, France	112	0.6	4.5	1971
	m	$10^9 m^3$		

Fig 1 Arch Dams experiencing reservoir induced seismicity (RIS)

REFERENCES		Reservoir Induced Seismiicity
1	USCOLD	Working Group on Induced Seismicity, 1977
2	Guha S K Padale J G Gosavi P D	Probable Risk Estimation Due to Reservoir Induced Seismicity. Chap 33, Dams and Earthquakes, Instn Civ Engrs 1981
3	Yague A G	Spanish Experiences on Earthquakes Induced by Reservoirs. ICOLD 13th Congress, New Delhi, Q51 R36
4	OKAMOTO et al	Static and Dynamic Behaviour of Kurobe Dam. University of Tokyo Press, 1988. ISBN 4-13-068139-7
5	Barros FP et al	Monitoring System for Seismic Activity Before and After Filling Binational Itaipu Reservoir. ICOLD 13th Congress, New Delhi, Q51 R5
6	Skipp B O Higgins M	The Potential for Induced Seismicity - Geological Approaches. Chap 32, Dams and Earthquakes, Instn Civ Engs 1981

DAM-RESERVOIR INTERACTION

1 The acceleration of a structure in a fluid causes pressure waves in the fluid, just like sound waves. This pressure field is additional to the hydrostatic pressure and has values on the accelerating surface itself, which when integrated over a representative area give rise to what in finite element terms is referred to as a nodal force. When divided by acceleration these nodal forces give rise to nodal masses, which are described either as 'added' or 'virtual' mass. They arise solely because of the acceleration of the structure in the fluid, and are additional to the structural mass. Whether the fluid causes additional damping or stiffness in the dam-reservoir system is still controversial (Ref. 1).

2 Historical Background. In 1933 Westergaard (Ref. 2) addressed for the first time the problem of a two-dimensional dam cross-section vibrating whilst in contact with a reservoir. This completely new theoretical approach dealt with a vertical upstream face, upstream/downstream ground motion and an infinite reservoir. By assuming that the natural frequencies of the dam would be much lower than the frequency content of the ground motion, he argued that the dam would vibrate as a rigid body. Westergaard's solution allowed the fluid to be compressible, the coefficient C_n, indicating the compressibility effect being given by

$$C_n^2 = 1 - \frac{4H^2\omega^2}{n^2\pi^2C^2} \qquad (1)$$

where H is the reservoir depth, ω is the circular frequency (r/s) and C is the speed of sound in water 1440 m/s. He noted that at reservoir resonance the impulsive pressure values would become infinite, which, with $C_n = 0$ in (1) gives $\omega_n = 2\pi nC/4H$. However, this resonance would be well below the earthquake range and so the mechanism would not be excited. Westergaard's proposed pressure distribution is shown in Fig 1; it is still the standard solution against which other theories and experimental assessments are compared, and is used commonly for design purposes, the argument here being that seismic design involves many ill-specified parameters, of which this is only one. Westergaard's mental image for dynamic reservoir effects considered a part of the reservoir to move in-phase with the structure, giving rise to the 'added-mass' concept.

Such a seminal paper as this, inevitably led to many developments, perhaps the most important was that by Bratz and Heilbron (Ref. 14) who addressed the issues of finite reservoir, incompressible fluid, and rigid body motion of the dam. The last of these led to an iterative scheme, assuming an acceleration profile for the upstream face, but their findings were that there was little change from the rigid body case. Later we shall see that structural flexibility greatly influences hydrodynamic pressures generated, and their deficiency probably lay in the assumed displacement shape. Another useful contribution was made by Perumalswami and Kar (Ref. 12)

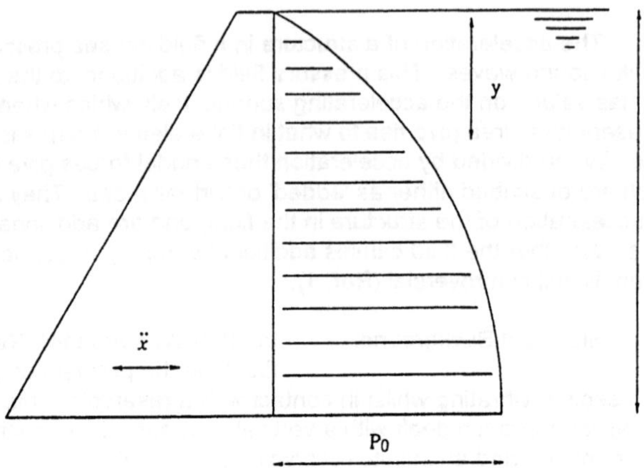

$\ddot{x}$ is lateral ground acceleration

$p(y) \approx 9.7\,\ddot{x}\,\sqrt{Hy}$ N/m^2

$P_0 \approx 9.7\,\ddot{x}\,H$ N/m^2

Fig. 1 Westergaard's pressure distribution.

3 Pressure (Eulerian) Formulation. The distribution of hydrodynamic pressure in a fluid is governed by the Helmholtz or modified wave equation, which, neglecting convective accelerations and viscous effects is

$$\frac{\partial^2 p}{\partial x^2} + \frac{\partial^2 p}{\partial y^2} + \frac{\partial^2 p}{\partial z^2} = \frac{1}{c^2}\frac{\partial^2 p}{\partial t^2} \qquad (2)$$

On a solid boundary and on a free surface the boundary conditions are, respectively,

$$\frac{\partial p}{\partial n} = -\rho a_n \quad \text{and} \quad p = \rho g W_z \qquad (3)$$

where n is the outward normal, ρ is the fluid mass density, a_n is the normal boundary acceleration and W_z is the surface wave height. This last is usually ignored, giving p = 0 on the free surface. These are the equations which Westergaard solved, with the simplifying assumptions alluded to. The removal of these assumptions, providing more valid solutions, required first an experimental approach which involved an electrical analogue, and second numerical methods usually involving the finite element method.

4 Electrical Analogue. For incompressible fluid the RHS of (2) is zero and if $\Phi = \lambda p$, where Φ is the electric potential (volts) and λ is a scalar multiplier, the LHS of (2) then governs Φ. The boundary conditions are

$$I_n = -k\frac{\partial \Phi}{\partial n} = k\lambda\frac{\partial p}{\partial n} \qquad (4)$$

where I_n is the normal current at the boundary and k is the material conductivity; (4) is then analogous to the first of (3), and where p is zero (on the fluid surface) Φ must be zero.

To use this analogy a physical model of the dam and reservoir must be made, although if one is not available for structural testing its construction need not be expensive in time or money. On the wet surface of the dam a series of discrete conductors is placed with the spaces in between made non-conducting with parrafin wax, for example Fig.2 By trial and error it was found that the conducting and non-conducting areas should be about equal. The reservoir is filled with a suitable electrolyte such as water. To model the boundary conditions, stationary fluid boundaries must be insulated (polythene sheet) and the free surface of the reservoir must have zero potential, achieved by floating earthed metal foil on it. Unit current is then applied to each of the conductors in turn, the voltages being measured at each discrete conductor on the wet surface, and these are used to build up a pressure influence matrix. Details are given in Refs. 3 and 4 but it will be seen that the wet surface acceleration distribution must be known.

The numerical solution of Poisson's equation had been made possible in 1950 by Southwell's finite difference based relaxation method (Ref. 7), and this removed the need for a physical model. The finite element method for the same purpose was developed in 1965 (Ref. 5) by Zienkiewicz and Cheung. However, the use of Poisson's equation, whether experimental or numerical, treats the fluid by one computational sequence and the structure by another; linking the two was tiresome and prone to error. In addition, a straight forward addition of the fluid mass (M_f) to the structure mass (M_s) considerably increases the bandwidth of the M_s matrix. A treatment of the coupled problem ab initio will now be discussed.

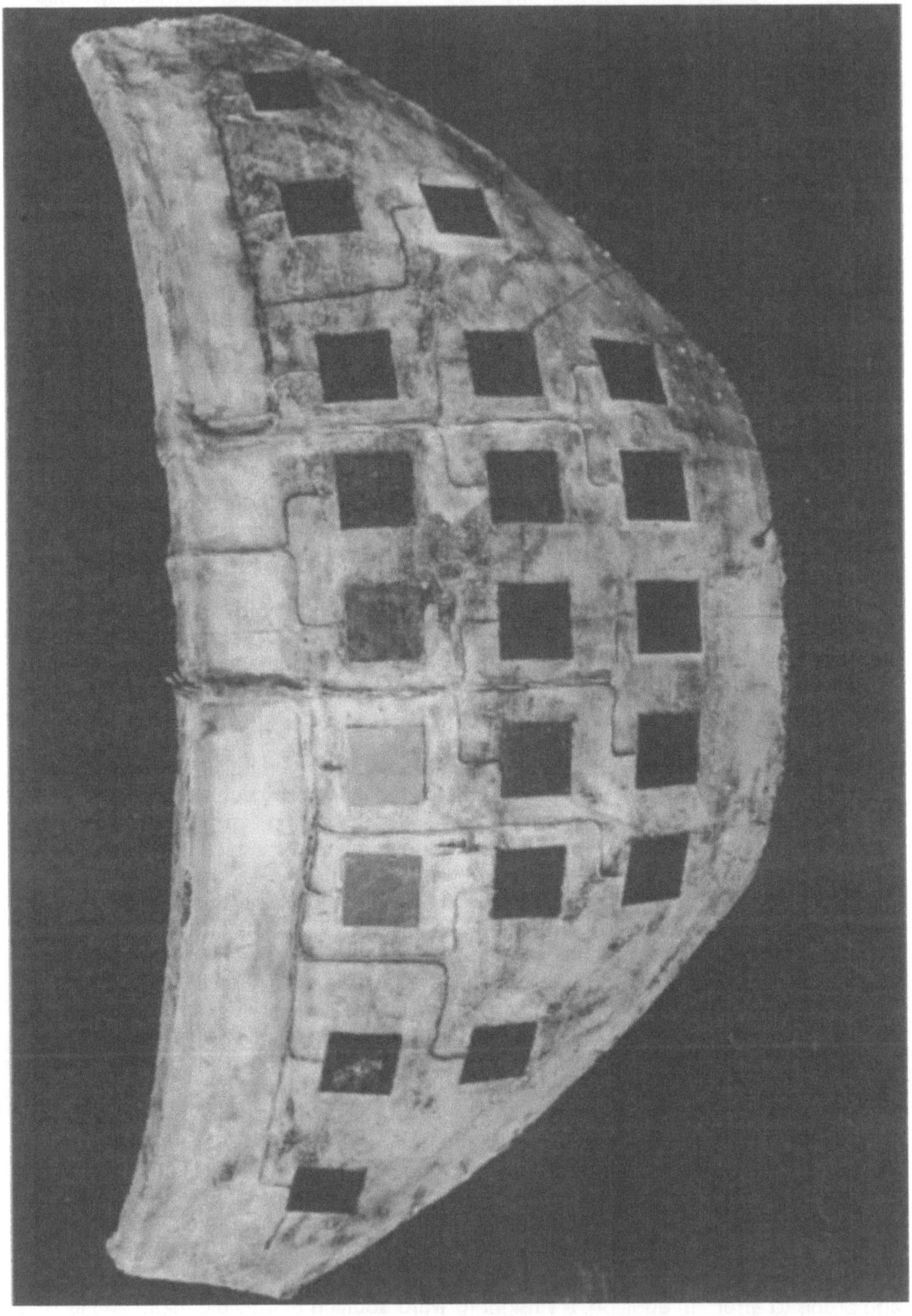

Fig. 2 Conducting plates on surface of arch dam model for electrical analogue.

5 The Coupled Problem

A review of the development of the coupled problem is given in Ref 1. Here, only the modal approach originated by Dungar and the present author for arch dams (Refs. 6 and 8) will be discussed. The dynamic equations of equilibrium are

$$M\ddot{x} + C\dot{x} + Kx = F(t) \tag{5}$$

where $M = M_s + M_f$. For the calculation of normal modes

$$K^{-1}MX_i = \lambda_i X_i \tag{6}$$

where $\lambda_i = 1/\omega_i^2$; ω_i and X_i being the ith natural frequency and mode shape respectively. If we knew the normal acceleration at every node on the wet surface, p could be calculated, followed by M_f , which with M_s would allow λ_i and X_i to be calculated from (6). Though this is not the case, in a normal mode the acceleration is proportional to the mode shape, which means that from a guessed structural mode shape, accelerations at the wet surface can be deduced apart from a multiplying factor. A standard way of calculating frequencies and mode shapes is by the so-called inverse iteration (or 'intensification') process starting with a guessed mode shape, and to this it is necessary to add only a few steps to incorporate the effect of the fluid. The complete sequence of computation for the first mode is as follows.

(i) Guess an initial vector X^0_1 for the first mode shape; a unit vector would serve.

(ii) For the moment assuming that M (= $M_s + M_f$) is completely known (see below), obtain the product $MX^0{}_1$, and call it $Y_1{}^1$, where the superscript indicates the first iteration cycle.

(iii) Obtain a new vector $V_1{}^1$ by solving the equation $KV_1{}^1 = Y_1{}^1$. [NB because this step is made a number of times, it is best if K is held in triangulated form in the computer]

(iv) From equation (6) above it is clear that $V_1{}^1 = \lambda^1{}_1 X_1{}^1$ Therefore, if $V_1{}^1$ is normalised so as to make its largest element unity, the factor required to do so will be $\lambda^1{}_1$ - a first approximation to $1/\omega_1{}^2$ - and the normalised $V_1{}^1$ will be the second approximation to the mode shape, ie $X_1{}^1$.

(v) Steps (i)-(iv) are repeated until two successive approximations to the mode shape are written an agreed tolerance. For arch dams it has been found that 10 iterations is sufficient.

It is useful to regard the guessed $X_1{}^0$ vector, and all subsequent X vectors, as being vectors of acceleration. To have such dimensions they would need to be multiplied by ω^2, but as X_i occurs on both sides of (6) this factor can be omitted. The product $M_fX_1{}^0$ will therefore be regarded as a force. This is a key statement, because at no time is M_f itself calculated, but rather the nodal forces arising from the hydrodynamic pressure field which the fluid exerts on the structure surface; these nodal forces only occur on the structure-fluid interface, and the $X_1{}^0$ vector used here has non-zero terms associated only with the nodal degrees of freedom on the interface.

A presentation of the necessary details of the calculation is given in Ref 6 for the interface between a variety of elements, the most complex of which is an 8-noded isoparametic representation of the doubly-curved surface of an arch dam. The essential common component is the solution of Poisson's equation in the reservoir, giving interface modal pressures from which the $M_f X_1^0$ forces are calculated.

Thus far, only X_1 and ω_1 have been considered, but higher modes can be found after deflation of previously found modes. It is to be noted, however, that because the complete mass matrix is not known, standard techniques such as shifting and Gram-Schmidt orthogonalisation, are no longer applicable. To overcome this difficulty a method devised by Simpson (Ref. 9) was used.

Further developments of this method are given in Ref 1, including that using implicit integration schemes based on Newmark's β-method or Wilson's θ - methods; whilst Ref 10 describes implementation of the method in SAP IV. Translations to other standard software packages have been made more recently.

6 Fluid Compressibility. It is noted that Westgaard included compressibility in his original work, but that other methods, including the one just discussed, ignore it by virtue of reliance on Laplace's equation. Porter and Chopra (Ref. 11), however, on the basis of analytical studies on very idealised arch dams, concludes that although compressibility has little influence at excitation frequencies much smaller than ω_1^r - the fundamental resonant frequency of the reservoir (c/4H), the response to upstream/downstream and vertical components of ground motion is reduced by compressibility effects when ω_1^r is exceeded. In real situations the reservoir is of a most irregular geometry with boundaries and base covered in silt, making particular resonances difficult to imagine.

7 Further Techniques In The Eulerian Formulation. The boundary element method (BEM) has been used as an alternative to FEM for solving the fluid domain equations. It has some advantages over FEM, but its formulation requires a difficult choice of weighting functions and its system matrices are full and do not necessarily have the banded structure of the FEM.

8 Displacement (Lagrangian) Formulation of Structure-Fluid Interaction.

When using the pressure (Eulerian) approach, the nodal variables for the fluid are pressures, whereas they are displacements for the structure. Because of this the two domains must be formulated separately and solved in parallel. The Lagrangian approach asks the obvious question, is it possible to use nodal displacements in discretising the fluid domain? Fluid and structural elements would then be directly compatible and could be joined at common nodes on the interface.

This attractive possibility would require only the development of the fluid element itself, with no special solution algorithms. In addition, because it would be formulated in the same manner as a structural element having a constitutive relationship, cavitation becomes a material non-linearity, and it would be possible to consider large displacement problems. This displacement fluid element would also allow the solution to 'non-grounded' fluid structure interaction problems; here, there exists rigid body motions which the fluid can support due to its stiffness. In view of these attractive facilities possessed by the displacement fluid element, the fact that it has not been universally adopted indicates that it also has some important disadvantages and these will now be considered.

9. Development of Displacement Fluid Elements.

Initially the approach was to use the standard solid element as a starting point, modifying the stress-strain matrix, D. separating the strain behaviour into bulk strain and shear strain.

$$D = K\begin{bmatrix} 1 & 1 & 1 & 0 & 0 & 0 \\ 1 & 1 & 1 & 0 & 0 & 0 \\ 1 & 1 & 1 & 0 & 0 & 0 \\ 0 & 0 & 0 & 0 & 0 & 0 \\ 0 & 0 & 0 & 0 & 0 & 0 \\ 0 & 0 & 0 & 0 & 0 & 0 \end{bmatrix} + G\begin{bmatrix} 4/3 & -2/3 & -2/3 & 0 & 0 & 0 \\ -2/3 & 4/3 & -2/3 & 0 & 0 & 0 \\ -2/3 & -2/3 & 4/3 & 0 & 0 & 0 \\ 0 & 0 & 0 & 1 & 0 & 0 \\ 0 & 0 & 0 & 0 & 1 & 0 \\ 0 & 0 & 0 & 0 & 0 & 1 \end{bmatrix}$$

where K is the bulk modulus and G the shear modulus. For non-viscous and irrotational behaviour G must be zero, but enforcing this results in the above statement containing zero diagonal terms, leads to singularity and the emergence of spurious zero energy modes. Early attempts to rectify this problem introduced small artificial shear stiffnesses, but this is clearly not satisfactory. Modelling incompressible behaviour, requiring no net volume change in the element, also produced difficulty because it requires use of 0.5 for Poisson's ratio, leading to serious inaccuracies in stress calculations.

An alternative approach is to introduce a penalty function approach directly to the rotational strain terms so as to produce irrotation. Illustrating this approach in 2D yz-space, the rotational strain is

$$L(u) = \frac{1}{2}\left(\frac{\partial v}{\delta z} - \frac{\partial w}{\partial y}\right)$$

The requirement that L(u) = 0 can be incorporated into the standard finite element formulation by introducing an additional term into the energy functional

$$\bar{\Pi} = \Pi + \alpha \int L(u)^T L(u)\, d\Gamma$$

where α is the penalty multiplier, the value of which must be judiciously chosen. The larger the better, but if too large the numerical stability of the assembled system stiffness matrix will be compromised. In addition, penalty formulations of this type require the use of reduced integration schemes for evaluating the second term above, and this brings other problems. (Here, 'reduced integration' means using fewer than the accepted number of Gauss points; further discussion will be given later.)

In view of the advantages and disadvantages possessed by both the Eulerian and Lagrangian formulations of fluid elements, it is not surprising that attempts have been made to produce a hybrid element where both fluid pressures and displacements were considered as primary modal variables. The resulting formulation ensures a symmetric coupled equation system, and the use of reduced integration is no longer necessary to ensure that incompressible fluid motions can be represented, thus giving a well-conditioned fluid mesh. The number of nodal variables is increased considerably, however, and this is a major disadvantage for large problems like arch dams.

Two types of displacement fluid element have been developed and used at the EERC Bristol for the dam-reservoir dynamics and these will now be discussed briefly. The first is the element originating from Wilson and Khalvati (Ref.13), which has been incorporated into SAP IV, but the second was entirely new in 1991, being based on what is referred to as the "Field Consistent" approach.

10. The Wilson-Khalvati Fluid Displacement Element.

Bulk stress (ie: pressure) and bulk strain are related by

$$P = K\epsilon_v \qquad \text{and} \qquad \epsilon_v = \epsilon_x + \epsilon_y + \epsilon_z$$

Rotational strains are

$$\bar{\epsilon}_x = \frac{1}{2}\left(\frac{\partial v}{\partial z} \;\; \frac{\partial w}{\partial y}\right) \qquad \bar{\epsilon}_y = \frac{1}{2}\left(\frac{\partial w}{\partial x} \;\; \frac{\partial u}{\partial z}\right) \qquad \bar{\epsilon}_z = \frac{1}{2}\left(\frac{\partial u}{\partial y} \;\; \frac{\partial v}{\partial x}\right)$$

Instead of the penalty formulation previously used to enforce irrotation, a physical stiffness parameter is introduced for each of the three strains above, and hence directly into the constitutive law for the element. These stiffnesses, C_{22}, C_{33}, C_{44} can be visualised as a watch-type spring, one end attached to the element at each integration point with the other end held fixed. The complete 3D constitutive law for the element is

$$\begin{Bmatrix} p \\ \bar{p}_x \\ \bar{p}_y \\ \bar{p}_z \end{Bmatrix} = \begin{bmatrix} K & 0 & 0 & 0 \\ 0 & C_{22} & 0 & 0 \\ 0 & 0 & C_{33} & 0 \\ 0 & 0 & 0 & C_{44} \end{bmatrix} \begin{Bmatrix} \epsilon_v \\ \bar{\epsilon}_x \\ \bar{\epsilon}_y \\ \bar{\epsilon}_z \end{Bmatrix} \qquad p = D\epsilon$$

As with the penalty number, the larger these stiffnesses are, the more irrotational the fluid element will be. If it is too large the matrix inversion routines will become unstable.

The formulation of the element is based on an energy method, there being 3 components:

(i). Strain energy $\quad = \frac{1}{2}\int_\Gamma \epsilon^T D \epsilon \, d\Gamma$

(ii). Kinetic energy $\quad = \frac{1}{2}\int_\Gamma v^T \rho v \, d\Gamma$

(iii) Free surface motion $\quad = \frac{1}{2}\int_S u_s^T \rho g u_s \, dS$

In (ii) above, v is the element velocity vector and ρ the density. In(iii) u_s is the vector of free surface displacements and g is gravitational acceleration.

The finite element discretisation now follows the standard procedure, usually for isoparametric elements, but some consideration must be given to the numerical Gaussian integration process - economy, convergence and singularity are the three key factors. If the integrals are evaluated exactly, the finite element solution is a lower bound, meaning that displacements and frequencies will be lower than the exact solution. As the mesh is refined and the C-values increased, convergence will be obtained, but the rate has proved to be unacceptably slow. Reduced integration is therefore introduced which decreases the stiffness of the elements. But the choice here is not governed solely by convergence requirements. In addition, the assembled system stiffness matrix must be non-singular, other than for geometrically acceptable rigid body motions. When meshes are 'grounded' these rigid body motions are removed, resulting in a positive-definite stiffness matrix. Individual element matrices are usually only positive semi-definite, since in their unrestrained state there exist several rigid body motions. To be avoided is a system matrix where there are more zero energy modes than possible rigid body modes, because these spurious modes will pollute the solution.

A very thorough evaluation of the Wilson-Khalvati element has been made by Greeves (Ref.15) for both 2D and 3D fluid elements, involving

(I) determining their stiffness performance under simple static loading;
(ii) determining their stiffness and mass properties for fluid meshes only;
(iii) determining their stiffness performance for complex dynamic fluid-structure conditions;
(iv) determining their stress performance under dynamic fluid-structure conditions.

In 2D, both 4-noded and 9-noded elements were used (Fig. 3), much effort being spent on the rotational stiffness parameter C_{22}. By increasing C_{22}, the frequency of the rotational or convective modes is increased; the aim being to select a value such that the frequency of these modes falls above the frequency range of interest. For (ii) above, Greeves agreed with Wilson-Khalvati's original suggestion that C_{22} should be 100 times the fluid bulk modulus, and also found that the 4-noded element was much inferior to the 9-noded element, probably because the former used a single-point integration grid, necessary to ensure mesh singularity. Here, the accuracy was measured by comparison with the first sloshing mode and the first compressive mode of fluid in a rectangular tank given by Lamb's closed form solutions. For (iii) and (iv) above, the Pine Flat gravity dam was used as a test case because results had previously been obtained using the Eulerian (pressure) element. The conclusions were the same: the 9-noded Lagrangian element was much superior to the 4-noded element. But something else was found, that the displacement element produces families of coupled system modes where the dam exhibits the same type of behaviour but the fluid does not. This was considered to be due to compressibility because all coupled modes had frequency close to the cut-off frequency (C/4H) of the reservoir. When this happens, the effect of compressibility is to introduce hydrodynamic damping, due to propagating fluid modes which radiate energy away from the structure; below the cut-off frequency the energy remains confined to the immediate vicinity of the vibrating structure. For

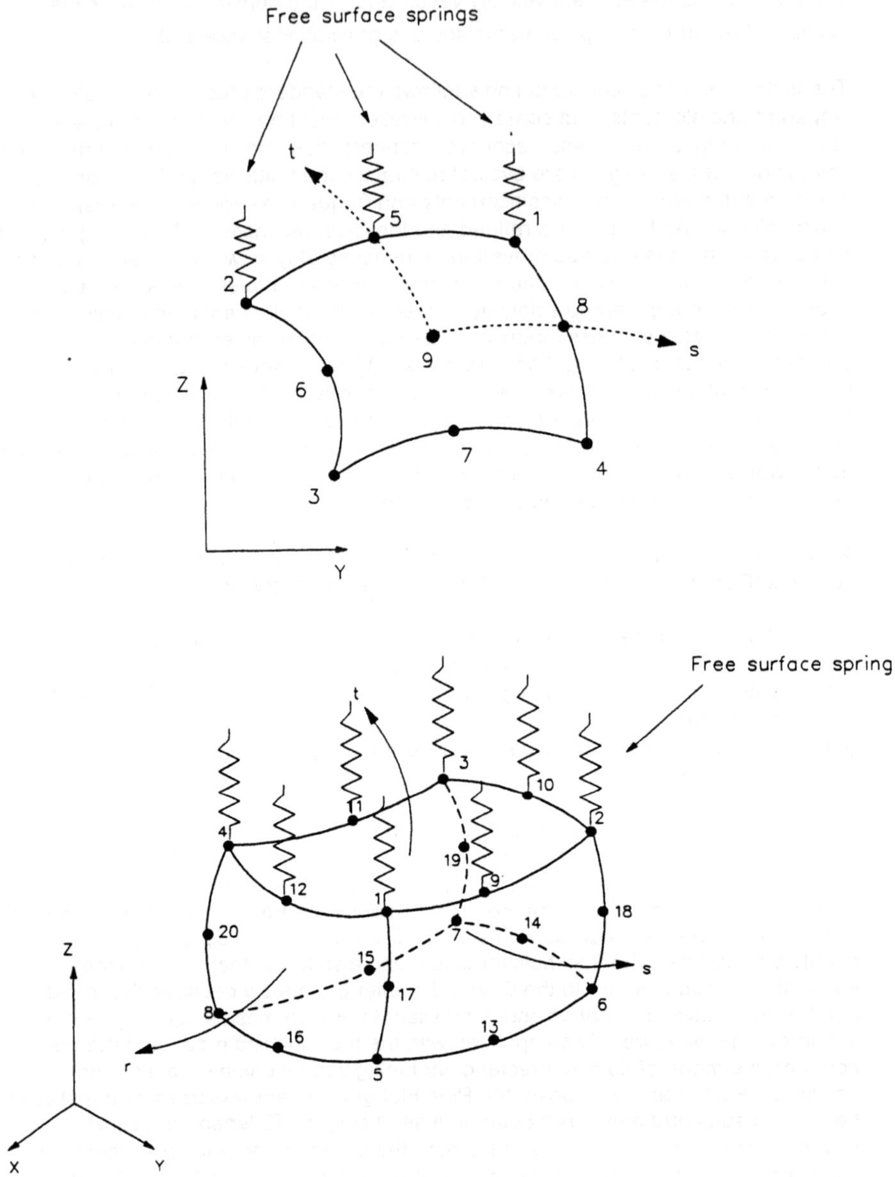

Fig. 3 2D and 3D fluid elements.

the 9-noded mesh, reference to all the coupled mode shapes when the dam is performing its first cantiliver bending mode shows the fluid to be participating fully throughout its extent. This was not the case for the 4-noded mesh because, as previously noted, it does not give an accurate value for the first compressive fluid frequency.

The Pine Flat gravity dam study reported above, using 2D fluid elements, was repeated using 8- and 27-noded 3D elements. Although the fluid element behaviour itself is polluted by the presence of spurious energy modes even after inclusion of free-surface stiffness terms the calculation of stress from these elements appears to be unaffected; in other words, they are accurate for this purpose.

11. Field Consistent (F.C) Formulation of the Fluid Element.

A key issue in the foregoing discussion of the Wilson-Khalvati fluid element was the required integration orders to ensure satisfactory performance. Reduced integration schemes were required to prevent 'locking' and an over stiff performance attributable to the penalty constraints imposed on the element strain operator, the element being required to behave both irrotationally and incompressibly. As a result the assembled fluid matrix is singular, but for 2D problems the free-surface stiffnesses remove any zero energy modes. To some, the use of reduced integration methods appears as a heuristic approach, rather than being logically based, because the explicit reason for over stiff performance is not identified.

The Field Consistent technique ensures that all stress and strain fields within the element are consistent, and was developed by Prathap (Ref. 17) for structural elements, where the locking and over-stiff behaviour occurs. It can be illustrated using the 2-noded beam bending element shown in Fig. 4. The strain energy, including shear, is

$$U = \frac{EI}{2}\int\left(\frac{\partial\theta}{\partial x}\right)^2 dx + \frac{GKA}{2}\int\left(\theta - \frac{\partial v}{\partial x}\right)^2 dx$$

where K is a shear correction factor. As the beam becomes thinner, K increases, and the shear strain energy may dominate. Thus, a satisfactory formulation must ensure that it tends to zero. This can be done by a penalty formulation. The shear strain is given by

$$\theta - \frac{\partial v}{\partial x} = b_0 + b_1 x - a_1$$

and in the limit this must be zero for all x, requiring that $(b_0 - a_1) = 0$ and $b_1 = 0$. The former is readily satisfied, but the latter is an un-attainable constraint which is responsible for locking of the element as the thickness reduces. It can be reduced by smoothing the shape functions used to describe the variation of rotation through the element. This effectively means that the shape function used in the formation of the shear strain must be reduced to a constant term from the original linear expression; this is achieved by a least square error fitting technique. If the required smoothed shape function is the constant C_0, the error function, X, is

$$X = \int (b_0 + b_1 x - c_0)^2 dx$$

and, by minimising in the standard manner, C_o is obtained from

$$\int b_1 (b_0 - b_1 x - c_0)\, dx$$

and is to be used <u>only</u> for the shear strain energy term, but with a full numerical integration scheme.

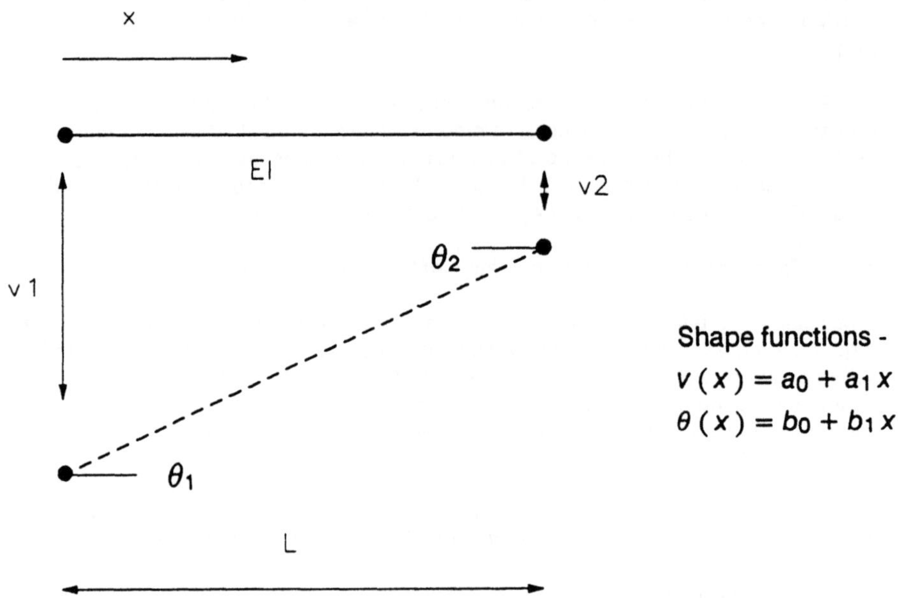

Fig. 4 2 - noded beam element and shape functions.

For 2D fluid elements in the yz-plane, the strains are

$$\epsilon_v = \frac{\partial v}{\partial y} + \frac{\partial w}{\partial z} \qquad \text{and} \qquad \bar{\epsilon}_x = \frac{1}{2}\left(\frac{\partial w}{\partial y} - \frac{\partial v}{\partial z}\right)$$

and for a 4-noded element

$$v = a_1 + a_2 y + a_3 z + a_4 yz \; ; \qquad w = b_1 + b_2 y + b_3 z + b_4 yz$$

For zero strains,

$$(a_2 + b_3) = (b_2 - a_3) = a_4 = b_4 = 0$$

and the final two conditions cannot be satisfied. They must therefore be smoothed in the manner described. 9-noded 2D elements and 8- and 27-noded 3D elements are treated in the same way.

The Field Consistent technique shows, in a rigorous way, the reason for the over-stiff behaviour of some penalised finite elements. It highlights the presence of spurious un-attainable numerical constraints in the strain operator which contributes to the strain energy of the element, making it too stiff. The smoothing of these spurious terms gives the same enhancement as that given by reduced integration techniques. This has been studied at some length by in Ref. 15 giving a more rigorous justification for the latter approach.

After several detailed numerical evaluations of these F.C fluid elements Greeves made the following conclusions.

1. The FC4 and FC8 elements are accurate for both fluid and fluid-structure response under dynamic loading conditions, particularly for stress predictions.

2. Both FC9 and FC27 elements show good results for fluid-only responses, but gave poor results for fluid-structure interaction. On investigation, it was found that the FC9 element possessed a variation in stiffness within its volume, indicating the element to be out of balance. A similar explanation is likely for the FC27 element, but detailed studies were not carried out.

12. Benchmark Analysis of Talvacchia Arch-Dam

As a result of the ICOLD meeting in Sydney (1990), a benchmark activity was organised to compare software used in the analysis of dams. Two standard problems were selected; non-linear dynamic response of Infiernillo embankment dam, and linear thermal and modal response of Talvacchia double curvature arch dam. The latter is considered here Ref. 16 since it allowed a comparison of use of the incompressible pressure fluid element (SAP IV) and the displacement fluid elements just discussed (SOLVIA).

Dry modal response. The first specified analysis was to find the first 6 modes for the dam on a rigid foundation and empty reservoir, using 20-noded and 15-noded isoparametric elements with the mesh shown in Fig. 5. A 27-point Gaussian integration was used, 84 elements representing the body of the dam. Both SAP IV and SOLVIA were used, with the results shown in Table 1 and Fig. 6. The mode shapes follow the standard pattern of increasing wave order along the crest, with the sixth mode being the first to show the second vertical cantilever motion. In Table 1 results are also given for the 21-node element because the 20-node element has supposedly suspect accuracy for dynamic studies, principally because of the mass distribution throughout the element. Such inaccuracy is not evident here. A useful modal parameter is the percentage of participating modal mass in each cartesian direction. This value indicates the amount of base shear which would arise for a unit acceleration in each direction, highlighting the significance of each mode in any subsequent response calculations. From Table 2 it is seems that for these first 6 modes, only 37% of the total dam mass is participating in the upstream-downstream direction, with even smaller percentages in other directions. In fact, it requires the use of 30 modes to involve 90% of the total mass.

Coupled modal analysis - incompressible fluid.

The second specified analysis was for the first 6 modes on a flexible but massless foundation with reservoir, using the mesh of Fig.7. There are 336 fluid elements (20 and 15-noded) of the pressure variety, 84 elements for the dam and 224 for the foundation block. The fluid substructure was linked to the dam by 8 and 6-noded interface fluid elements. At the free surface, the nodal pressures were deflated, but at the far boundary they were retained as a check on truncation errors. Table 3 gives the results; for comparison the 'dry' results are repeated from Table 1 together with 'dry' results for a flexible (but massless) foundation. The effect of the reservoir is seen to reduce modal frequencies as expected, bearing in mind that the reservoir depth was 56 m for a crest height of 76m. This particular reservoir level was determined by corresponding conditions during prototype tests on the dam. It was, of course, unfortunate that a top reservoir level could not have been used because it is known from many studies that the top 20 m of water have a major effect on a thin arch dam, and a full reservoir would have provided a better test of the different fluid elements. The mode shapes are the same as those of Fig. 6 except for an interchange between fifth and sixth modes. This reflects the increased interaction effects for the modal response containing the second vertical bending motion of the dam, resulting in a greater percentage reduction in modal frequency.

Modal hydrodynamic pressures were also found as part of the solution; selected values at the dam face are given in Fig.8. They follow the expected pattern, with the effect of the second vertical bending motion in mode 5 resulting in peak hydrodynamic pressure occurring at greater depth than for other modes. Along the centreline of the reservoir these pressures were shown to decay quickly, being insignificant at a distance from the dam of twice the reservoir depth.

Coupled modal analysis - compressible fluid.

The reservoir depth of 56 m resulted in the fluid-structure interaction effect being modest; this also reduces the possible influence of compressibility. The fundamental compressive mode for the trapezoidal reservoir used here is 8.54 Hz, and Table 3 shows that the first few modes fall below this. Modes 4, 5 and 6 may

Mode	SAPIVP:20-noded (Hz)	SOLVIA:20-noded (Hz)	SAPIVP:21-noded (Hz)
1 - A1 B1	4.193	4.175	4.180
2 - S2 B1	4.478	4.463	4.461
3 - S1 B1	5.534	5.511	5.506
4 - A2 B1	6.530	6.502	6.537
5 - S3 B1	8.266	8.229	8.258
6 - S2 B2	8.984	8.956	8.966

Notes A:antisymmetric plan shape S:symmetric plan shape B:vertical bending mode

Table 1 Talvacchia; thesis Table 8.2

Mode	X - mass (%)	Y - mass (%)	Z - mass (%)
1 - A1 B1	0.05	8.40	0.01
2 - S2 B1	10.42	0.01	0.41
3 - S1 B1	13.01	0.04	2.63
4 - A2 B1	0.03	3.91	0.01
5 - S3 B1	0.91	0.02	2.61
6 - S2 B2	12.69	0.01	5.01
Sum	37.11	12.39	10.69

Table 2 Talvacchia; thesis Table 8.3

Mode	Rigid base (Hz)	Flexible bsae (Hz)	Flexible base & fluid effects (Hz)
1 - A1 B1	4.193	3.806	3.712
2 - S2 B1	4.478	4.142	3.976
3 - S1 B1	5.534	5.331	5.130
4 - A2 B1	6.530	6.287	6.052
5 - S2 B2	8.266	7.947	7.509
6 - S3 B1	8.984	8.405	7.810

Table 3 Talvacchia; Table 8.4

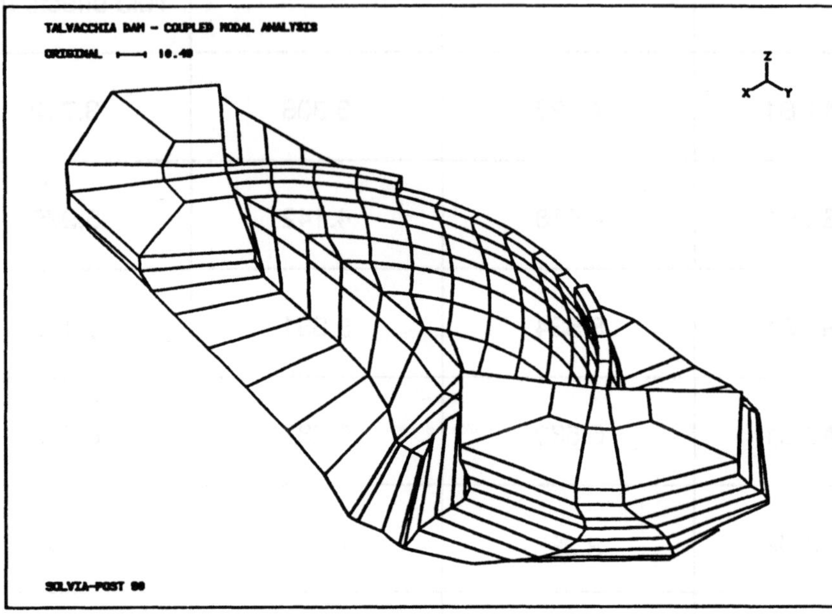

Fig. 5 Talvacchia; thesis 8.19

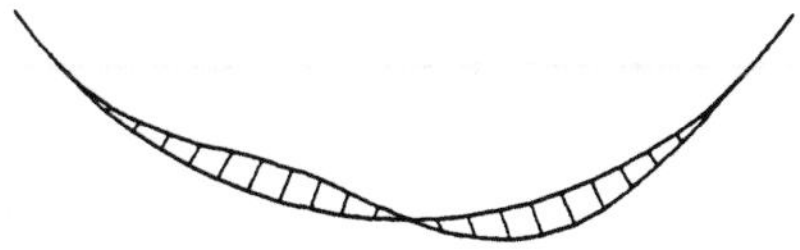

Mode 1 - A1 B1 - 4.193Hz

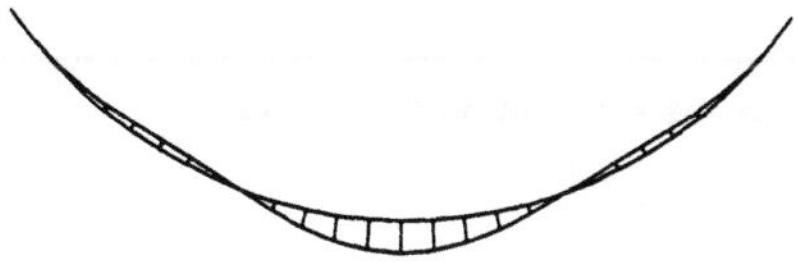

Mode 2 - S2 B1 - 4.478Hz

Mode 3 - S1 B1 - 5.534Hz

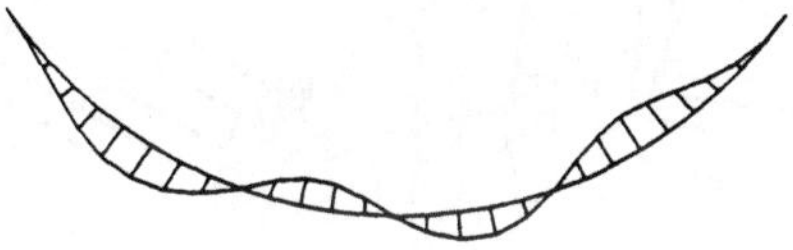

Mode 4 - A2 B1 - 6.530Hz

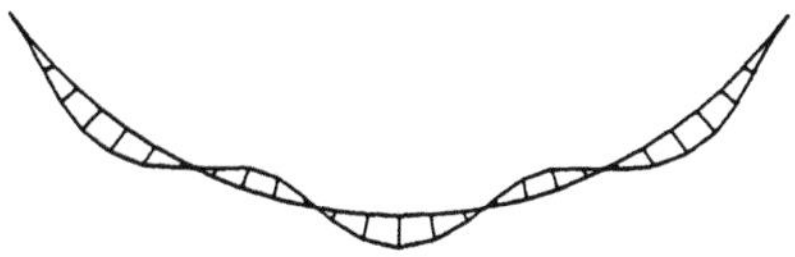

Mode 5 - S3 B1 - 8.266Hz

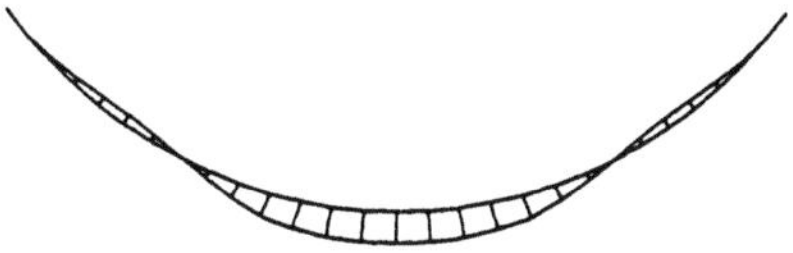

Mode 6 - S2 B2 - 8.984Hz

Notes - S:symmetric mode A:antisymmetric mode B:vertical cantilever mode

Fig. 6 Talvacchia thesis Table 8.20

8.55

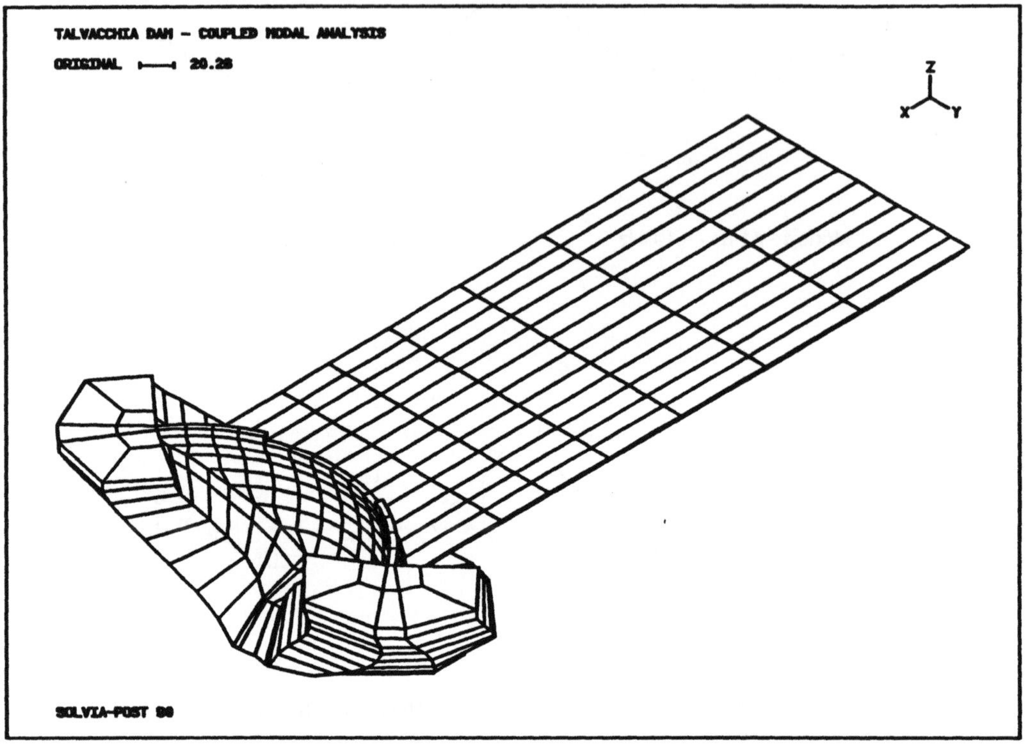

Fig. 7 Talvacchia thesis Table 8.21

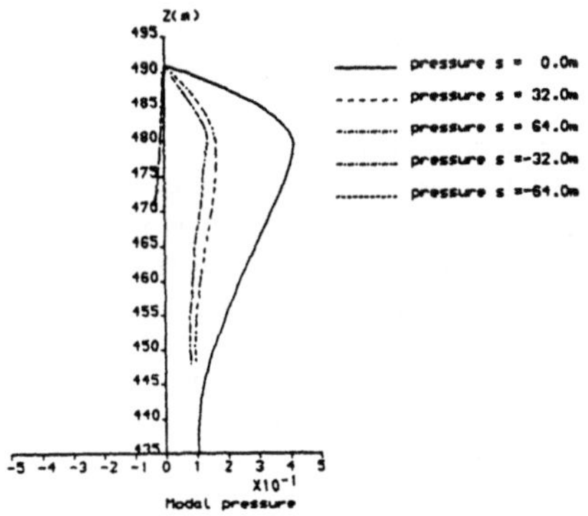

Interface pressures - mode 2

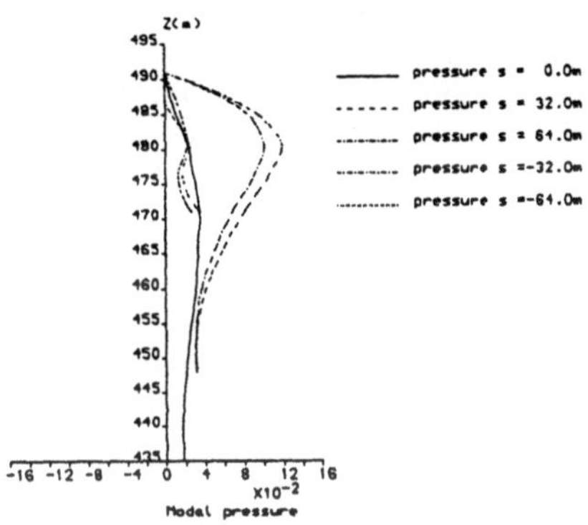

Interface pressures - mode 3

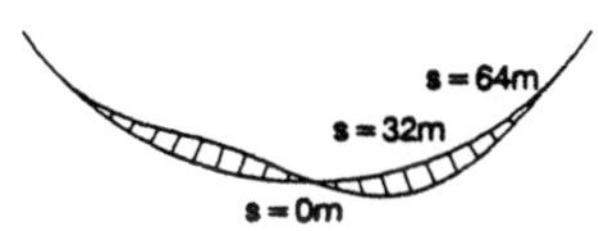

Positions for pressure distributions

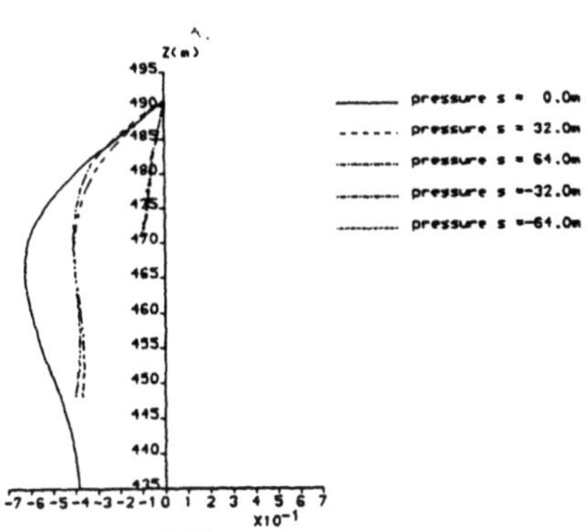

Interface pressures - mode 5

Fig. 8 Talvacchia; Fig. 8.23

however be influenced, their dimensionless frequency ratios being 0.83, 1.05 and 1.11.

The dam and foundation mesh remain unchanged, but 20 and 15-noded displacement fluid elements are now used. Because the fluid is only able to transmit normal forces at any boundary, special link elements were required; for the fluid/dam interface these were rigid links, but for the fluid/canyon boundaries the inclusion of a set of 'grounded' nodes was required. Sets of node points were automatically generated at a unit distance perpendicular from the reservoir boundary which were completely restrained, and rigid link elements introduced between them and other corresponding fluid nodes. In this way, tangential movement between fluid, canyon and dam was possible.

For previous meshes of displacement type fluid elements, a stability evaluation was performed to determine the possible presence of low frequency sloshing modes. The response of such modes is controlled by introduction of free surface springs. Such behaviour arises because of a nett singularity introduced by either the use of reduced integration (R.I.) schemes, or by modified shape function derivatives in the Field Consistent elements. For these studies on the Talvacchia dam, the RI20 fluid element was used, and in the fluid mesh there were 1710 nodes, resulting in 5130 degrees of freedom (1710 X 3). The 363 fluid elements each used a 2x2x2 (8 point) Gauss integration. For the 3D fluid element there are four element strains, one bulk and three rotational; this gives a total of 363 X 8 X 4 = 11616 independent strain relations which can be introduced by the fluid elements. This is greater than the number of degrees of freedom in the fluid, and so there is no nett singularity. This introduces an important point - that the effectiveness of R.I. schemes cannot necessarily be discussed by singularity requirements. Here, mesh singularity has not occurred, but the fluid will still behave incompressibly and inrotationally.

We note here one major disadvantage of using displacement fluid elements compared to the incompressible pressure model; there are now a total of 8925 equations with an optimised bandwidth of 1188 - a much larger problem size than that produced by the pressure model.

Table 4 gives frequencies obtained for this combined dam-foundation-fluid system, where it will be seen that there is now no uniqueness between frequency and mode shape, the differences between the modes being accounted for by fluid behaviour. (NB: This is similar to the results found for the Koyna gravity dam, but there such behaviour only occurred above the compressible limit, C/4H. For the arch dam, it is found for all modes.)

The differences between the two sets of results given in Table 4- incompressible pressure model (SAP IV) and the displacement fluid element (SOLVIA) are unlikely to be attributable to compressibility effects because of the low reservoir level. Furthermore, such differences were not found by other participants in the benchmark programme who were using different types of element. Various studies have been made (Ref. 15)) to explain these differences, but none are conclusive - many more analyses of this type are required by independent researchers. One useful study by the Bristol group was to determine the significance of these multiple modes, some may be dominant and others too weak to participate in the response of the dam, by calculating the percentage participation of total mass for each mode. Table 5 presents these factors for both dam and fluid mass, for each of the multiple modes found, showing that they are all significant. This would appear to indicate

Mode	SAPIVP (Hz)	SOLVIA (Hz)
1 - A1 B1	3.792	3.183 3.324 3.618 4.438
2 - S2 B1	3.976	4.291 5.015
3 - S1 B1	5.130	3.892 5.730 6.986
4 - A2 B1	6.052	4.826 7.061 7.361
5 - S3 B1	7.810	9.527 9.644 9.829
6 - S2 B2	7.509	7.591 8.291 10.100

Table 4 Table 8.5

Mode	DAM			RESERVOIR		
	X-mass (%)	Y - mass (%)	Z - mass (%)	X - mass (%)	Y - mass (%)	Z - mass (%)
A1 B1	1.8	2.1	0.0	1.4	0.4	0.0
	2.2	2.6	0.1	1.9	2.6	1.2
	1.5	4.1	0.0	1.1	0.0	0.5
	0.0	3.8	0.0	0.6	5.3	0.2
S1 B1	7.5	0.4	0.0	7.3	2.8	0.3
	9.5	2.0	0.4	1.6	3.2	1.0
	1.6	1.5	1.3	0.3	2.3	1.7
S2 B1	8.4	0.3	0.3	2.4	1.2	3.9
	2.2	0.3	0.7	2.9	2.6	1.4
A2 B1	1.5	2.6	0.3	0.1	1.9	0.0
	2.9	0.7	3.8	2.0	0.2	2.4
	1.5	1.3	1.2	1.1	0.6	3.0
S3 B1	0.8	0.3	0.4	1.0	0.4	0.5
	1.0	0.1	0.1	0.6	0.4	0.8
	1.5	1.2	1.4	0.6	0.9	1.4
S2 B2	0.7	4.4	0.2	0.8	2.2	2.0
	2.2	0.9	2.0	3.5	0.7	11.3
	0.6	2.4	0.4	0.2	0.5	0.4

Table 5 Table 8.6

that these multiple modes are not spurious, but should be included in the dynamic response analysis. Table 6 summarises the participation total for each group of modes and compares these with the dry modal values. The increase in dam response due to fluid effects is clear, in the X direction from 37% to 47% and a much larger increase in the Y direction from 12% to 31%.

A further contribution towards explaining the above behaviour can be found by noting that an eigen-solution to the dam-foundation-fluid system is obtained using the iterative procedure

$$K Y_j^{i+1} = M X_j^i$$

where K and M are global stiffness and mass matrices, respectively; X_j^i is the current approximation to the jth eigen-vector, and Y_j^{i+1} provides a better approximation. Here the right-hand side load vector is applied to the complete system, fluid as well as dam. In contrast, when the pressure model (SAP IV) is used, the fluid is loaded only at the interface by the acceleration of the dam. Clearly, for the displacement fluid model the fluid has greater potential to drive the response of the dam.

The foregoing research study of Talvacchia arch dam is incomplete, and leaves many 'loose-ends'. Its findings of multiple modes is, however, supported by limited experimental evidence. In 1987 the Maentwrog arch-gravity dam in Wales was subjected to excitation by rotating-mass machines and modal responses found. The first 3 modes were the first antisymmetric response but at different frequencies. These tests were carried out because the concrete exhibited Alkaline Aggregate Reaction, and it was assumed that this was the reason for the measurement of multiple modes; fluid effects were not considered. A similar series of tests were carried out on the Hermitage gravity dam in Jamaica, where the same modes were found to occur at different frequencies.

Mode	DRY			COUPLED		
	X - mass (%)	Y - mass (%)	Z - mass (%)	X - mass (%)	Y - mass (%)	Z - mass (%)
A1 B1	0.05	8.40	0.01	5.51	12.60	0.09
S2 B1	10.42	0.01	0.41	10.61	0.53	1.01
S1 B1	13.01	0.04	2.63	18.60	3.90	1.71
A2 B1	0.03	3.91	0.01	5.88	4.55	5.22
S3 B1	0.91	0.02	2.61	3.26	1.63	1.90
S2 B2	12.69	0.01	5.01	3.42	7.69	2.55
SUM	37.11	12.39	10.69	47.28	30.90	12.48

Table 6 Table 8.7

REFERENCES

1	Greeves E J Severn R T	A State of the Art Review of the Analysis ad Modelling of Fluid-Structure Interaction. Chap 7 Structures Subject to Dynamic Loads, 1991, 233-270, Elsevier, ISBN I-85166-582-X
2	Westergaard H M	Water Pressure on Dams during Earthquakes ASCE EM Div Vol 98, 1933
3	Zinkiewicz O C Nath B	Analogue Procedure for Determination of Virtual Mass. ASCE Sept, 1964, HY5
4	Back P A A et al	The Seismic Study of a Double Curvature Arch Dam Proc Instn Civ Engs 1969 (June), 217-248.
5	Zienkiewicz O C Cheung Y K	Finite Elements in the Solution of Field Problems. Engineering, Lond, 1965, 24 Sept, 507-510
6	Altinisik D Karadeniz H Severn R T	Theoretical and Experimental Studies on Dynamic Structure - Fluid Coupling Proc Instn Civ Engrs. Part 2, 1981, 71, Sept 675-704
7	Allen D N de G	Relaxation Methods McGraw Hill Book Co 1954
8	Dungar R	An efficient Method of Fluid Structure Coupling in the Dynamic Analysis of Structures. In J Num Meth Engng 1978, 13, 93-107
9	Simpson A Collar A R	Matrices in Engineering Dynamics Ellis Horwood, Chicherster, UK
10	Greeves E J Dumanoglu	The Implementation of an Efficient Computer Analysis for Fluid-Structure Systems using the Eulerian Approach within SAP IV Univ Bristol, Civ Eng Int Report UBCEE-EE-89-10 Sept 1989
11	Porter C S Chopra A K	Hydrodynamic Effects in the Dynamic Response of Simple Arch Dams J E'quake Eng Struct Dyn, Vol 10. 1982
12	Perumalswami P R Kar L	Earthquake Hydrodynamic Forces on Arch Dams ASCE EM Dw 99, 1973, 965-977
13	Wilson E L Khalvati M	Finite Elements for the Dynamic Analysis of Fluid-Solid Systems Int J Num Meth Eng Vol 19 1983
14	Bratz Heilbron	Discussion on ref 2 ASCE EM Div Vol 98 1933

15.	Greeves E.J.	The modelling and analysis of linear and non-linear fluid-structure systems with particular reference to concrete dams. PhD thesis, Univ. of Bristol, July 1992.
16.	Greeves E.J. Severn R.T. Taylor C.A.	The linear dynamic analysis of an arch dam including both incompressible and compressible fluid interaction Benchmark Workshop on Numerical Analysis of Dams. ISMES, Bergamo, Italy 28-29 May 1991.
17.	Prathap G. Ramesh Babu C.	Field Consistency and violent stress oscillations in the finite element method. Int. J. Num. Meth in Eng., 24, 1987.

MODEL STUDIES FOR EARTHQUAKE LOADING

1 The Arch Dams Committee of the Institution of Civil Engineers (1)

The contents of this lecture are based largely on my own experience, which began in 1958 when, as a member of the above committee, I was asked to look at the effect of earthquakes on arch dams. The remit included both theoretical and experimental studies; the former has been dealt with in other lectures of this course, but it is worth noting that during its 10 years of work, the newly emerging finite element method was developed at a great rate in the need to produce solutions for the arch dam problem. These theoretical studies are reported in Ref 1, and form the basis of much subsequent development in dam analysis generally.

With regard to earthquake loading, it should be noted that prior to the late 1950s, any consideration of such loads would be restricted to their application as an equivalent static load applied in the trial load method. Computational facilities were such that only the very simplest structure could be analysed for real dynamic loads by classical methods. In truth, seismic loading was not considered to be a serious problem, even in such areas as California. The Pacoima arch dam, completed in 1928, being a case in point and this was certainly true of the Dokan dam, designed in 1952-54 by British Consultants for construction in Iraq (2), for which the three-dimensional finite-difference equations were solved for the first time for gravity, water and temperature loads.

The terms of reference of the Arch Dams Committee were:-

> "To examine critically the various methods of designing arch dams of different profiles commencing with a comparison, by mathematical calculation, of specified idealised dams which should be further analysed by model studies."

Five different designs were produced, and the various groups of researchers each used their own analytical methods and computer software, which at that time was written by the researchers themselves. Regrettably, only two groups addressed the dynamic problem, and only the author's group attempted model studies (1), and even here, the approach was that of checking out the theoretical studies rather than attempting to produce similitude between model and prototype.

2 Tang - E Soleyman Dam (3)

Seismic model studies on this arch dam are reported here because they were of a pioneering nature which necessitated the installation of new testing equipment and the development of new measuring techniques allowing the models to be subjected to random as well as sinusoidal input. They were carried out in 1960-61 by LNEC Lisbon for Sir Alexander Gibb and Partner; a UK firm of Consulting Engineers, the Senior Partner of which was the chairman of the Arch Dams Committee referred to above. It is noted that this arch dam is only now (1995) under construction in Iran.

The general arrangements for the tests are indicated in Fig.1, from which it is seen that the 1:200 model of dam and foundation block is cradled in a steel frame which is suspended (about 10 cm from the floor) from an overhead gantry by chains. The electro-magnetic vibrator, whose maximum force is given as 1360 kg, is attached to the model through the foundation block. Provision is made for the simulation of

1. Tang-E-Soleyman dam: vibration tests on models at LNEC, Lisbon.

reservoir in the model but no details are given, and tests were carried out both with, and without, this reservoir, the not surprising conclusion being reached that "earthquake action do not seem to be much affected by water level in the reservoir. The filling of this, giving rise to compressive stresses, will improve the behaviour by reducing the tensile stresses developed".

Two concrete and one plastic models were made and tested using sinusoidal and random input, with 8 strain-gauges on each of the upstream and downstream faces measuring response; their recordings were processed to produce plots of spectral density against frequency. From the strains, so-called "apparent stresses" were produced because they were not corrected for the influence of Poisson's ratio. At each of the 16 points, each of the 3 models produced such a stress value, which though of the same order of magnitude, differed considerably in value. These were, however, novel tests breaking new ground in both equipment and methodology.

Pioneering tests were also carried out on a bell-mouth spillway model for this dam, but here, as with the dam itself, the results were too inconclusive to be of much value to the designers.

3 Hendrik Verwood Dam (4)

This dam was designed in the mid 1960s for construction on the Orange river in S Africa, and by this time computer software had been written which could be used with some confidence for calculation of response to earthquakes, either by use of spectral curves or by time-history direct integration methods; for the former, however, it was usually necessary to re-calculate such spectral curves as had been produced primarily for building structures, to obtain values at the frequencies appropriate for dams (5,6). Even so, response calculations using a modal approach depended on the prior calculation of natural frequencies and modal shapes, which in turn depended on assumptions as to the dynamic effects of the reservoir. Here, doubts could be resolved by model studies, and such was the case for the Hendrik Verwood 100 m high double curvature arch dam where the 1:200 model shown in Fig.2 was produced, having a reservoir length extending upstream three times the height of the dam. The ratio of model to prototype natural frequencies is given by the linear scale, and from accumulated knowledge of frequencies of similar prototype arch dams, the 1:200 scale was chosen to be within the 0-1100 Hz range of the measuring equipment. Concrete and water were used to avoid problems in dynamic similarity, and since geophysical data gave an elastic modulus for the valley similar to that for concrete, the same concrete was used in the model for both dam and valley block.

In deciding how to measure the natural frequencies and mode shapes of the model, it was necessary to bear in mind that the purpose of the model testing was comparison with modes calculated on the assumption of zero damping. It has been shown that for a structure in which the damping is small and viscous, the excitation of an undamped mode occurs when, with a sinusoidal force input at a number of points, the outputs are all over the structure are also in phase, but lag behind the force inputs by 90 degrees. The measuring system was designed to satisfy these criteria, being based on a modified version of that devised for testing the Concorde supersonic jet (12). In theory, it is necessary to use as many exciters as there are modes in the structure, but with 5 exciters it is usually possible to excite undamped modes with acceptable accuracy - this being defined by the 90 degrees phase lag requirement.

2. Hendrik Verwoerd dam: 1:200 scale model for 5-point excitation modal analysis.

For the H Verwood model the first seven modes were investigated with reservoir empty and full. For each mode the procedure was first to establish an approximate resonance frequency from vector response plots. These plots were made with an arbitrary force distribution using two or three exciters. An attempt was then made to arrive at a force distribution and frequency appropriate to a pure mode (ie quadrature force-response relationship at the excitation points). A limited check was then made on the responses at other points on the structure, and if all were in phase the mode shape was measured in detail. But if the check showed that some parts of the structure were significantly out of phase with others, then further exciters (up to 5) were deployed. If the responses were still unsatisfactory the exciters were deployed at different positions and the process repeated. When the mode had been finally obtained and its responses measured in detail over the structure, a vector response plot was made for a single point by varying the frequency through the resonance bandwidth whilst maintaining the force distribution appropriate to the mode. Modal damping was obtained from this plot and its circularity and position relative to the origin were indicative of the purity of the mode (Fig 3). The calculated and measured mode shapes for the fundamental antisymmetric mode, reservoir empty and full, are given in Fig 4. The calculations for the reservoir full condition used the electrical analogue method to determine the hydrodynamic pressure involving 42 brass plates on the upstream face of the dam, polythene covering the valley block, and metal foil floating on the water surface. This analogue method will be described in another lecture.

The frequencies and dampings of all measured modes are given in Table 1. Although these damping values are not relevant to the prototype, it is of interest to note that damping values are greater for the reservoir full condition, which in view of other comment in this course about structure/fluid coupling, could be taken as indicative of water compressibility.

There is little doubt that use of multipoint excitation gave results of an accuracy not possible with single point excitation. Even so, several of the modes presented difficulties, a few in finding the proper force distribution and some in obtaining acceptable vector response plots over the resonance bandwidth. If a pure undamped mode of a damped structure is excited in a resonance test, and there is no cross-damping between the modes, then variation of the excitation frequency on either side of the resonance (whilst maintaining the same force distribution) should result in the tip of the displacement response vector at any point on the structure describing a circle which is tangential at the origin to the force input reference. This circular vector response curve is also characteristic of a single degree of freedom system, and the aim in using multipoint excitation is to 'isolate' each mode so that it behaves as a single degree of freedom. Pure, and impure modes are shown in Fig 3.

Two subsidiary tests were made; the first to determine resonance frequencies of concrete test specimens from which an average E-value was obtained ($32.2.10^3$ kN/mm^2), and the second a transient input applied to the valley block with a hammer, a form of dynamic testing which was developed later on a number of arch dam models as data collection and processing equipment became available (7). This second test was devised to provide a check on the modal frequencies already obtained, by analysing the waveform of responses measured on magnetic tape at the base of the dam and the overspill such as is shown in Fig 5, which indicates clearly that it contains contributions from a number of modes. Such data measured

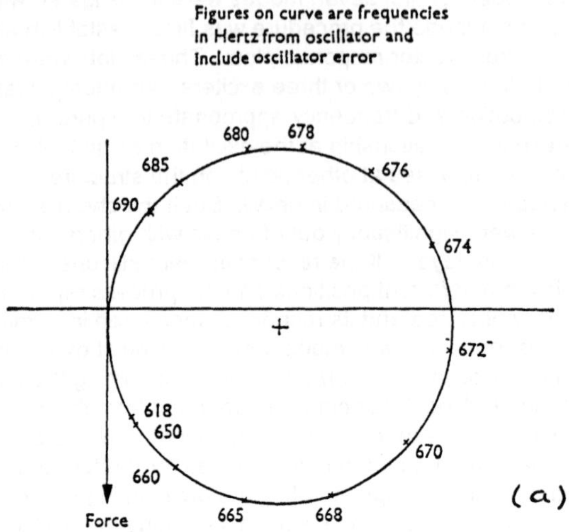

SINGLE POINT RESPONSE VECTOR-MODE AT 672 Hz RESERVOIR EMPTY

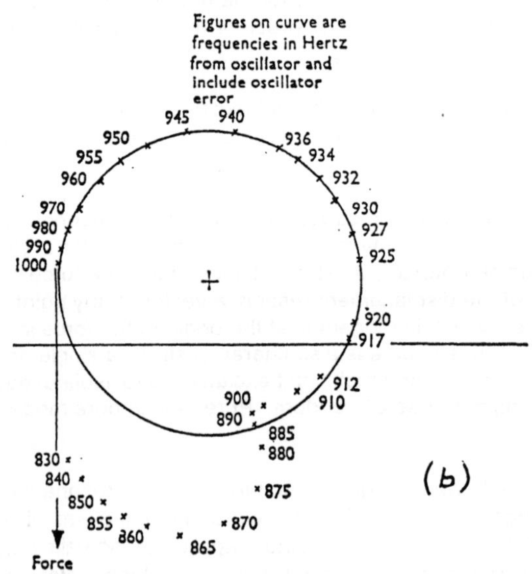

SINGLE POINT RESPONSE VECTOR-MODE AT 916 Hz RESERVOIR EMPTY

3. Hendrik Verwoerd dam: 1:200 scale model. Single point response (a) pure mode; (b) impure mode.

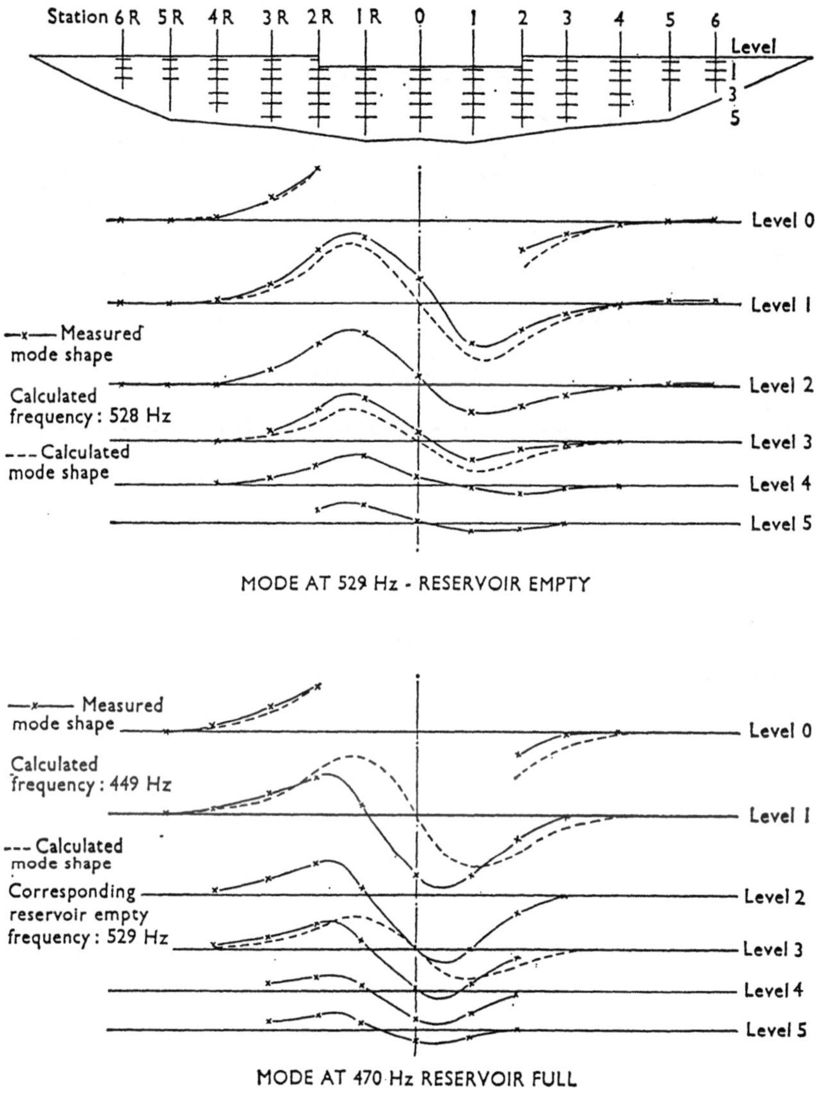

4. **Hendrik Verwoerd dam: 1:200 scale model. Fundamental antisymmetric mode; reservoir empty and full.**

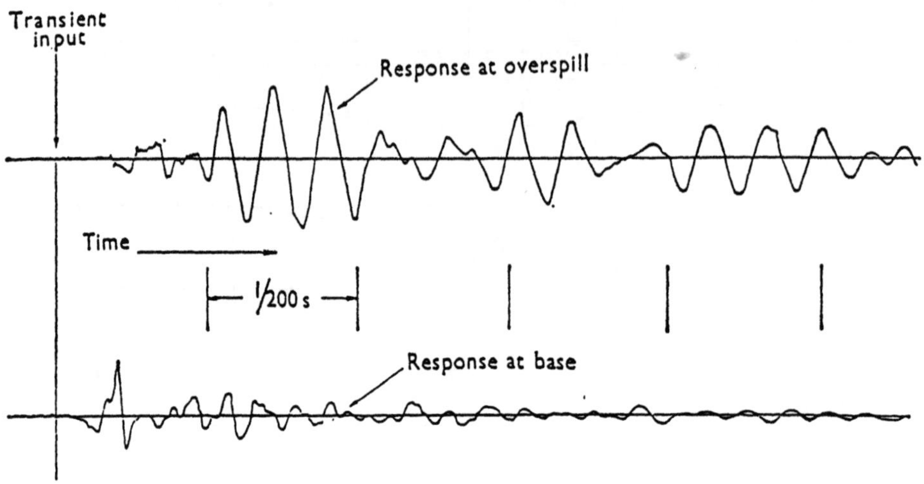

5. Hendrik Verwoerd dam: 1:200 scale model. Response to a hammer blow on base block.

Mode description	Reservoir empty			Reservoir full		
	Calculated frequency,* *Hz*	Measured frequency, *Hz*	Measured damping, c/c_c	Calculated frequency, *Hz*	Measured frequency, *Hz*	Measured damping, c/c_c
Fundamental antisymmetric	528	529	0·009	449	470	0·016
Fundamental symmetric	548	573	0·013	488	434	0·015
First harmonic symmetric	673	672	0·010	611	573	0·040
First harmonic antisymmetric	840	770	0·010	734	683	0·016
Second harmonic symmetric	1021	916	0·024	941	799	0·025
Second harmonic antisymmetric	1220	1006	0·011	1076	919	0·055
Third harmonic symmetric	1334	1093	0·011	1237	1016	0·026

* Calculated frequencies are based on $E=4{\cdot}67\times10^6$ lbf/sq. in. and $p=141{\cdot}2$ lb/cu. ft from test beams (Table 2).

Mode	1	2	3	4	5	6	7
Measured frequency ratio: (Reservoir full/ Reservoir empty)	0·89	0·76	0·85	0·89	0·87	0·91	0·92
Calculated frequency ratio: (Reservoir full/ Reservoir empty)	0·85	0·89	0·91	0·87	0·92	0·88	0·93

Table 1. Hendrik Verwoerd dam: 1:200 scale model. Mode frequencies and dampings.

at many points throughout the dam can be used (7) to determine model shapes, but this development came later, for prototype arch dams and for models. Table 2 gives the three frequencies obtained from this hammer test on the Hendrik Verwood model compared with those obtained from the multipoint resonance tests; agreement is achieved to within 1%.

These model tests gave confidence that the in-house software specially written for calculating natural frequencies and modal shapes was correct, and could be used in response calculation. For this purpose the Santa Barbara (Calif.) earthquake of 30 June 1941 was thought to be most relevant to the S African conditions, and using it, the stresses produced were nowhere such as to cause considerations of re-design.

4 Model Studies for Reservoir Effects (8,13)

Following the H Verwood dam studies, a research programme using models was launched in the mid 1960s to actually measure the hydro dynamic reservoir pressure against a vibrating dam. This was made possible by the developments of the equipment used there for isolating reasonably pure modes of complex structures; it came to be known as MAMA - Manual Automatic Multipoint Apparatus, a picture of which is shown in Fig.6. It was somewhat easier to use than the earlier version, though still required manual adjustment of force levels. The model 'dams' were a steel plate and a ferrocement cylinder which was later replaced by a ferro-cement wall set in either end of a massive concrete block of dimensions 3 x 2 x 1.5 m, so that a reservoir was formed between them (Fiig.7). Only one sufficiently sensitive pressure transducer was available at the time, and so an array of 20 holes was drilled through each dam to take the transducer in turn, which was 2 mm from the water face. The lack of equipment necessitated a lengthy measuring process which would not have been feasible had the MAMA equipment not been able to re-locate the same mode precisely each time the location of the transducer was changed.

Fig 8 shows hydrodynamic pressure profiles for the first symmetric and first anti-symmetric modes of the ferrocement wall and also summarises the measured and calculated resonance frequencies. The finite element solutions used here are those for Laplace's equation and it is seen that the 3 D version is in good agreement with the measurements in all cases. The Chopra equation mentioned here is a 2D solution, essentially the same as Westergaard's original solution. Similar conclusions were arrived at for the steel plate, that the 3D nature of the dam-reservoir system must be taken into account. From the fact that measurements were very close to the 3D solution of Laplace's equation, it was concluded that water compressibility was not evident at this model scale. In fact, a series of tests was performed in an attempt to produce hydrodynamic pressure resonance within the reservoir. In each test the reservoir was partly filled to a depth of 30 cm and the dam excited in a normal mode at a frequency somewhat above the 590 Hz calculated (c/4H) reservoir resonance frequency. The reservoir was then slowly filled to its ful depth of 61 cm, and as this filling progressed the phase of the hydrodynamic pressure and its value were observed at a single point. By this process it was expected that the natural frequency of the reservoir would at some instant be equal to the excitation frequency, resulting in a very large pressure and a quadrature phase relationship with the acceleration of the wall. But no such phenomenon was observed.

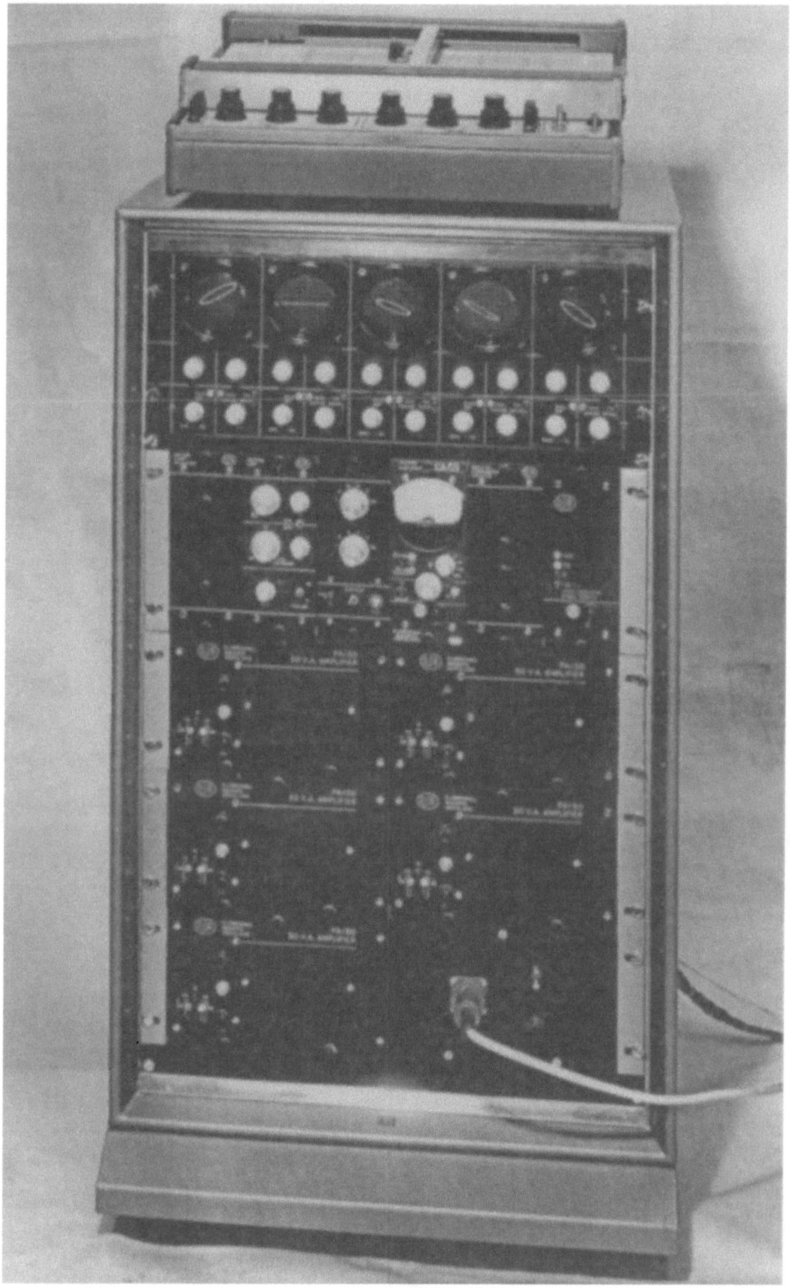

6. Manual Automatic Multipoint Apparatus (MAMA).

7. Model study of hydrodynamic pressure: (a) ferrocement cylinder.
(b) steel plate.

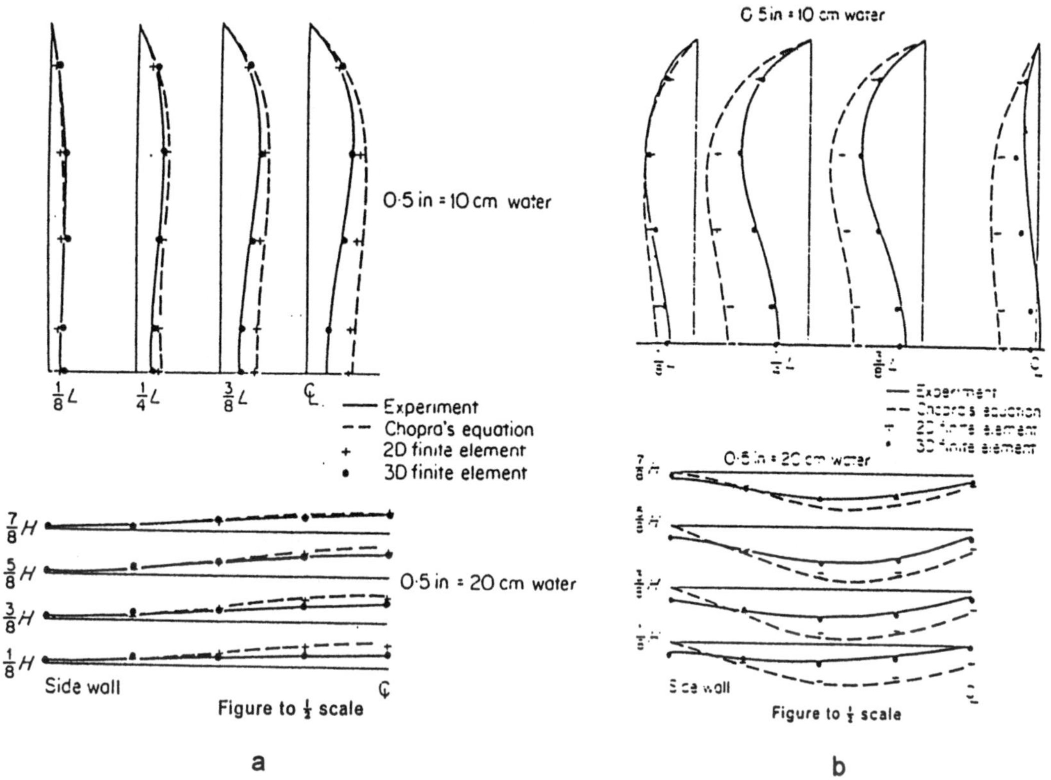

Mode	Reservoir empty: Experimental	Reservoir empty: Theoretical	Reservoir full: Experimental	Reservoir full: Theoretical	Percentage reduction in frequency: Experimental	Percentage reduction in frequency: Theoretical
1	122	115	75	73	39	37
2	264	258	157	154	41	40
3	444	462	285	263	36	43
4	557	520	369	341	34	35

c

8. Hydrodyanmic pressure profiles measured on ferrocement wall
(a) first symmetric mode, (b) first anti-symmetric mode
(c) experimental and theoretical frequencies.

Reservoir empty			
Resonance test		Transient test	
Frequency, *Hz*	Damping, c/C_c	Frequency, *Hz*	Damping, c/C_c
529	0·009	531	0·01–0·02
573	0·013	569	0·01–0·02
672	0·010	677	0·01–0·02

Table 2. Hendrik Verwoerd dam: 1:200 scale model. Comparison of frequencies obtained by resonance and transient testing.

The foregoing studies were continued in 1978-80 by Altinisik (13) using a very thin cylindrical ferrocement arch mounted in a steel tank (which was made as rigid as possible) filled with water, the whole being attached to a uniaxial shaking table for harmonic and earthquake input.

5 Victoria and Contra Arch Dams (7)

The final phase of the H Verwood tests was the use of transient input by means of a hammer to the base of the valley block. This idea was taken up by Trevelyan (7) in 1982-84 using models of the Contra (Switzerland) and Victoria (Sri Lanka) arch dams. Fig. 9 shows the male mould for a 1.150 model of Contra, whilst Fig. 10 is an actual view of the Victoria dam. The intervening years had seen many and substantial improvements in signal capture and processing techniques, as well as in finite element capability, and it was therefore possible to make comparison between modal properties obtained by a variety of methods, including MAMA, instrumented hammer impact, finite element calculation and, for Contra, measurements on the actual prototype dam. Table 3 gives a selection of these results for Contra.

6 Further LNEC Studies (9,10,11)

The pioneering studies at LNEC on the Tang-E Soleyman dam have already been mentioned; it is therefore necessary to comment briefly on impressive developments there since 1962 although Prof J O Pedro will probably already have referred to Ref 10, where the point is made that it is difficult to satisfy similitude requirements for both dam and reservoir, so that model studies are uncoupled. There is another reason too, that the mass of water required to represent the reservoir is large, requiring very high input forces to produce the necessary response in the model dam. Thus scale model testing is most often carried out without reservoir to validate special features which involve uncertainties in finite element models. For example, Ref II describes an LNEC model study of considerable complexity of a 185 m high Greek arch dam which was to be built across a well defined fault line where the risk had been estimated as movement of several centimetres for the design earthquake, and several tens of centimetres for the maximum credible earthquake. Two loading systems were used, one to represent gravity and static water load, and the other to impose step-by-step horizontal displacement along the fault. One very important conclusion of this model study for the safety of arch dams in seismic regions, is that for fault displacement of 10 cm the stresses remain within the limits normally accepted for operating conditions at arch dams. And even when the fault movement reaches 50 cm, local failures occur but the dam does not collapse.

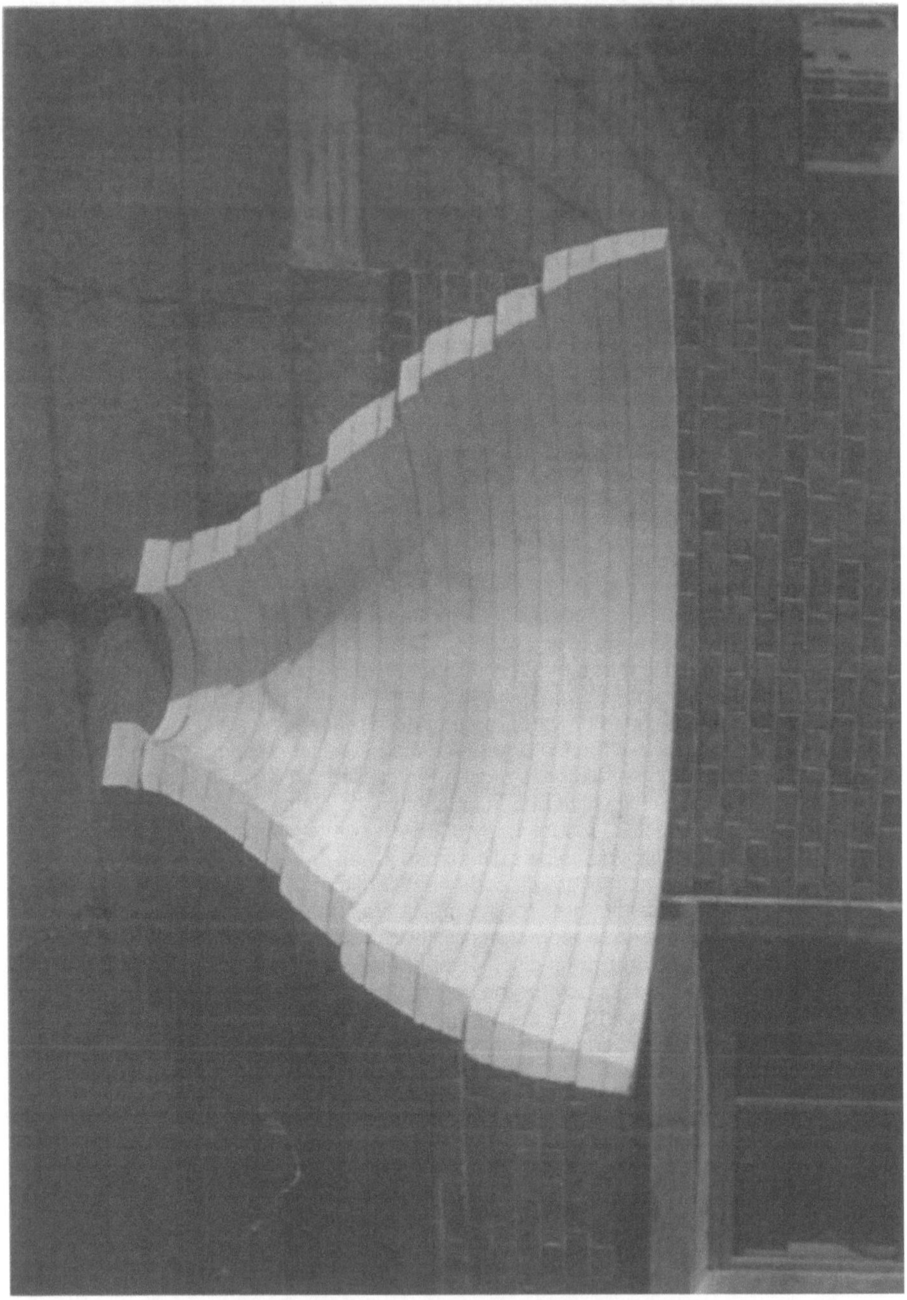

9. Polystyrene male mould for the 1:150 scale model of Contra dam.

10. Victoria arch dam, Sri Lanka.

	Mode number					
Method of analysis	1	2	3	4	5	6
Finite element	1.99	2.32	3.04	3.77	4.05	4.71
MAMA	1.74	2.29	3.01	3.42	4.12	4.59
Impulse SDOF	1.84	2.46	3.36	3.70	4.48	
Impulse MDOF	1.83	2.46	3.36	3.69	4.48	4.71
Multisine SDOF	1.74	2.34	3.04	3.46	4.18	4.64
Multisine MDOF	1.76	2.33	3.04	3.45	4.20	4.63
Prototype test	1.74	1.94	3.00			

Table 3. Contra arch dam. Natural frequencies (Hz) wiith reservoir empty, obtained by various model and prototype tests.

REFERENCES		Arch Dam Model Studies
1		Arch Dams - A Review of British Research and Development. Proc Symp 20-21 March 1968, Instn Civ Engs 1968
2	Allen D N de G Chitty L Pippard A J S Severn R T	The Experimental and Mathematical Analysis of Arch Dams, with Special Reference to Dokan Paper No.6113, Proc. Instn Civ. Engrs. 1955/56
3	Borges J F Pereira J J	Experimental Study of Tang-E Soleyman Dam (Seismic Tests) LNEC Lisbon, Proc 680-11, May 1961
4	Back P A A et al	The Seismic Study of a Double Curvature Arch Dam Proc Instn Civ Engrs 1969 (June), 217-248
5	Severn R T Taylor C A	Earthquake Effects on an Arch Dam by Response Spectra Methods Symp on Vibrations in Civil Engineering, ICE April 1965
6	Taylor P R	The Response of Arch Dams to Earthquakes PhD Thesis, Bristol University 1965
7	Trevelyan J	The Assessment of the Dynamic Properties of Concrete Arch Dams Using Transfer Function Techniques. Ph.D. Thesis, Univ of Bristol, 1984.
8	Selby A Severn R T	An Experimental Assessment of the Added Mass of Some Plates Vibrating in Water E'quake Eng and Struct Dyn Vol 1, 1972, 189-200
9	Serafim J L Pedro J O	Seismic Study of Arch Dams Proc ASCE, ST2, 1965 (April)
10	Pedro J O Pereira J M S	Models for Seismic Analysis of Arch Dam ICOLD 13th Congress, New Delhi, 1979, Q51 R20
11	Pedro J O Azevedo M C	Arch Dam on Active Fault-Model Analysis of Special Construction Arrangements. LNEC Lisbon Mem No 722, 1988
12	Taylor G A Gaukroger D R Skingle C W	MAMA - A Semi-Automatic Technique for Exciting the Principal Modes of Vibration of Complex Structures Roy Aircraft Estab Farnborough UK TR 67211 (1967)
13	Altinisik D	A Seismic Design of Concrete Dams PhD Thesis, Dept of Civ Eng, Univ of Bristol, 1980

FULL-SCALE DYNAMIC TESTING AND MATHEMATICAL VALIDATION

1 Summary The mutual support between dynamic testing and mathematical modelling is the theme of this lecture, in which the point is made that such studies can inform dam behaviour when subject to static loads as well as dynamic loads. The different types of dynamic force input are covered, from rotating eccentric-mass exciters to ambient forces caused by wind or water passing through power station intakes, and occasionally explosions in the reservoir or actual earthquakes. But whichever type of input is employed, the essential task is the measurement of a transfer function between input force and output response. The processing of these transfer functions to produce resonance frequencies, modal shapes and damping values is considered, with some possible sources of error discussed, and an awareness that these dam characteristics are obtained in the linear range at low strain levels.

Illustrations are presented of dynamic testing of arch dams based on the lecturer's own experience (Refs. 1, 2, 3) for the most part, with appropriate reference to other studies. One study presented is that of an intake tower (Ref. 4), an appurtenant structure of vital importance in the ability to drawdown the reservoir for inspection after an earthquake.

2 Introduction Dams are designed on the basis of theoretical models of their assumed behaviour, ranging in the case of arch dams from independent arch analysis, through trial load to non-linear finite element methods . All these methods have one aspect in common; they are tools used by the designer to form an opinion on the safety of the dam, and in view of this, all methods of physical validation are valuable.

Every dam of reasonable size is instrumented in some way, and its continuing existence over a large number of cycles of loading is in itself a full-scale test of, and support for, the theoretical design model. Such tests are too wide in scope to be covered by this lecture, which is restricted to dynamic testing procedures which impose forces on the dam which are small in comparison to those which it normally sustains. Although the essential thrust of such testing is an understanding of the way in which dams will respond to an earthquake, valuable information about static behaviour (eg stiffness) is often obtained. Very occasionally a well-instrumented dam experiences a real earthquake (Refs. 5,6) of sufficient magnitude to produce high-level response, but this must be regarded as an infrequent bonus of particular importance here is the Pacoima dam, which not only suffered damage in the 1971 San Fernando earthquake (Ref. 17), but was affected again in the 1994 Northridge event.

One of the major interests in dynamic studies on thin arch dams, and to a lesser extent on other concrete dams, is the dynamic interaction between dam and reservoir. This manifests itself principally as a hydrodynamic pressure, additional to the static water pressure, on the upstream face, which can be interpreted in terms of a mass which is additional to the mass of the dam itself. This being so, the natural frequencies of the dam fall as the reservoir level rises, and frequency measurements are therefore an indirect assessment of hydrodynamic pressure.

However, sufficiently sensitive transducers have become available during the last 10 years to make possible the direct measurement of this pressure on the reservoir-dam interface (Refs. 4,7), allowing problems such as cavitation and the effects of water compressibility to be studied. A particularly valuable contribution here (Ref. 7) indicates the crucial factor as being the ratio of the first resonant frequency of the water only to the first resonant frequency of the dam-foundation system without water. At a value less than 1.1 compressibility effects are significant, but fall to zero as this ratio rises above 1.5.

3 Aims of Full-Scale Testing The overall aim of full-scale testing can be summarised as that of structural identification, in which all aspects of the structural problem, from loading through material behaviour to structural mechanisms, can be characterised. Integrity monitoring is a phrase which is often used, including defect location, covering a technique which has found ready application in mechanical and aerospace engineering. Its application to dams is more difficult, due to size, the relatively coarse tolerances to which they are built, the variability of properties within the same material, notably concrete, and the harsh environment in which the structure often exists. For some simple structures. such as piles, fault location through dynamic testing is now used commercially, but for dams, precise fault location would seem to be a distant hope.

To come to the details of full-scale dynamic testing, it may be said that the essential aim is to determine the structural response to one of a number of dynamic force inputs, from which the fundamental dynamic characteristics namely natural frequencies, modal shapes and damping may be determined. The response referred to here can be displacement, velocity, acceleration, strain or stress, but whatever it is, the essential objective of dynamic testing is that of measuring a 'transfer function' between some excitation quantity and some response quantity, at various locations on the dam. From this transfer function it will normally be possible to extract the modal properties previously referred to.

4 Excitation Methods in Dynamic Testing The available methods can be divided into two main categories, the first involving excitation with a force of known magnitude and frequency content, whilst the second uses ambient forces arising from wind, wave, traffic and water passing through the dam, which are not precisely defined. The former is generally to be preferred since a true transfer function between input force and response can be obtained. For ambient vibration some assumption must be made as to the frequency content of the input force in order that the transfer function, which is the response normalised to unit input, can be estimated. Usually, an attempt will be made to measure the ambient input, such as wind velocity, to determine its frequency content; however, this is often statistically non-stationary and some average value has to be taken, a process which involves considerable numerical effort.

5 Forced Excitation (i) Controlled The most common form of controlled forced excitation uses rotating eccentric mass (REM) exciters, which are suitable only for steady state excitation. Refined REM systems with accurate speed control and contra-rotating masses were introduced first in 1961 in California (Ref. 8). The system shown in Fig. 1 was built by the present lecturer (Ref. 9) in 1975. It consists of four machines, operating in the 0-20 Hz range, producing a sinusoidal force in the horizontal plane which depends on the masses attached to the contra-rotating spigots, which at a

1. Bristol rotating eccentric-mass exciter in gallery of Emosson dam

maximum reaches 10 KN for each machine. The machine is attached to a steel ring by clamps, which in turn is attached to the concrete of the dam by six bolts. The direction of the sinusoidal force can be changed in azimuth by rotating the machine on the steel ring. The digital electronic control system ensures that all four machines operate at the same frequency to within 0.001 Hz and 0.01 radian in phase. The phase can be varied between exciters however, this being an essential feature in exciting antisymmetrical nodes in the dam.

An alternative form of controlled excitation can be produced when masses reciprocate along an axis, rather than rotate around a shaft. In principle, any kind of force time-history can be produced by such a device, which is normally a hydraulic actuator, but digital computer control is necessary for accuracy. The low frequency force output is dependent on the maximum stroke of the actuator, and with a long stroke the low-frequency performance can be superior to REM output. As an illustration, a 1m-stroke actuator driving a 26-tonne trolley proved capable of destroying an 11-storey block of flats! REM machines are not capable of such forces, but they have the advantages of relatively small size, simple electrical power requirements (240V, 13 amp) and low cost. In contrast, hydraulic actuators require a large pump of considerable weight and power consumption.

6 Forced Excitation (ii) Uncontrolled As far as dams are concerned we include here only explosions in the reservoir or natural earthquakes, but for general dynamic testing we add impulses produced by a dynamically tuned hammer equipped with a transducer which measures the input force, or by vehicle impact. For impulse testing the nature of the impulse function implies that energy is distributed over a broad bandwidth, the sharper the impulse, the wider the bandwith.

7 Ambient Excitation Ambient forces are those over which the tester has no control, examples for dam studies being wind, and forces caused by reservoir water passing through the dam. The deep narrow valley and warm airface of an arch dam produces air currents which are sufficiently strong to produce measurable response in modern transducers (Ref. 9) even though it is several orders of magnitude less than that caused by forced vibration. For ambient forces the characteristics of the excitation are usually difficult to obtain, only a statistical description being possible. However, ambient vibration surveys are becoming increasingly popular due to the availability of multi-channel signal processors and compact data acquisition and storage equipment. Experience shows (Ref. 1) the reliability of the frequency and mode-shape data obtained. The main difficulty associated with ambient excitation is that of accuracy of damping values. Several algorithms have been proposed (Ref. 10) for obtaining damping estimates from ambient response data, but these do not necessarily circumvent the problem of obtaining data from non-deterministic signals. However, since damping can only be obtained from full-scale tests of some sort, all information is valuable.

8 Measurement of Response In general, response can mean any measurable parameter, but for dams acceleration in the body of the dam is normally measured, with pressure measurements at the dam reservoir interface when possible, discriminations of 10^{-6}g and 1.5N/m^2 being achievable.

Using REM, frequency sweeps with the machines located appropriately will serve to determine natural frequencies as peaks of the transfer function. For this purpose it is strictly only necessary to use one accelerometer, so long as it is not located at a node point of the modal shape, but deployment of several provides necessary corroboration. Having isolated a natural frequency its corresponding mode shape can be mapped, with the method used depending on the number of accelerometers available. If this is large, they can be distributed throughout the dam at the beginning of the test, measuring amplitudes of response simultaneously as the machines rotate at the resonance frequency. But since mode shapes represent relative amplitude, it is vital that all accelerometers have been calibrated to give the same performance.

If the available accelerometers are small in number, two as a minimum, a different procedure is required, in which one accelerometer remains stationary, preferably at an anti-node, whilst the other is moved from point to point throughout the dam. During such a test, which might last for several hours, the machines are rotating at the resonance frequency. As before, the accelerometers must be calibrated, but this is now a simpler process which can be carried out frequently during the test. An advantage of this 'travelling accelerometer' approach is its flexibility in selecting measuring points as the tests proceeds, using radio contact.

9 Transfer Functions Dynamic testing is essentially that of measuring the transfer function (or frequency response function) between an input quantity and a response quantity. For a single degree of freedom oscillator the undamped and damped natural frequencies are

where β is the viscous damping ratio equal to $C\ (M/K)^{1/2}/2$. Fig 2 shows the transfer function for forced input, being the ratio of the displacement (velocity or acceleration) response to a unit force plotted against f/fn where f is the frequency of the input force. Fig 2 gives the transfer function for unit ground acceleration. The transfer functions peak at the natural frequency of the structure and the amplitude of response at this resonance is governed by the damping. Real structures are multi-degree -of-freedom systems of course, and the measured response normally consists of overlapping contribution from several modes, superposed on which is measurement noise and other unwanted contributions. The skill of the experimenter consists of abstracting the maximum amount of useful information from transfer functions of this type.

10 Measurement of Transfer Functions Transfer functions can be measured in a number of ways. The simplest is to excite the structure continuously with a force of known amplitude and frequency, measuring the amplitude of response at that frequency. This yields one point on the transfer function curve, other points being found by changing the excitation frequency. Very simple data acquisition equipment is required, normally a digital storage oscilloscope will suffice. It is, however, a laborious operation.

When a digital spectrum analyser is available, a more sophisticated approach can be employed in which the response to a wide variety of excitations can be measured. Multiple response signals are recorded and processed to give the Fourier spectrum of each signal, a combination of which produces the required transfer function. If x(t) and y(t) are the input and output signals, the Fourier spectra of the former, X(f), is given by

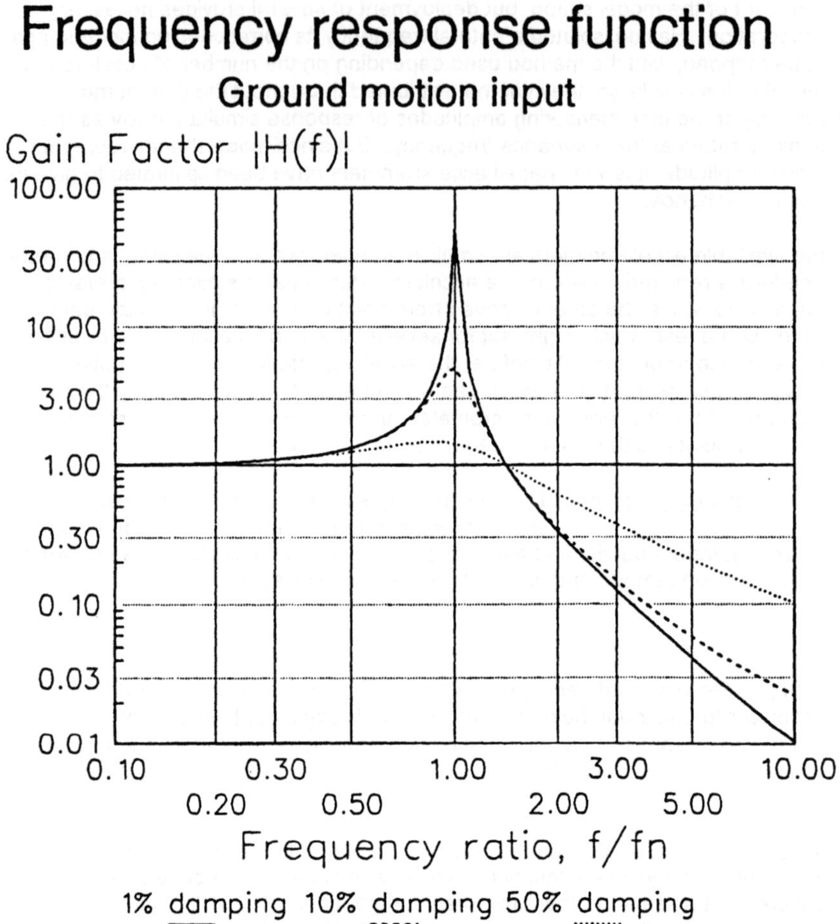

2. **Transfer function for unit ground acceleration**

and similarly for y(f), from which the autopower spectra, $S_x(f)$ and $S_y(f)$, and crosspower spectra, $S_{xy}(f)$, can be defined as follows.

where the asterisk denotes complex conjugate. The crosspower spectrum is complex valued and is a measure of the degree of causality between the input and response signals. Its modulus is the product of the moduli of X(f) and Y(f), whilst its phase is the difference in phase between these two functions. The transfer function required, H(f), is formed from

$$H(f) = S_{xy}(f)/S_x(f)$$

which gives the magnitude of the response for a unit input at each frequency. Peaks in the modulus of H(f) correspond to the natural frequencies of the dam, whilst the spread of each peak is a measure of the damping in that node.

11 Errors in Transfer Function Estimation The two most important sources of error are bias and variance of the spectral estimates. In practice, spectral estimates have a finite frequency resolution; this must be fine enough to define accurately the true peaks corresponding to natural frequencies. Fig 3 illustrates the problem, the most significant effect being an overestimate of damping. The solution is to increase the frequency resolution.

If a statistically stationary signal x(t) is divided into a number N of equal length segments and the S(f) of each is calculated, the normalised error of the mean spectrum is $1/N^{½}$. Thus, to improve the accuracy of the spectral estimate, effectively eliminating noise, many segments must be averaged. This implies many hours of testing and the processing of large volumes of data. Over such a long period stationarity is called into question, so that a compromise must be reached.

12 Emosson The Emosson Arch Dam is in Switzerland, although the pumping station is in France. It was not designed for seismic loads but experienced minor (M<3) reservoir induced earthquakes in 1973-74. Post-construction calculations using a finite element model sought corroboration from dynamics tests. Fig 4 shows that it is 180 m high with a right wing dam of gravity type, making it acutely asymmetrical. Vertical blocks are numbered 1 to 26 and there are horizontal galleries at 21 m intervals, which facilitate the mapping of mode shapes.

The study of Emosson was a joint project of the UK Building Research Establishment (BRE), Bristol University and Motor Columbus. The first was responsible for decisions on site as to the occurrence of natural frequencies and mode shapes, while Motor Columbus carried out the finite calculations three months before the site visit. The University's role was that of co-ordinator. This information is given for two reasons. First, BRE did not know of the computed predictions until after the completion of the site work; in other words, the computed modal characteristics were in no way allowed to prejudice what was measured. Second, the computer program used a coupled reservoir/dam formulation, for which it was necessary to estimate the water level. This was assumed to be 26.5 m below crest-

Predicted frequency			Measured frequency	
Before testing	'A' adjusted	'B' adjusted	Induced vibration	Wind excited
2·09	1·85	1·918	1·85	1·918
2·22	1·96	2·03	—	2·11
—	—	—	—	2·7
3·33	2·94	3·05	2·96	3·0
—	—	—	—	3·2
4·03	3·57	3·70	—	3·5
4·52	4·00	4·15	4·07	4·115
5·05	4·47	4·63	—	4·66
5·99	5·30	5·50	5·32	5·3
6·58	5·82	6·04	—	5·8
7·09	6·27	6·51	—	6·2
7·30	6·46	6·70	—	6·7
—	—	—	—	7·3
—	—	—	8·26	7·9

'A' adjusted values are obtained by multiplying the pre-testing predictions by 1·85/2·09.
'B' adjusted values are obtained by multiplying the pre-testing predictions by 1·918/2·09.

Table 1 Emosson Dam: Comparison of calculated and measured frequencies (Hz)

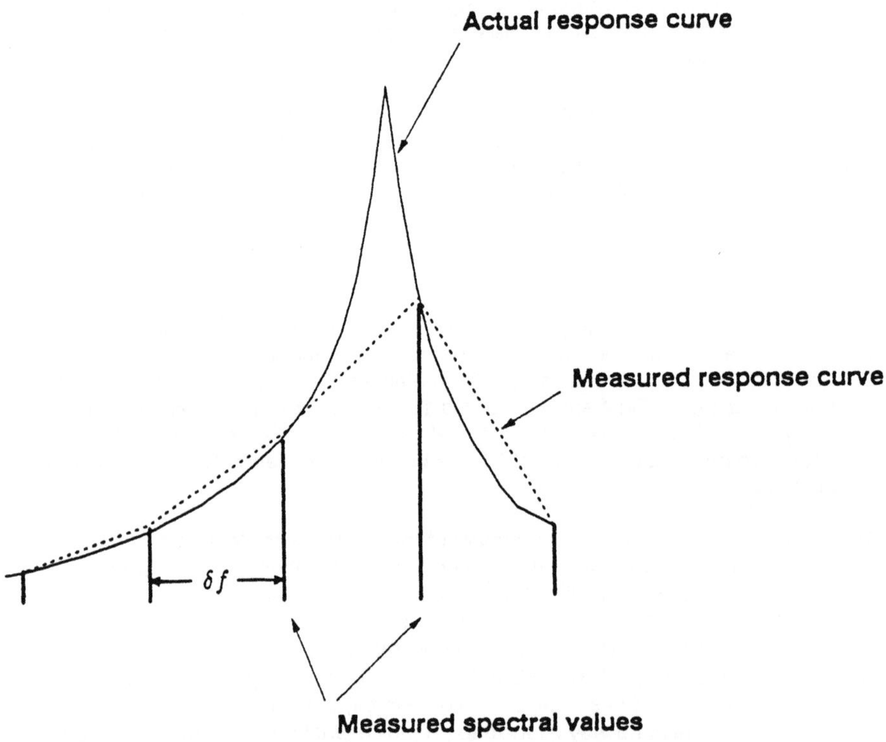

3. Increased damping estimate caused by too large a spectral interval

level, whereas during testing it was actually 7 m below. This difference is referred to again when comparisons are considered.

13 Emosson: first visit (1980) For all the forced vibration tests, the exciters were placed in the highest gallery (1926 m). Twelve positions were drilled at which an exciter could be located, four of them close together in central block 15 (Fig 4) and therefore the position at which maximum force should be deployed to excite symmetrical modes. The remaining eight positions were distributed uniformly across the gallery and were used as appropriate in the exploration of mode shapes.

Throughout the forced vibration tests a background vibration wax observed, unaffected by heavy wagons crossing the crest and therefore assumed to be due to the strong wind which blew steadily up the downstream face. When the exciter tests had been concluded, the opportunity was taken to record this ambient vibration overnight on two accelerometers placed at the 1926 m level in blocks 10 and 14. This action was speculative, since no spectral analysis facilities were available on site and anything of value which these records contained would only appear after processing at BRE. In fact, they were of great value, producing a range of natural frequencies to compare with those calculated and those measured. The two measurements did not, however, permit mode shapes to be deduced.

14 Discussion of Results Table 1 is a comparison of calculated and measured frequencies. It must be noted the calculated values were obtained first, for which a dynamic modulus of concrete was taken as 3.75×10^6 t/m^2. This value was also used for most of the rock, but in the upper and lower parts of the right abutment 2.34×10^6 and 1.87×10^6 t/m^2 were used. These were estimates, as was the reservoir level previously referred to . The first column of Table 1 gives the calculated frequencies based on these assumptions, whereas column 4 gives the frequencies measured using the exciters and column 5 those from the spectral analysis of wind excitation. If the disparity between the first entries in columns 1 and 4 is assumed to be due to incorrect assumptions in the calculations, that is, the measured value is correct, the factor 1.85/2.09 applied to the first column gives the second column, of corrected values for comparison with other measured frequencies in column 4. The agreement is then seen to be acceptable. In a similar way, it might be assumed that the first wind -excited frequency is correct, in which case the factor 1.918/2.09 applied to column 1 gives column 3 for comparison with column 5.

Comparisons for the first two measured and calculated mode shapes is given in Figures 5 and 6. Reference 1 gives the remainder and all are remarkably close.

A significant feature of the forced vibration results is that it was not possible to excite the first symmetrical mode, with any disposition of the exciters. But it was obtained from the wind excitation (2.11HZ) and was calculated at 2.22 HZ. An explanation of this failure is the lack of symmetry of the dam, coupled with the fact that in model tests on symmetrical arch dams, the first symmetrical and asymmetrical modes are closely spaced in frequency and the model vibrates preferentially in the latter unless the multiple exciters are carefully positioned. Figure 7 shows a spectral analysis of response due to wind in an accelerometer in block 10 at level 1926 m. There are clearly two peaks, at frequencies corresponding to the first symmetrical and first asymmetrical frequencies. But, in the absence of other information, it is possible to distinguish between them only

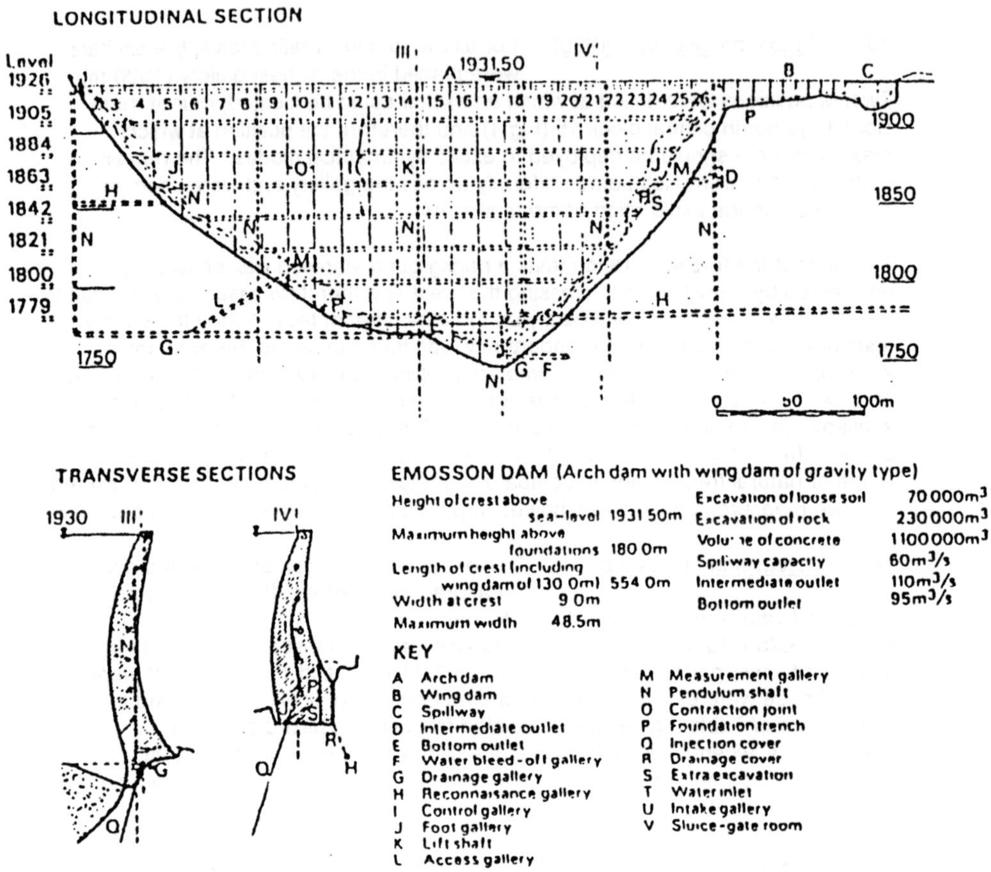

4. Emosson arch dam

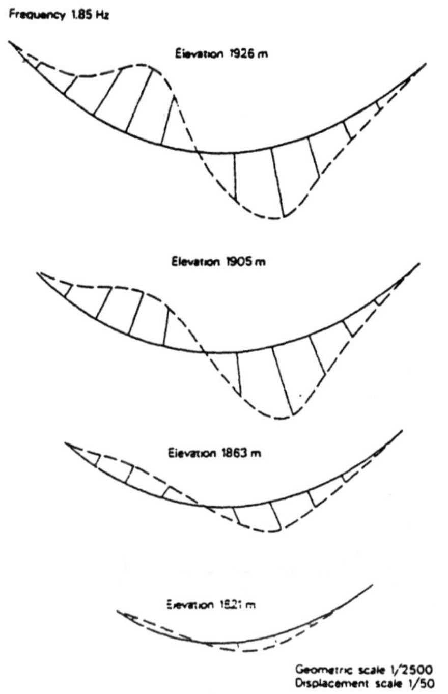

Figure 5(a). Emosson dam—test results August. 1980: period 0·541 s

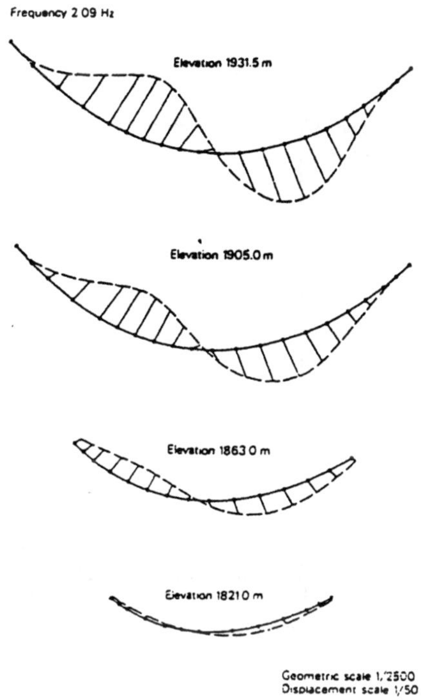

Figure 5(b). Computed displacements for mode 1 period 0·478 s

5. Emosson: Comparison of measured/calculated first antisymetrical mode

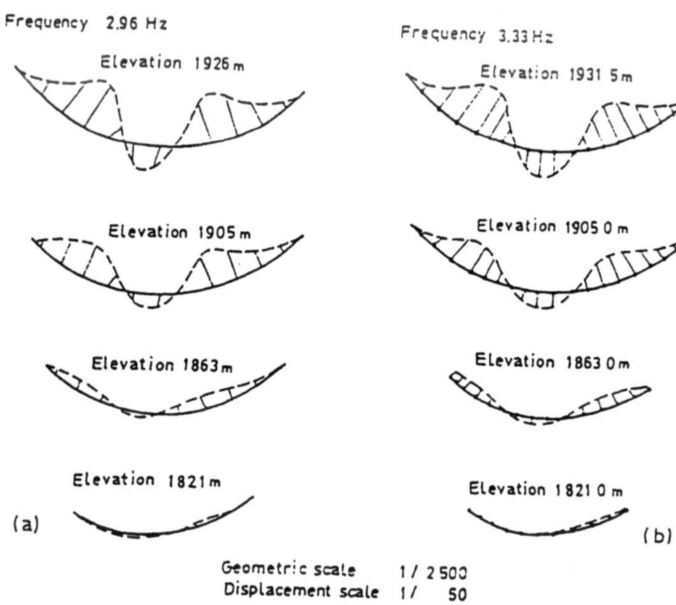

6. Emosson: Comparison of measured/calculated first symmetrical mode

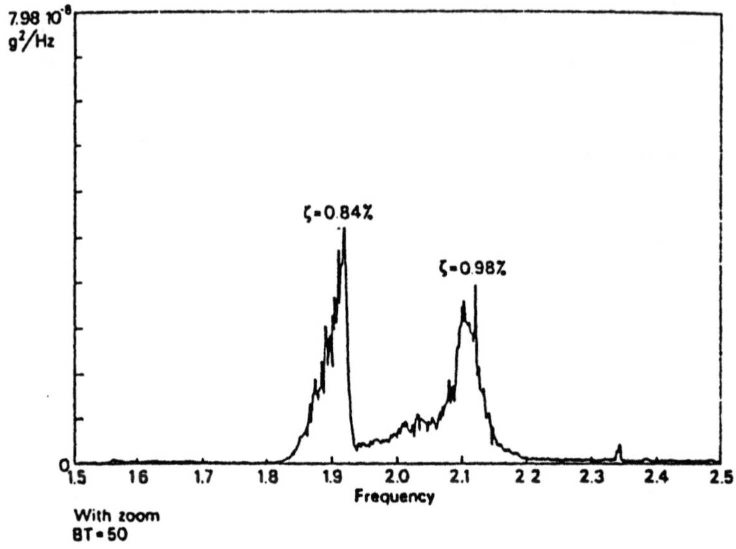

7. Emosson: Spectral analysis of response due to wind

from a knowledge of phase and two accelerometers do not give sufficient information. It was therefore decided to make a second visit to Emosson, concentrating on a detailed study of wind excitation.

15 Emosson: second visit (1982) One interesting and surprising feature of the wind response data on the 1980 visit to Emosson was its stationarity, and a more protracted assessment of data from the second visit showed it to be so again, which means the spectral analysis can be pursued with confidence, particularly relating to mode shapes.

On this second visit, accelerometers were deployed at level 1926 m in blocks 6, 12, 15, 18 and 24, and in block 18 at levels 1884 and 1842 m. These seven positions were specially selected to resolve the ambiguity in the first two modes mentioned above and, although space precludes a full discussion of results, it was clearly established from phase comparisons that the first frequency was that of an asymmetrical mode, with the seven accelerometers giving a value of between 1.85 and 1.90 Hz, and that the first symmetrical mode was at 2.1Hz 2. Table 1 gives other results from the spectral analysis, for frequency only, because the limit of seven accelerometers imposed by the tape recorder made it impossible to discern mode shapes for all frequencies.

16 Contra Contra Dam is situated in Ticino, Switzerland. It also experienced reservoir induced seismicity in October 1965 during first filling; the reservoir was drawn down but no damage was found. No further RIS has been reported but an assessment of possible seismic loading was part of normal safety considerations. It is of double curvature form, with height 220 m and crest length 380 m. Its thickness varies with height, from 25 m at the base to 7 m at the crest. The horizontal galleries and the vertical block numbers are indicated in Figure8. A 1:150 model of Contra was built, including foundation, and tested dynamically with various reservoir level. Finite element calculations were made for comparison with both the model and prototype measurements.

17 Contra: first visit (1982) A first visit to Contra was made immediately following the second visit to Emosson and a similar dynamic survey was intended, using both forced and ambient excitation. Unfortunately, during the first night, with the exciters and cabling already in position, a violent storm with much lightning damaged all the control electronics. As a result, only two exciters could be used, and this led to difficulty in determining higher mode shapes. The storm did however have a compensating benefit: the reservoir level rose from 445.15 m to overflow level at 470.0 m, and beyond, in 36 hours, enabling ambient measurements to determine reduction in frequency with increasing water level.

During the forced vibration tests, 18 modes were located but only the ten lowest were studied in detail. The characteristics of these are given in Table2, together with two calculated frequencies. Because the water level was changing during the testing, it is necessary to record its value in column 2. All except the first two frequencies were also measured at full reservoir and these are given in column 3. The values here for the first two frequencies are estimates; their measured mode shapes appear in Figure 9.

For ambient vibration testing at Contra, seven accelerometers were deployed along the gallery at 440 m, and on in block 10 of each of the galleries at 414, 336 and 315

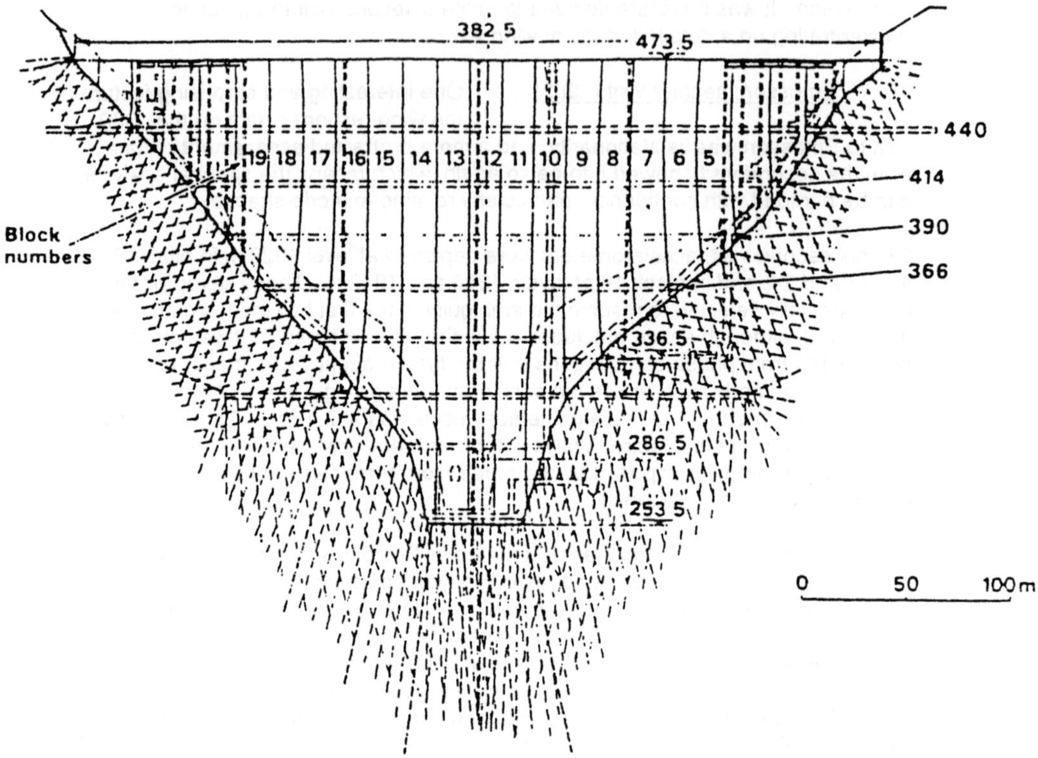

8. Contra: profile showing galleries and vertical blocks

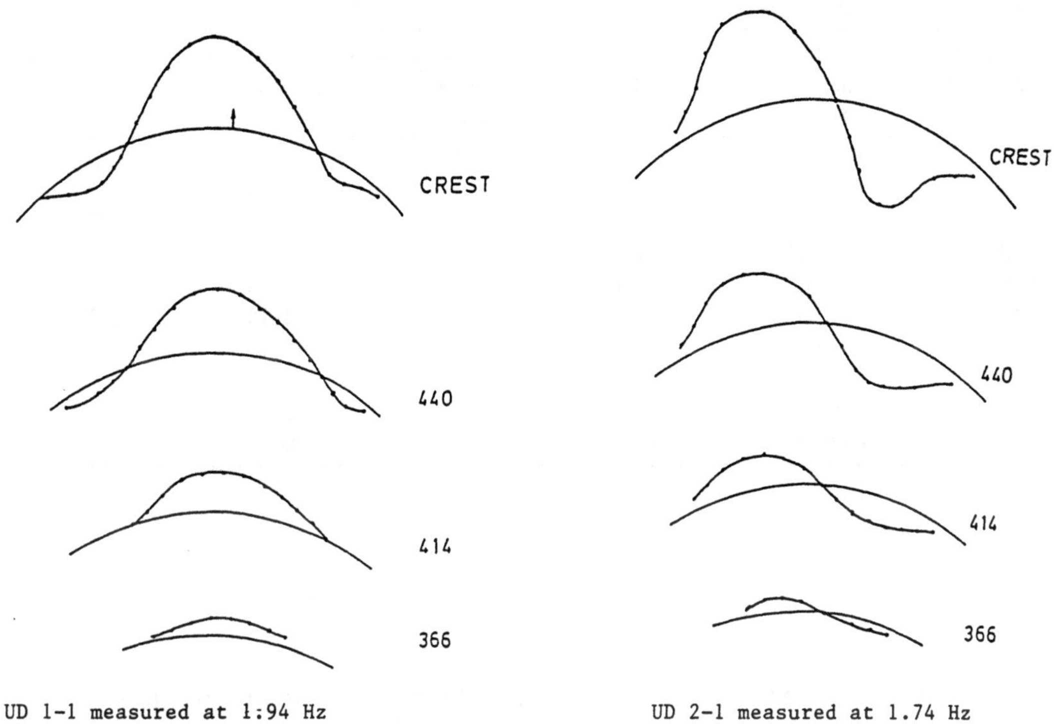

9. Contra: the first two measured mode shapes

Table 2·Contra Dam. Characteristics of the first ten forced vibration modes.

Measured res. frequency Hz	Res. water height m	Frequency at full res. Hz	Calculated frequency (full res.) Hz	Damping % crit	Mode class
1.94	454.05	1.77E	1.79	1.72	UD 1-1
1.74	454.05	1.58E	–	1.76	UD 2-1
2.50	454.75	2.28	2.42	4.18	UD 3-1
3.68	470.00	3.68	–	–	UD 5-1
4.60	470.00	4.60	–	1.08	UD 6-1
3.10	470.00	3.10	–	1.45	UD 2-2
4.00	470.00	4.00	–	1.05	UD 3-2
3.00	455.30	2.62	–	2.69	–
3.94	454.75	3.24	–	2.68	–
4.28	470.00	4.28	–	0.91	–

Table 2 Contra Dam: Characteristics of the first ten forced vibration nodes

m, Unlike the data obtained at Emosson, it proved to be non-stationary (probably because of the changing water level during the tests) and was not used for mode shapes. Natural frequencies can be obtained from non-stationary data and these were found to be consistent with corresponding frequencies from the forced vibrations tests, if allowance is made for different water levels. One particularly valuable and unique result is presented in Figure10, showing the change in frequency of the first asymmetrical mode (UD 2-1) as the water level changed.

18 Contra: second visit (1985) Between 1982 and 1985, improvements had been made in our equipment; in particular, it had become possible to analyse ambient data on site (Ref. 11). The second visit to Contra, therefore, concentrated entirely on ambient response. This was expected to derive from the wind but nothing was measurable above noise level. Instead, a clearly measurable intermittent response was being recorded which eventually proved to coincide with water flowing through the dam for power generation. From the 33 recordings made, at stations throughout the dam, eight upstream-downstream modes were clearly identified. Three were symmetrical modes at 2.03, 2.63 and 3.1 Hz for a reservoir level of 444.5 m, which are consistent with previous results. A mode at 1.83 Hz was probably the first asymmetric mode, but four others could not be classified simply.

19 Tests on a Reservoir Intake Tower Damage to an intake tower during an earthquake could result in failure to draw down the reservoir, with possible consequential loss of the dam itself. Rational seismic design methods are therefore important. One major issue here, as with the dam itself, is the part played by the surrounding water in the dynamic behaviour. Analytical solutions (Ref. 12) have shown that for compressible water the hydrodynamic interaction can be represented by an added mass plus an added damping of the tower. The latter comes from propagating pressure waves which carry energy away from the tower. Such waves can only occur at excitation frequencies higher that the fundamental resonance of the reservoir, which for constant depth, H, and infinite length is given by (C/4H) Hz, where C is the speed of sound in water (1440 m/s). For incompressible water C is infinite and so radiation damping does not occur.

Few attempts have been made at experimental validation of these theories. For the San Bernadino (Ref. 13) tower the study was limited to the fundamental bending mode of the tower which had a natural frequency well below the critical frequency referred to above, at which compressibility becomes theoretically important. Laboratory tests also fall into this category (Ref. 12). The study reported by Bristol colleagues in Ref 4 is therefore particularly valuable, the purpose of these dynamic tests on the Wimbleball (45 m) intake tower being to produce data which could be used to validate the theory that the compressibility of water was not important in predicting the seismic response of such towers. This was done by comparison of measurement and finite element, calculation (without compressibility) for the two bending modes which would dominate the response of the tower during an earthquake. The natural frequencies of these modes were below, and close to, the so-called 'cut-off frequency' at which compressibility becomes important. The results are givern in Figs. 11 and 12. During the tests the reservoir level was at 37.5m, the cut-off frequency (C/4H) for which is 9.67 Hz. Fig. 12 shows the first and second binding frequencies of the tower to be somewhat less than 2 and 8 Hz, respectively But Fig. 12 does also show that beyond the cut-off frequency the peaks are less well defined, this being a characteristic of either closely coupled

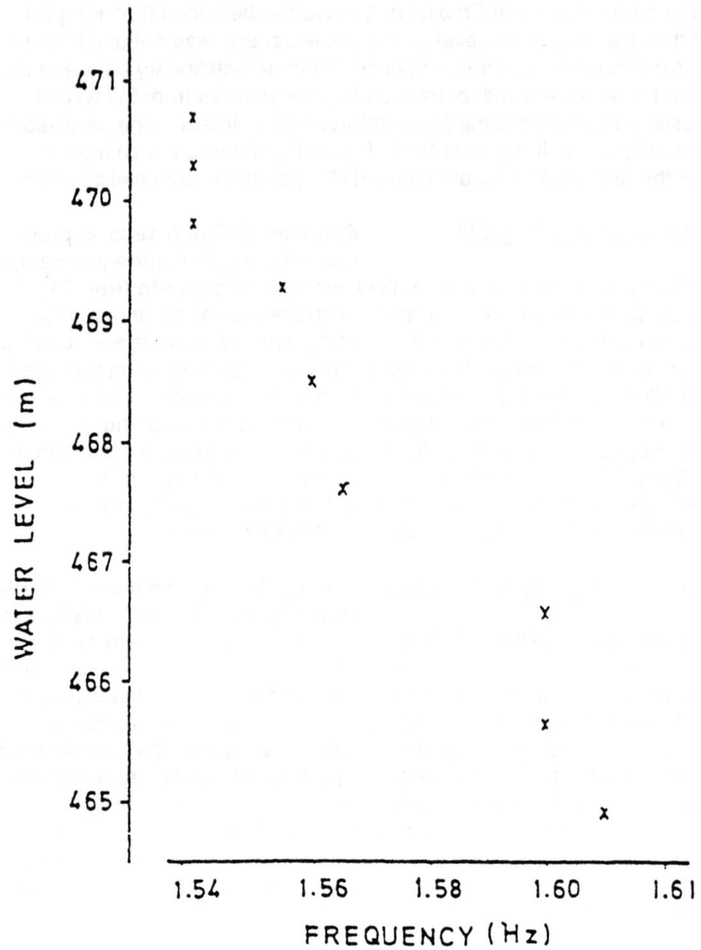

10. Contra: Frequency for UD 2-1 mode versus reservoir level

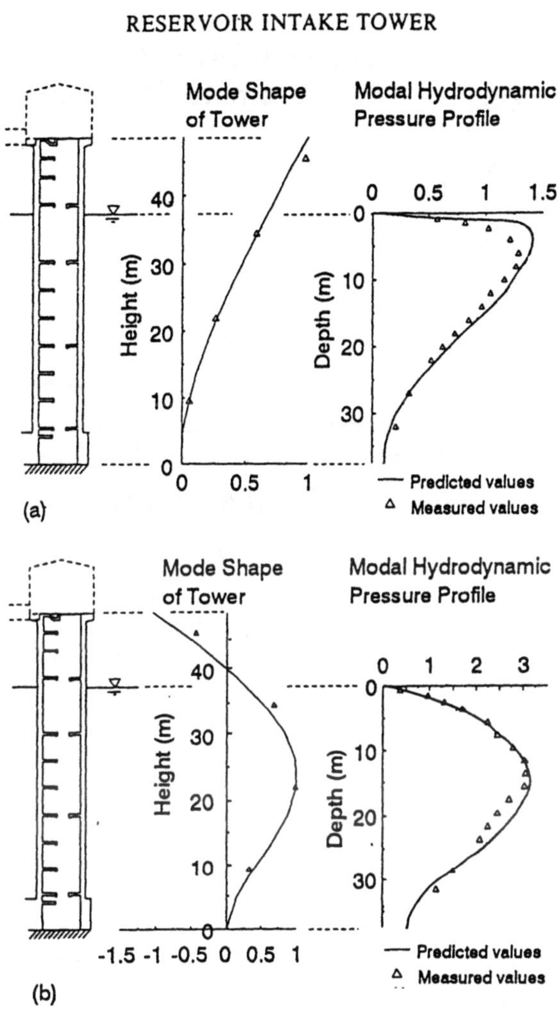

11. Reservoir intake tower: first two bending mode shapes and hydrodynamic pressures

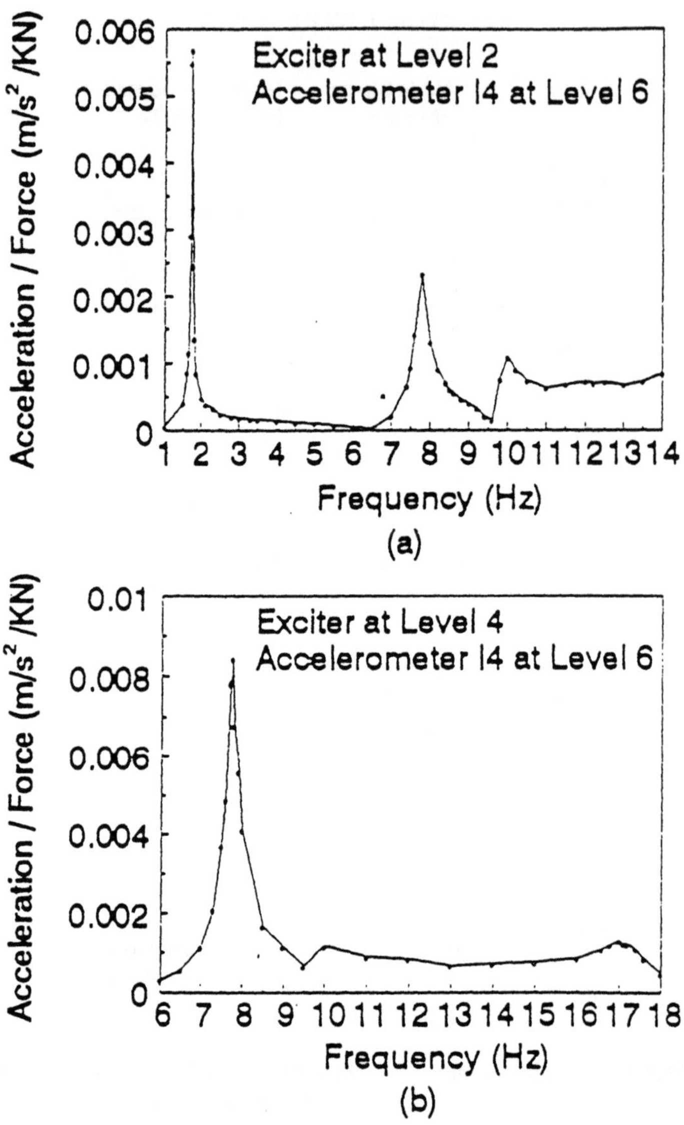

12. Reservoir intake tower: frequency response curves form forced vibration tests

modes or high levels of damping. The latter could have been caused by radiation damping through the reservoir, which has been shown theoretically to occur at frequencies above the cut-off frequency.

.20 Pacoima Arch Dam (Ref. 17). This dam is of special interest because it experienced the M=6.6 San Fernando earthquake of 9 February 1971, being only 6 km from the epicentre of the earthquake, which occurred at a focal depth of 12.8 km. The dam is a thin double curvature arch, 113 m high with crest length 180 m; its crest and base thickness are 3.2 and 30.2 m, respectively. Construction was completed in 1928, the stress analysis being based on a crown-cantilever trial-load approach, but without seismic effects. In 1967-68 a major review was carried out using the radial deflection trial load method with gravity, full reservoir and a 0.15 g horizontal acceleration in the upstream-downstream direction to represent earthquake loading. The only concern raised by these studies was the stability of the left abutment if the earthquake acceleration was increased beyond 0.15 g.

Following this work, two strong-motion instruments were installed, one at the crest centre and one on a rock ridge above the left abutment. The 8 second duration earthquake of 1971 produced the strongest accelerations ever recorded in these instruments, being 1.25 g for both horizontal components and 0.7 g for the vertical component. Subsequent studies concluded that the base rock acceleration might have been in the range 0.6-0.8 g. Precise survey after the event showed a narrowing of Pacoima gorge by 23.9 mm and a lowering of the right abutment by 17.3 mm relative to the left. However, the only visible damage was the opening of a radial contraction joint between the arch, and the left abutment thrust block and a crack in this block. The joint opening was 6.10 mm and extended from the crest to a depth of 13.7 m where it stopped at a horizontal lift joint. At the time of the earthquake the reservoir level was 45 m below crest. All six piezometers responded initially with a sudden increase in water level, followed by an abrupt decline, but the new stabilised levels were higher in three cases, and lower in the other three, that the pre-earthquake levels. This was attributed to pressure effects and random re-arrangement of joint openings and seepage passages.

Immediately after the 1971 earthquake the Pacoima dam was re-analysed using both the full trial-load approach and the 3D FEM, with the 0.15 g horizontal acceleration as in 1968. It is interesting that the stress patterns given by the two methods were in close agreement, and for both tension and compression within the allowable limits for Pacoima concrete.

Special dynamic response analyses were carried out for two scenarios:

(i) Full reservoir and an M=8 event on the San Andreas fault 30 km from the dam. Peak ground acceleration 0.28 g.
(ii) Full reservoir and an M=6.5 event on the same fault as the 1971 event but 20 km from the dam. Peak ground acceleration 0.32 g.

All three components of ground acceleration were applied simultaneously at the foundation, and Westergaard's added mass approach was used to represent the dynamic effect of the reservoir. The stresses calculated for both scenarios were within the dynamic strength of the concrete, and they were also less than those calculated for the situation obtaining at the time of the 1971 event with the 3-coordinate input motion scaled by two-thirds to allow for the influence of the siting of the recording instruments as previously discussed.

In association with the remedial works, an extensive monitoring system was installed which included an array of 11 accelerometers distributed throughout the dam, its foundations and abutments, 6 of which are of the triaxial type. A water pressure transducer was installed on the upstream face at about mid-height of the crown section. During the controlled reservoir re-filling, no untoward observations were recorded.

It is important to note that Pacoima arch dam came safety through the 1971 San Fernando event, which caused severe damage to two old hydraulic earthfill dams which form part of the Van Norman reservoir complex, roughly twice as far as Pacoima from the epicentre. The near failure of these two dams has caused strenuous activity within California dam engineering during the last 20 years, with the Lower Crystal Springs dam study being noteworthy. This curved concrete gravity dam pioneered concrete dam design in the USA, and safety withstood the M=8.3 event of 1906 originating on the San Andreas fault. The conclusion of recent studies is that no remedial work is required for the dam or its foundation for either the maximum expectable earthquake (50 sec, M = 8.3, maximum acceleration 0.6 g) or the maximum credible earthquake (60 sec, M = 8.5, maximum acceleration 0.7 g).

The Pacoima dam was attacked again in 1994 by the Northridge earthquake, resulting in repeat cracking between the arch and the left abutment thrust block. Major rock slides damaged ancilliary structures, but the dam itself remained intact.

21 Italian Arch Dams (Ref. 15) Three, thin double-curvature arch dams were in the area affected by the Friuli earthquake of 1976, in which the acceleration exceed 0.3 g and M = 6.2. The Ambiesta arch dam is particularly interesting because a geophone network, installed immediately after the main event, was able to record a number of aftershocks, and actual response could be compared with FEM calculations using modal response analysis. Fig shows such a comparison for crest acceleration using 2% damping. The good agreement is remarkable, if only for the fact that no account was taken of the dynamic effect of the reservoir, not even by Westergaard's approach. It may have been that the reservoir was at a low level during the aftershocks, but no information is given.

Two valuable conclusions are drawn from these Italian studies:

(i) Displacements and stresses in the three arch dams studied are approximately the same for an earthquake like Fruili as they are for a full reservoir head.

(ii) Arch dams, being designed to take horizontal load by arch action, are better able than other types of dam to resist earthquakes which produce predominately horizontal motion.

22 Morrow Point Dam (Ref 7) This double-curvature thin arch dam in Colorado USA is 465 ft high, with crest and base thickness of 12 and 52 ft respectively. Its almost perfect symmetry was considered to allow considerable economy of computational effort by modelling only

one half of the dam, and imposing either symmetrical or antisymmetrical boundary conditions on the centreline.

Forced vibration tests were carried out with two large REM exciters of 5000 lb force capacity located on the crest at the crown cantilever, with force orientation upstream-downstream for symmetric modes, and cross valley for antisymmetric modes. The reservoir throughout the testing was only 9 ft below crest level and the principal aim of the study was to examine hydrodynamic interaction between dam and reservoir. It is suggested that although many previous forced vibration tests have produced agreement between measurement and calculation with water compressibility neglected, they have often done so by adjusting material parameters of dam and foundation, and damping, in order to do so (Ref. 1), whereas some of the disagreement might be due to neglect of compressibility of water. In this Marrow point study it is argued that because antisymmetric modes are not greatly affected by compressibility, the adjustment of material properties should be based on them; the symmetric modes can then be used for a detailed analysis of compressibility. Using f_w to represent the first resonance frequency of the water only, and f_d for the first resonance frequency of the dam-foundation system without water, if $f_w/f_d > 1.5$ then water compressibility is not important, but it becomes so if this ratio is less than 1.1. Such criteria are to be applied separately to symmetric and antisymmetric responses.

The following conclusions are taken from Ref 7.

(a) An attempt to isolate the response of the fundamental water mode appeared to be successful, and a strong resonance was observed.

(b) Damping values determined by the half-power method from the experimental data at full reservoir were approximately 1.5% for the first two antisymmetric resonances and approximately 4% for the first three symmetric resonances. A possible explanation for the higher damping values in the symmetric response is the added damping associated with water compressibility. It is also possible that radiation damping associated with dam-foundation interaction could be greater for symmetric vibration.

(c) The fit of the numerical results to experimental data for the antisymmetric frequency responses at full reservoir resulted in a slightly better match at the observed second antisymmetric resonant frequency when water compressibility was included in the analysis. This frequency is close enough to the first antisymmetric resonance of the water to be affected by the increased added mass from the water resonance.

(d) Prediction of the symmetric frequency response data at full reservoir after calibrating the numerical model with the antisymmetric data yielded poor results with the incompressible water model. Inclusion of compressibility improved the correlation, which is not surprising since the first symmetric frequency of the water is below the first symmetric frequency of the dam-foundation system, indicating the potential for strong water compressibility effects.

(e) The first two measured symmetric resonating shapes of the system, being similar for the dam, differed primarily in the response of the water domain, which is possible only with compressible water. This feature was captured qualitatively by the compressible water model but not by the incompressible water mode.

(f) Prediction of the symmetric frequency response data obtained for the partially full reservoir (158 ft below crest) also improved when compressibility was included. Here the water resonance frequency is low enough to affect significantly the second symmetric resonant frequency of the system.

23 Kurobe Dam (Ref. 6) This 180 m high arch dam in a seismically active zone was completed in 1963 and is unique from a seismic point of view because of the comprehensive array of instruments for dynamic measurement installed in the dam, and the volume of data which has been recorded during the past 30 years. Forced vibration tests have also been carried out, giving from the total study the most comprehensive picture of the dynamic behaviour of any arch dam. Only a brief selection of the findings can be given here.

Many earthquakes were recorded at the dam from 1970 onwards, the most significant of which had M=3.9 with epicentre only 8 Km from the dam with focal depth of 10 Km. Because seismographs had been installed at base and crest, magnification factors could be deduced. Power spectrum analysis was carried out for the most significant earthquakes to obtain resonance frequency, and because they occurred at different water levels. Table 3 has some interest from the structure/fluid interaction aspect. It will be seen that for some components increase in water level does not appear to cause reduction in frequency. Such data was augmented by forced vibration tests at two different water levels with results shown in Fig 13 which do show the normally expected behaviour. From the earthquake records damping was deduced using the half-power method, and for the first symmetrical mode 10% was obtained, contrasting with 3% from the forced vibration tests; it is argued that the larger strain levels produced by the earthquakes is the reason for this, it being known of course from laboratory tests that damping does increase with strain.

The Kurobe observations also provides us with one of the very few measurements of excitation of the two abutments separately. Looking at acceleration first, Fig 14 compares maximum acceleration recorded on the two abutments for many earthquake. There is concentration of data along the 1:1 slope line, but closer observation shows many values being in a band of from half to twice that of the opposite abutment. The total analysis of this data also includes power spectra of acceleration in the base rock of each abutment, and the coherence between these records, the general conclusion being that there is no close relation between records on the two abutments.

24 Portuguese (LNEC) Experience It is assumed that this has been dealt with by Prof J O Pedro, but it will be mentioned here that a resume of activity to 1986 has been presented in Ref 19. These activities include the measurement of reservoir induced seismicity (RIS) at Cahora Bassa arch dam on the Zambesi river in Africa and at Portuguese dams, seismic monitoring and vibrating testing using mechanical vibrators. The studies on the multiple-arch Aguieira dam being particularly interesting. These LNEC studies form part of a systematic on-going programme.

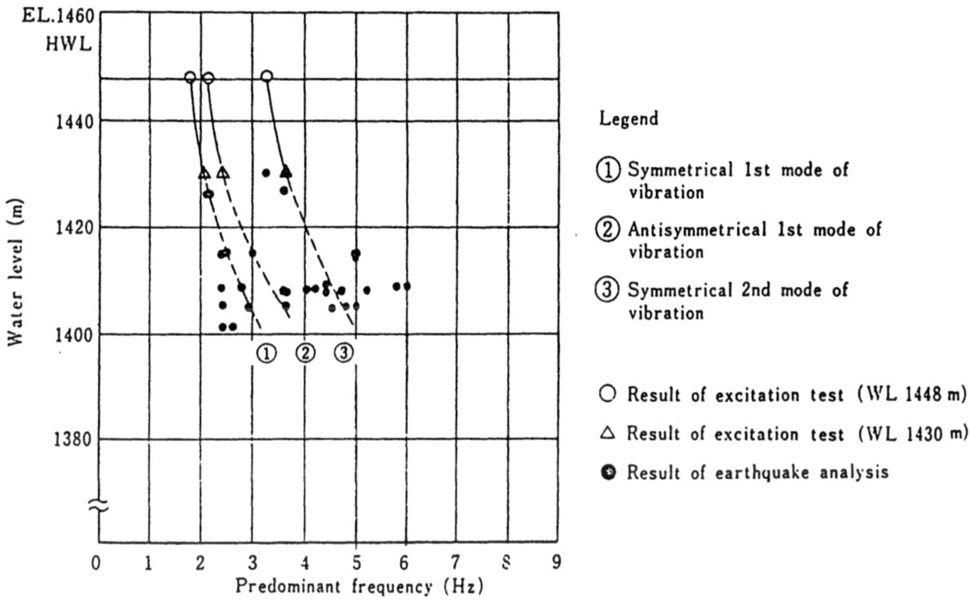

13. Kurobe dam: predominant frequencies in the radial direction versus water level

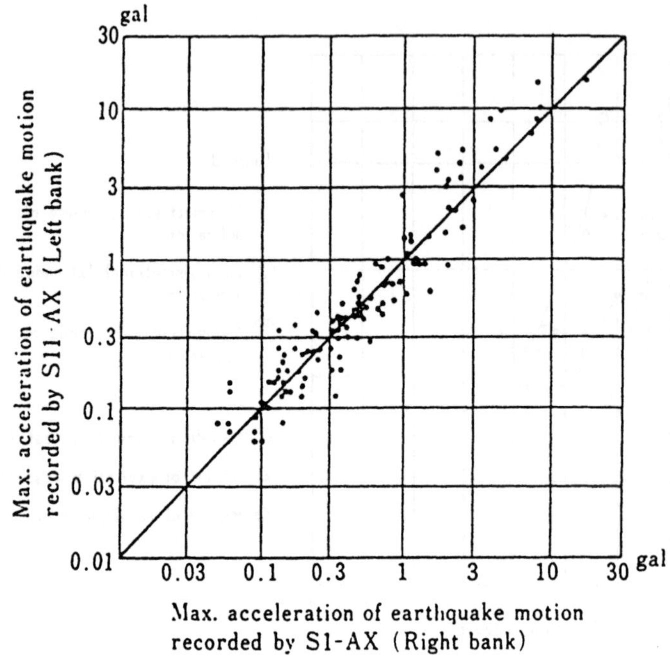

14. Kurobe dam: relationship between maximum acceleration at the left and right foundation rock in the direction of the dam axis

Date	Component	Max. acceleration (gal)	Water level (m)	Predominant frequency (Hz)
	S 5-AX	19.1		4.2 7.4
	S 6-AX	28.9		2.4 3.5 5.2
1970. 2. 27	S 7-AX	53.1	1408	3.6
(19°13′)	S 9-AX	31.8		4.4
	S 10-AX	24.3		2.8 4.7 6.4 7.8 9.0
	S 5-AX	–		—
	S 6-AX	117.9		1.0 2.5 4.5 6.0 7.2
1972. 9. 8	S 7-AX	116.3	1415	2.3 5.0 6.8
(23°02′)	S 9-AX	–		—
	S 10-AX	20.0		3.0 5.5 7.5
	S 5-AX	–		—
	S 6-AX	20.4		2.5 5.0 7.3 9.0
1972. 9. 8	S 7-AX	16.6	1415	2.5 5.0 7.5 9.0
(23°04′)	S 9-AX	–		—
	S 10-AX	2.3		5.0 7.5 15.3
	S 5-AX	–		—
	S 6-AX	7.2		2.60
1974. 5. 9	2 7-AX	10.7	1401	2.40
(8°34′)	S 9-AX	4.2		2.40
	S 10-AX	–		—
	S 5-AX	8.3		5.0
	S 6-AX	22.9		3.6
1976. 4. 28	S 7-AX	24.5	1405	2.9 4.6
(21°31′)	S 9-AX	9.1		2.4 4.6
	S 10-AX	11.3		4.8
	S 5-AX	7.8		6.0 8.3 9.7 11.8 13.6 17.0
	S 6-AX	12.2		4.0 5.8
1977. 10.19	S 7-AX	–	1409	—
(5°53′)	S 9-AX	5.9		4.4 6.0 13.0
	S 10-AX	5.9		5.8 7.8 12.4 13.7
	S 5-AX	–		—
	S 6-AX	4.2		2.2 3.6
1978. 1.14	S 7-AX	6.1	1427	2.1
(12°25′)	S 9-AX	–		—
	S 10-AX	–		—
	S 5-AX	19.1		5.2 8.2 10.2
	S 6-AX	16.1		3.6 9.5
1978. 11. 8	S 7-AX	17.1	1430	3.2 5.3 9.8
(13°42′)	S 9-AX	5.6		5.0
	S 10-AX	8.5		7.0 8.8

Table 3. Kurobe Dam: predominant frequencies in the radial direction

REFERENCES		ARCH DAM PROTOTYPE STUDIES
1	Deinum P J et al	Vibration Tests on Emmosson Arch Dam E'quake Eng Struct Dyn, Vol 10, 1982
2	Brownjohn JMW Severn R T Taylor C A	Observations of the Dynamic Behaviour of Two Arch Dams. Int Workshop on Arch Dams, Coimbra Portugal 5-9 April 1987, Balkema Publishers, Rotterdam
3	Ellis B R Jeary A P Severn R T	Lessons From Prototype Dynamic Measurements and Corresponding Analyses ICOLD 15th Congress, Lausanne, Q56 R70, 1985
4	Daniell W E Taylor C A	Full-Scale Dynamic Testing and Analysis of a Reservoir Intake Tower E'quake Eng Struct Dyn Vol 23, 1219-1237, 1994
5	Okamoto S et al	Dynamic Behaviour of an Arch Dam during Earthquakes. Univ of Tokyo Press, 1965 DUC 627 825 7 042 7
6	Okamoto S et al	Static and Dynamic Behaviour of Kurobe Dam Univ of Tokyo Press, 1988, ISBN 4-13-068139-7
7	Duron Z H Hall J F	Experimental and Finite Element Studies of the Forced Vibration Response of Morrow Point Dam E'quake Eng and Struct Dyn Vol 16, 1988, 1012-1029
8	Keightley W O Housner S W Hudson D E	Vibration Tests of the Encino Dam Intake Tower EERI Report, Cal Tech, Pasadena, Calif July 1961
9	Severn R T Jeary A P Ellis B R	Forced Vibration Tests and Theoretical Studies on Dams. Proc ICE Pt 2, 1980, 69, Sept, 605-634 Proc ICE Pt 2, 1981, 71 , June 575-595
10	Brownjohn J M W	Assessment of Structural Integrity by Dynamic Measurements. PhD Thesis, Dept of Civ Eng, Univ of Bristol 1988
11	Brownjohn J M W Severn R T Taylor C A	Ambient Vibration Survey of Contra Dam Univ of Bristol, Dept of Civ Eng, Int Report 1986
12	Tanaka Y Hudspeth R T	Restoring Forces on vertical Cylinders Forced by Earthquakes. E'quake Eng Struct Dyn 16, 99-119, 1988
13	Rea D Liaw C Y Chopra A K	Dynamic Properties of the San Bernadino Intake Tower. Report UCB/EERC 75/07, Univ of Calif, Berkeley, 1975

14	Mau S T Wang S	Arch Dam Identification using Vibration Test Data. E'quake Eng Struct Dyn, Vol 18 No 4, May 1989, 491-
15	Calciati F et al	Experience Gained During In Situ Artificial and Natural Dynamic Excitation of Large Concrete Dams in Italy. ICOLD 13th Congress, New Delhi 1979, Q51 R32
16	Widmann R	The Dynamic Behaviour of Arch Dams Investigations by Means of Calculations and Measurements ICOLD 13th Congress, New Delhi 1979, Q51 R9
17	Swanson A A Sharma R P	Effects of the 1971 San Fernando Earthquake on Pacoina Arch Dam ICOLD 13th Congress, New Delhi 1979, Q51 R3
18	Wulff J G Van Orden R C	Evaluation of the Earthquake Stability of Lower Crystal Springs Dam, California USA. ICOLD 13th Congress, New Delhi 1979, Q51 R1
19	Pedro J O et al	Seismic Monitoring of Concrete Dams and Reservoirs, Portuguese Experience 8th Euro Conf on E'quake Eng 8-12 Sept 1986
20	Morrison J et al	Earthquake Recordings on or Near Dams U.S.C.O.L.D, Nov. 1977
21	Darbre G R	Strong-Motion Instrumentation of Dams Earthquake Eng. and Struct. Dyn. Vol. 24 1101-1111 (1995)
22	Rouse G C Bouwkempt J G	Vibration Studies of Monticello Dam U.S. Dept. of Interior, Bur. of Rec. Res. Report No. 9, 1967
23	Mau S T Wang S	Arch Dam System Identification Using Vibration Test Data Earthquake Eng. and Struct. Dyn. Vol. 18, No.4 May 1989

DIFFERENTIAL GROUND MOTIONS

1 In seismic analysis of structures it is customary to assume that the ground input is common to all parts of the base of the structure. This is clearly not correct, because it assumes that the excitation is travelling with infinite velocity. However unless the structure is large the error is of no significance. For example, if the velocity of wave propagation is 1830 m/s, with a predominant frequency of 3 Hz and therefore a wavelength of 610 m, a building with a base dimension of 30 m would be subjected to essentially the same ground motion throughout. However, a dam spanning 500 m would experience different ground motions at different support points. If the same signal, unaffected by the ground through which it passes, is considered to be imposed on the structure, different base points will experience the same input, but out of phase due to the finite travel time; such an input is described as ***asynchronous***. Usually ground conditions will modulate the signal causing different base points to experience totally different inputs; the structure is then said to experience ***multiple support excitation*** (Ref. 1, 2).

2 The foregoing is written with a single component of ground input in mind, whereas in an actual earthquake motion takes place in all six degrees of freedom, which clearly gives rise to multiple support excitation. The common design practice of allowing for vertical input by taking a proportion (two-thirds?) of the horizontal input acting synchronously with it is clearly of doubtful validity.

3 There is not much recorded evidence of differential ground motion relating to arch dams, possibly the most significant being that from the Italian Ambiesta double curvature arch dam during aftershocks associated with the Friuli earthquake of 1976 (Ref. 3) where geophone measurements were taken on the left and right abutments 10m below the crest, and also in the lowest region of the crown cantilever where it meets the foundation, Fig. 1. The response spectra from these three stations, though similar in shape, have maximum spectral acceleration values of around 65, 26 and 13 cm/s^2 respectively, Fig. 2. For this dam, measured response was compared with calculated responses based first on the three different boundary accelerations just discussed, second on constant acceleration of 65 cm/s^2, and third on constant acceleration of 13 cm/s^2. The first of these three gave much the best agreement between calculation and measurement, thereby emphasising the importance of differential ground motion. Other measurements were made at the Kurobe (Japan) dam (Ref. 6), where seismographs were installed at 1380m on the right bank and 1395m on the left bank to measure upstream-downstream acceleration. Many small shocks of magnitude less than 5.1 were recorded, as shown in Fig. 3. The majority of acceleration points are distributed on either side of the 1:1 line in a region ranging from half to twice the acceleration on the opposite bank. Thus, it is reasonable to conclude that a factor of 2 covers the difference between the two banks.

4 It is not easy to predict the effect on all response characteristics of these differential base inputs, whether multiple support or asynchronous. However, if a response spectrum approach is used for the calculation it is evident that response of the structure will be lower for the asynchronous case then for the synchronous case where all base points have common input. This is because the definition of response spectrum characteristics (displacement, velocity or acceleration) requires them to be maximum values for the earthquake being considered. All other values are less than the response spectrum values. For synchronous excitation <u>all</u> support points experience the spectral acceleration, whereas in the asynchronous case only one point has this value; all others have lower acceleration. For multiple support

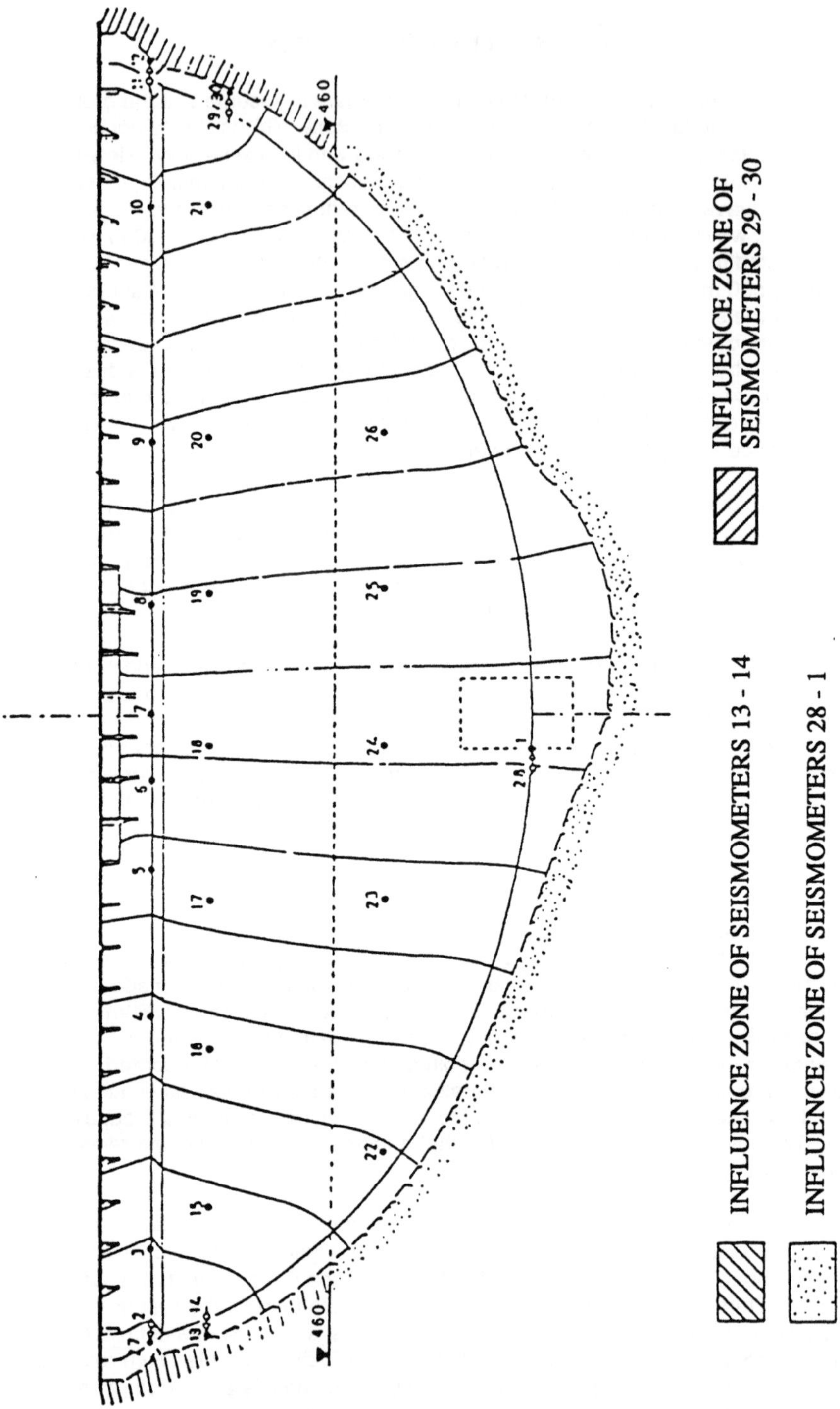

Fig.1 Network of seismometers on the Ambresta Dam downstream face

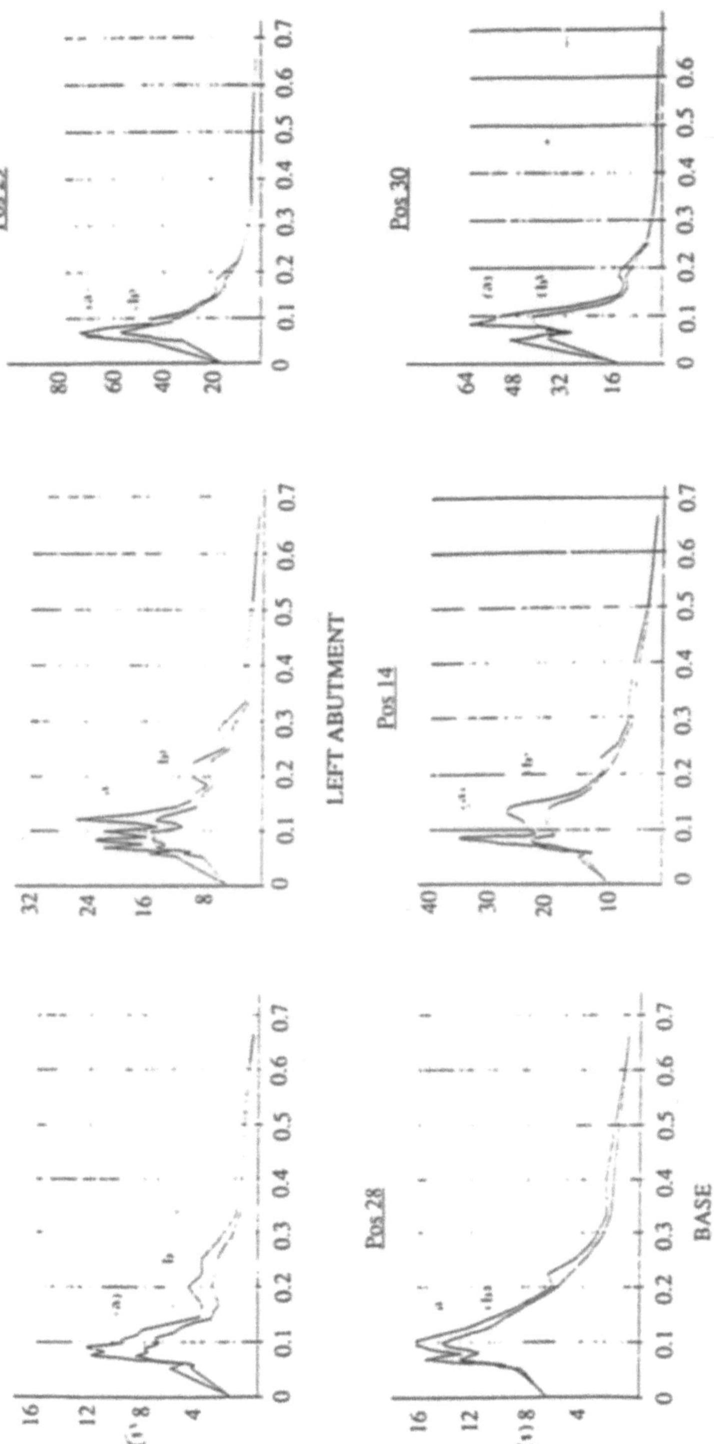

Fig. 2 Response spectra of an aftershock recorded at different dam-foundation points (see Fig. 1)

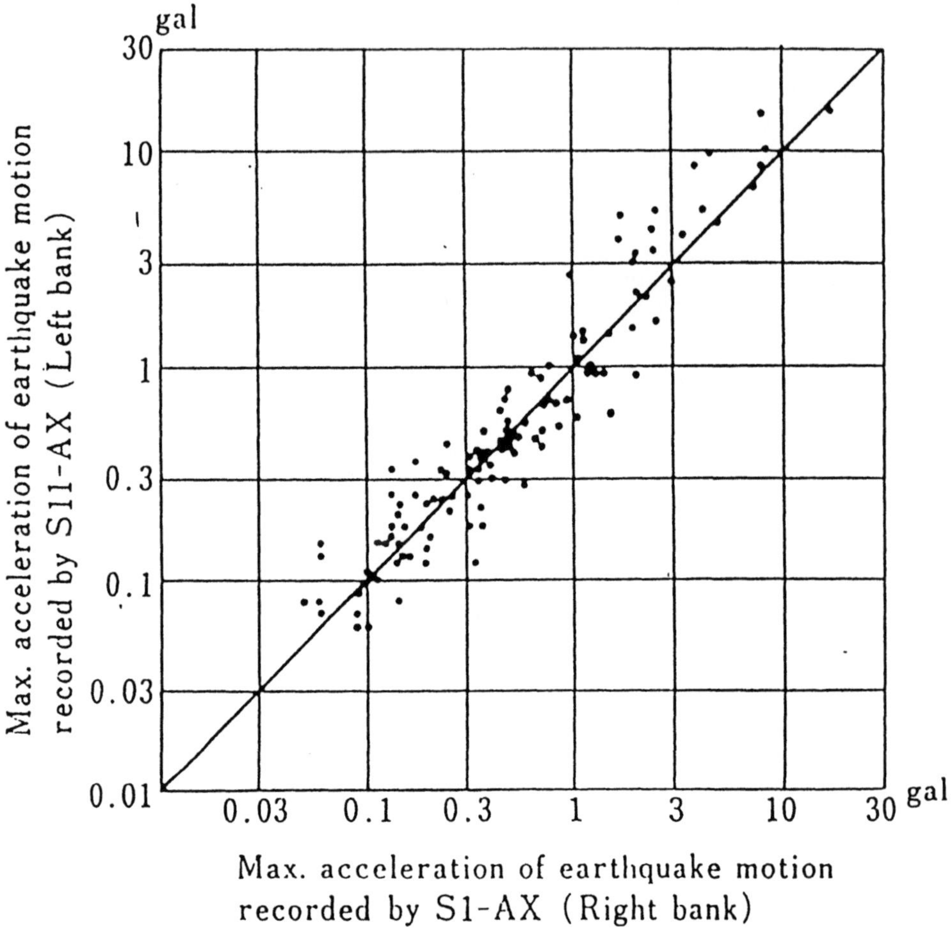

Fig. 3 Kurobe dam. Relationship between maximum acceleration at the left and right foundation rock in the direction of the dam axis

excitation the same conclusion is likely to be valid, because it is unlikely that the separate contributions to response at a certain point from all the different ground inputs will be at their maximum values simultaneously.

5 For synchronous support motion response stresses arise solely from the inertia of the structure. If the structure were to possess no mass, no dynamic stresses would be generated. This is not so far asynchronous or multiple support motion because the differential movements between support points cause stresses which depend on the elasticity of the structure and not its mass. Consider the portal frame shown in Fig. 4. If this is first assumed to be without mass and the foot A is given a time-varying movement while foot B is stationary, at any instant during the movement of A, the frame will be distorted and stresses generated. Given the movement of A these stresses can be calculated by a static analysis. They are therefore referred to as 'pseudo-static' because although they are time-dependent, it would be misleading to refer to them as dynamic, because dynamic stresses are those arising from inertia. Returning to the movement of A, it is clear that all that we need, assuming linear behaviour, is the shape taken up by the frame for a unit movement of A in the appropriate direction. If A has 6 degrees of possible movement, there would be 6 such shapes to calculate. The pseudo-static stresses at any time would then be calculated from this unit shape (later to be called the r-vector) multiplied by the actual displacement at that time taken from a record of displacement, or the integration of a velocity or acceleration record.

Fig. 4 Portal frame

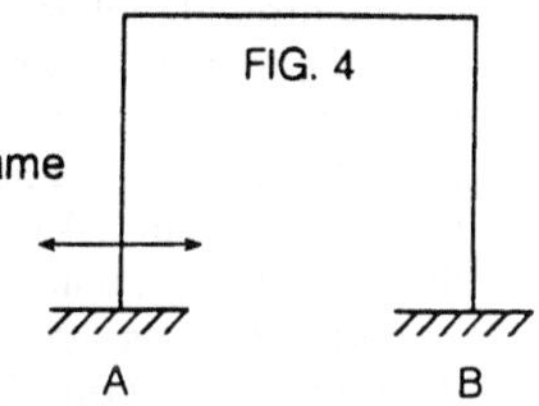

If the portal frame is now allowed its mass, the movement of A will have generated dynamic movement in the whole frame, thereby causing true dynamic stress. Thus, when the base of a structure is given differential movement, whether multiple support or asynchronous, displacement and stress derive from two sources, dynamic and pseudo-static.

6 For a mathematical analysis it will be helpful to refer to the simplest possible structure, a simple shear-type cantilever shown in Fig. 5.. In the absence of external forces applied directly to the structure, the equilibrium condition is

Fig. 5

Simple cantilever

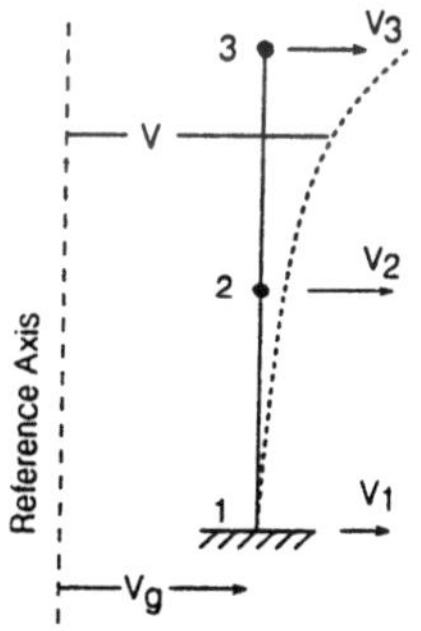

Inertia force + damping force + stiffness force = 0.

Because it is intended to use the finite element, method with response expressed by nodal degrees of freedom, the term 'response degree of freedom, (RDOF) is used to define those structural degrees of freedom not attached directly to the ground; the latter are called 'ground degrees of freedom' (GDOF). In the shearing cantilever, V_2 and V_3 are RDOF, whereas V_1 is GDOF. V is used to represent total displacement, that is, ground displacement plus dynamic displacement due to inertia.

With M, C and K being the conventional 3 x 3 matrices, (1) is written

$$\overline{M} \begin{Bmatrix} \ddot{V}_1 \\ \ddot{V}_2 \\ \ddot{V}_3 \end{Bmatrix} + \overline{C} \begin{Bmatrix} \dot{V}_1 \\ \dot{V}_2 \\ \dot{V}_3 \end{Bmatrix} + \overline{K} \begin{Bmatrix} V_1 \\ V_2 \\ V_3 \end{Bmatrix} = 0 \qquad (2)$$

If the cantilever were either infinitely stiff, or without mass, the total displacement at any point would be the same as the ground displacement, but in the normal condition the RDOF have a displacement additional to the ground displacement. Thus

$$V(t) = \nu(t) + \nu^s(t) = \nu(t) + r\nu_g(t) \qquad (3)$$

where $\upsilon(t)$ is the vector of dynamic displacement (due to inertia) and $\upsilon^s(t)$, depends on the ground motion; it is convenient to write it as $r\upsilon_g(t)$, with r being a 3 x 1 vector (in this case) multiplying the ground displacement. It may be referred to as the 'ground displacement shape vector' because it determines the deformed shape of the structure when a particular GDOF is given unit displacement with all other GDOF held fixed. A different r-vector is therefore calculable for each GDOF. [For this cantilever, a unit displacement of the GDOF υ_1 causes unit displacement in both υ_2 and υ_3 because there is no other ground point to be held fixed; the r-vector is then a column of ones].

Substitution of (3) into (2) gives

$$\overline{M}\begin{Bmatrix} o+\ddot{\nu}_g \\ \ddot{\nu}_2+r_2\ddot{\nu}_g \\ \ddot{\nu}_3+r_3\ddot{\nu}_g \end{Bmatrix} + \overline{C}\begin{Bmatrix} o+\dot{\nu}_g \\ \dot{\nu}_2+r_2\dot{\nu}_g \\ \dot{\nu}_3+r_3\dot{\nu}_g \end{Bmatrix} + \overline{K}\begin{Bmatrix} o+\nu_g \\ \nu_2+r_2\nu_g \\ \nu_3+r_3\nu_g \end{Bmatrix} = 0 \qquad (4)$$

$$M\ddot{\nu} + C\dot{\nu} + K\nu = -\left(Mr + m_g\right)\ddot{\nu}_g - \left(Cr + C_g\right)\dot{\nu}_g - \left(Kr + k_g\right)\nu_g \qquad (5)$$

$$M = \begin{bmatrix} m_{22} & m_{23} \\ m_{32} & m_{33} \end{bmatrix} \quad C = \begin{bmatrix} c_{22} & c_{23} \\ c_{32} & c_{33} \end{bmatrix} \quad K = \begin{bmatrix} k_{22} & k_{23} \\ k_{32} & k_{33} \end{bmatrix} \qquad (6)$$

$$m_g = \begin{Bmatrix} m_{21} \\ m_{31} \end{Bmatrix} \quad c_g = \begin{Bmatrix} c_{21} \\ c_{31} \end{Bmatrix} \quad k_g = \begin{Bmatrix} k_{21} \\ k_{31} \end{Bmatrix} \qquad (7)$$

In (6) the matrices relate RDOF to each other; in (7) the vectors relate RDOF to the single GDOF in this case.

The dynamic equilibrium equation (5) necessarily contains the static equilibrium equation, obtained by removing all dynamic terms. Thus

$$\left(Kr + k_g\right)\nu_g = 0, \quad giving \;\; r = -K^{-1}k_g \qquad (8)$$

For this shear-type cantilever

$$\overline{K} = \frac{12EI}{h^3}\begin{bmatrix} 1 & -1 & 0 \\ -1 & 2 & -1 \\ 0 & -1 & 1 \end{bmatrix}, \quad K = \frac{12EI}{h^3}\begin{bmatrix} 2 & -1 \\ -1 & 0 \end{bmatrix}, \quad k_g = \frac{12EI}{h^3}\begin{Bmatrix} -1 \\ 0 \end{Bmatrix} \tag{8a}$$

which with use of (8) shows that r is indeed the unit vector. The terms on the right-hand side of (5) constitute the effective forces due to the ground motion, and (8) shows that the last term is zero. As the second and third terms are of the same form, if damping is directly proportional to stiffness, the second term also makes zero contribution. Although this is not true if damping is proportional to mass, the damping term makes only a small contribution relative to the mass term and may be ignored. Thus

$$M\ddot{v} + C\dot{v} + Kv = -\left(Mr + m_g\right) v_g \tag{9}$$

For a diagonal mass matrix m_g is zero, and even for a consistent mass matrix there will be only a few non-zero terms in the coupling matrix m_g. Thus, with little loss of accuracy, bearing in mind other parameters in seismic analysis,

$$M\ddot{v} + C\dot{v} + Kv = -Mr\,\ddot{v}_g \tag{10}$$

7 To extend the analysis to multiple support excitation consider the dam-like structure shown in Fig. 6, which, to avoid algebra, but without loss of generality, has only one degree of freedom at each node. RDOF are 2 and 4; GDOF are 1,3 and 5.

The matrix statement relating pseudo-static displacements and corresponding forces is

$$\begin{bmatrix} k_{11} & & & & sym \\ k_{21} & k_{22} & & & \\ k_{31} & k_{32} & k_{33} & & \\ k_{41} & k_{42} & k_{34} & k_{44} & \\ k_{51} & k_{52} & k_{54} & k_{54} & k_{55} \end{bmatrix} \begin{Bmatrix} v_1 \\ v_2 \\ v_3 \\ v_4 \\ v_5 \end{Bmatrix}^s = \begin{Bmatrix} f_1 \\ f_2 \\ f_3 \\ f_4 \\ f_5 \end{Bmatrix} \tag{11}$$

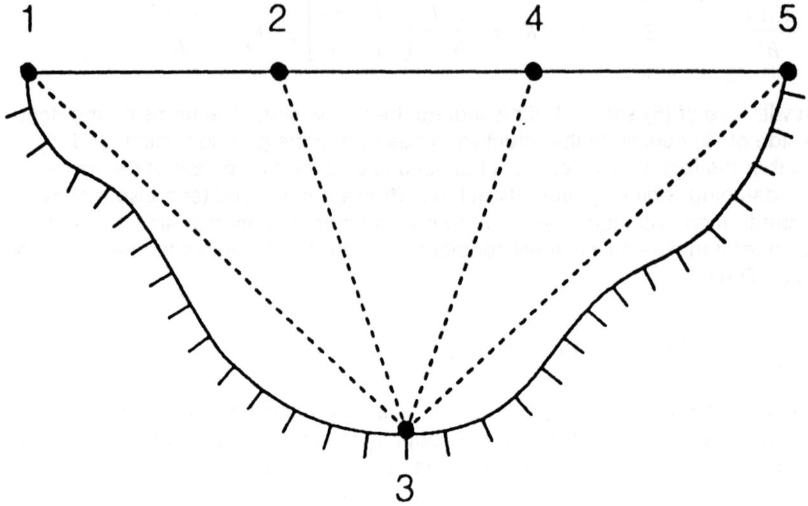

Fig. 6 Membrane dam

Putting $v_5^s = 1, \quad with \ v_1^s = v_3^s = 0 \quad and \quad f_2 = f_4 = 0$

$$\begin{bmatrix} k_{22} & k_{24} \\ k_{42} & k_{44} \end{bmatrix} \begin{Bmatrix} v_2 \\ v_4 \end{Bmatrix}^s = - \begin{Bmatrix} k_{25} \\ k_{45} \end{Bmatrix} . \tag{12}$$

The square matrix is K and the right-hand side vector is k_g; because the r-vector is defined as $K^{-1}k_g$, (12) shows that it is the vector of displacements in the RDOF produced by a unit displacement in a corresponding GDOF, here V^s_5

8 If all three GDOF are given unit displacement together,

$$k_g = - \begin{Bmatrix} k_{21} + k_{23} + k_{25} \\ k_{41} + k_{43} + k_{45} \end{Bmatrix} .1 \tag{13}$$

Because the sum of terms in any row of a stiffness matrix is zero, it can be shown that the r-vector is now the unit vector, as it should be for this case of synchronous input. However, such input in a certain direction produces a unit r-vector only if there are no coupling terms in the stiffness matrix between the degrees of freedom in the direction of this synchronous ground motion and the other degrees of freedom. This is unlikely to be so, and so the use of a simple 'direction vector' of noughts and ones which is sometimes used instead of r in (10) is at best only a good approximation.

9 The r-vector describes the shape which the structure takes up when one of the GDOF is given a unit value; an r-vector has to be calculated for each GDOF which experiences ground movement. Equation (8) can always be used for this, but some computer programming is required which involves partitioning the system stiffness matrix and this takes considerable computer time, particularly when many different r-vectors are required. An alternative method is available, which though theoretically inexact, involves errors which are trivial when one recalls all other uncertainties in seismic analysis. This method consists simply of carrying out a static analysis with the addition of a relatively large stiffness to the diagonal term corresponding to the GDOF for which the r-vector is being found, with a load numerically equal to that large stiffness applied corresponding to this degree of freedom. No other loads are applied, and so the displacement of the GDOF can be made as close to unity as is desired by increasing the ratio of the spurious large stiffness to the real diagonal stiffness terms. As will be seen, the necessary additional computer programming involved in this new approach to r-vector calculation is trivial.

10 Design against dynamic load normally requires calculation of stresses as well as displacements, and for this it is necessary first to calculate dynamic nodal forces associated first with RDOF and then with GDOF. When these are known, the internal dynamic stresses can be found by either a time-domain or frequency domain analysis depending on the type of structure being considered.

As illustration consider the 'membrane-dam' again (Fig 6)

$$\begin{bmatrix} k_{11} & & & & \\ k_{21} & k_{22} & & & \\ k_{31} & k_{32} & k_{33} & & \\ k_{41} & k_{42} & k_{43} & k_{44} & \\ k_{51} & k_{52} & k_{53} & k_{54} & k_{55} \end{bmatrix} \begin{Bmatrix} V_1 \\ V_2 \\ V_3 \\ V_4 \\ V_5 \end{Bmatrix} = \begin{Bmatrix} f_1 \\ f_2 \\ f_3 \\ f_4 \\ f_5 \end{Bmatrix} \tag{14}$$

where the displacements are total and the forces are dynamic. Considering the two RDOF first, and supposing $V_1 = V_5 = 0$ with $V_3 = V_g$. Then

$$\begin{Bmatrix} f_2 \\ f_4 \end{Bmatrix} = \begin{bmatrix} k_{22} & k_{24} \\ k_{42} & k_{44} \end{bmatrix} \begin{Bmatrix} V_2 \\ V_4 \end{Bmatrix} + \begin{Bmatrix} k_{23} \\ k_{43} \end{Bmatrix} V_g \tag{15}$$

or

$$f_R = KV + k_g V_g = K\left(v + rv_g\right) + k_g V_g = Kv + \left(Kr + k_g\right) v_g \tag{16}$$

because $V_g = \upsilon_g$. The last term is zero by definition of r and so the dynamic forces acting on the RDOF are obtained from dynamic displacements only. In this illustration only one GDOF has been allowed to move, but the result is clearly valid for multiple movement of GDOF.

For the forces associated with the GDOF

$$\begin{Bmatrix} f_1 \\ f_3 \\ f_5 \end{Bmatrix} = \begin{Bmatrix} k_{12} \\ k_{32} \\ k_{52} \end{Bmatrix} \left(v + v^s\right)_2 + \begin{Bmatrix} k_{13} \\ k_{33} \\ k_{53} \end{Bmatrix} v_g + \begin{Bmatrix} k_{14} \\ k_{34} \\ k_{54} \end{Bmatrix} \left(v + v^s\right)_4 \tag{17}$$

which, with $\upsilon^s = r\,\upsilon_g$, can be written

$$\begin{Bmatrix} f_1 \\ f_3 \\ f_5 \end{Bmatrix} = \begin{bmatrix} k_{12} & k_{14} \\ k_{32} & k_{34} \\ k_{52} & k_{54} \end{bmatrix} \begin{Bmatrix} v_2 \\ v_4 \end{Bmatrix} + \begin{bmatrix} k_{12} & k_{14} \\ k_{32} & k_{34} \\ k_{52} & k_{54} \end{bmatrix} \begin{Bmatrix} r_2 \\ r_4 \end{Bmatrix} v_g + \begin{Bmatrix} k_{13} \\ k_{33} \\ k_{53} \end{Bmatrix} v_g \tag{18}$$

Here each row of the 3x2 matrix consists of stiffnesses which link the two RDOF with a single GDOF and this vector has previously been called k_g; we therefore call the matrix K_g^T. The final 3x1 vector consists of stiffnesses linking the GDOF which has been allowed to move (ie node 3) with all other GDOF; this is now referred to as k_{gg}. [NB Not all the stiffnesses in (16) are non-zero in this illustration, but they have been written to show the pattern, from which generalisations can be made].
Because $r = -K^{-1} k_g$,

$$f_g = K_g^T v + \left[k_{gg} - K_g^T \left(K^{-1} k_g\right)\right] v_g \tag{19}$$

The first term here gives GDOF forces due to the RDOF, whereas the second gives forces due to the differential movements of the supports. The generalisation for multi-support movement is straightforward.

11 The dynamic displacement is of course found from

$$M\ddot{v} + C\dot{v} + Kv = -MR\ddot{V}_g \tag{20}$$

where M, C and K are the n x n RDOF; the number of GDOF is m, so that R is n x m (the assemblage of r-vectors) and $\ddot{V}g$ is m x 1, representing the m different ground accelerations. When some of these last are common a useful reduction in effort is possible.

If the modal superposition approach is used for (20), a prior calculation of natural frequencies, w_i, and nodal shapes phi - ϕ , is assumed.

If Y_i is used to denote modal amplitude

$$v = \sum_i \phi_i Y_i \tag{21}$$

and

$$\ddot{Y}_i + 2\beta_i \dot{Y}_i + w_i^2 Y_i = \left[\frac{\phi_i M}{\phi_i^T M \phi_i}\right] \left[r_1 \ddot{v}_{g1}\left(\tau_1, t\right) + r_2 \ddot{v}_{g2}\left(\tau_2, t\right) + \quad\right] \tag{22}$$

Here, the motion is assumed asynchronous, but the computational process, being digital, is the same for truly multiple support excitation.

The total pseudo-static displacement is calculated from

$$v^s = r_1 v_{g1}\left(\tau_1, t\right) + r_2 v_{g2}\left(\tau_2, t\right) + \ldots\ldots\ldots\ldots \tag{22a}$$

which means that the given ground accelerations must be integrated twice to give ground displacements. This causes error, which when added to the errors in the original record could cause spurious residual displacements which would affect the results obtained. In nearly all cases, zero residual ground displacement is required and this must be checked for.

In view of the very large amount of response data which can be generated, the Bristol software retains only the maximum value of each component of the displacement and stress vectors.

As an alternative to such response calculations the response spectrum approach can be considered, and this has been studied by Der Kuireghian and Neuenhofer in Ref. 6, but it has not been applied to arch dams.

12 A study has been made (Ref. 2) of the effects of differential ground motion on three types of dam, but only that on the 110m high Victoria arch dam in Sri Lanka is discussed here. The S16E component of the San Fernando (1971)

earthquake record was used, and the results presented compare this record applied in the upstream/downstream direction to the whole of the dam, with the same record applied to the left-hand half of the foundation only, the right-hand half being held fixed.

The digitised record contained 6000 time/acceleration pairs at an interval of $2.25.10^{-3}$ s. The displacements were also calculated at this interval, but in a first approximation only every 100th value was used to calculate stresses. A second approximation used every 50th value and this gave significantly different stresses, but using 10th and 5th values gave the same stresses and so in the results presented the sampling was for every 5th displacement value, that is, at an interval of $11.25.10^{-3}$s. Five natural frequencies and modal shapes were employed, with damping at 5%. In the calculations 108 twenty-node 3D elements were used in the dam itself and 114 eight-node 3D elements in the foundation block , this gave 2694 degrees of freedom in total. Fig 7 gives a digital map of the xx stress-component on the developed downstream face for a common base motion; the decimal point in each number indicates the point at which the stress has been calculated. These stresses are referred to Cartesian axes with origin at the base of the crown cantilever. Fig 8 gives the corresponding values of xx when only the left-hand half of the base is excited; the stress pattern is now quite different. In particular, xx in the crown cantilever region now has its maximum value at the base, rather than at the crest as in Fig.7. Other stress maps are given in Ref 2 from which Fig 9 is reproduced; it shows a direct comparison of the crown cantilever direct stresses for the two cases of ground motion.

Some earlier studies on the Victoria dam are given in Ref 1, but these were made with an inferior computer programme which allowed only twenty 20-node isoparametric elements to represent the whole dam. The principal value here, was a study of the use of the δ -vector approach compared to the r-vector approach. The former consists of ones (in the excitation direction) and zeros only, ignoring the stiffness coupling between the degrees of freedom, which the r-vector includes. Displacements obtained using the δ -vector are, as expected, smaller than those given by the r-vector because the former imposes unwarranted constraints.

The general conclusion of this and other work on dams (Ref. 3), is that the pseudo-static stresses are significant in magnitude and should certainly be considered in any design studies for seismic actions. The most recent ICOLD publication on the subject, Bulletin 52 - Earthquake Analysis Procedures for Dams (Ref. 4), does give the general theory in very formal terms, but gives no illustrative examples and makes no clear recommendations.

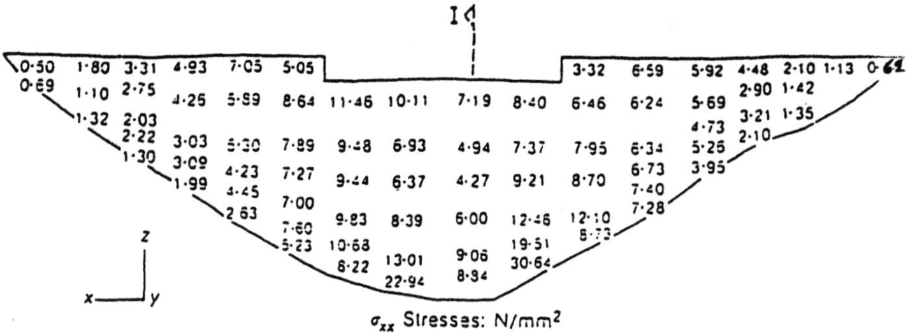

Fig. 7 Victoria Arch Dam. xx-stresses on downstream face due to common ground motion

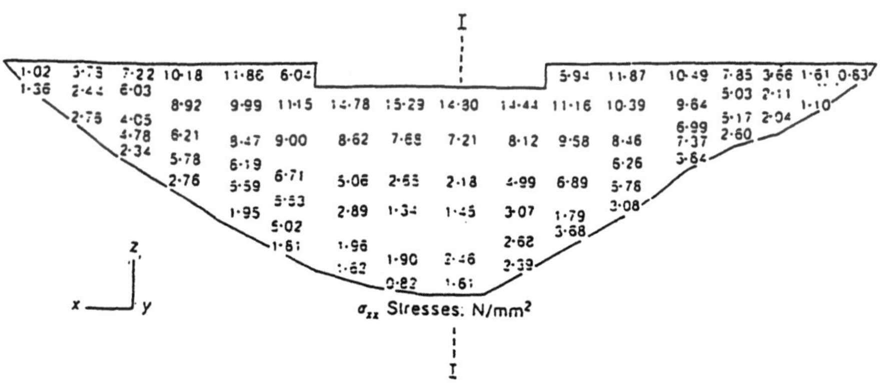

Fig. 8 Victoria Arch Dam. Stresses on downstream face due to half-base motion

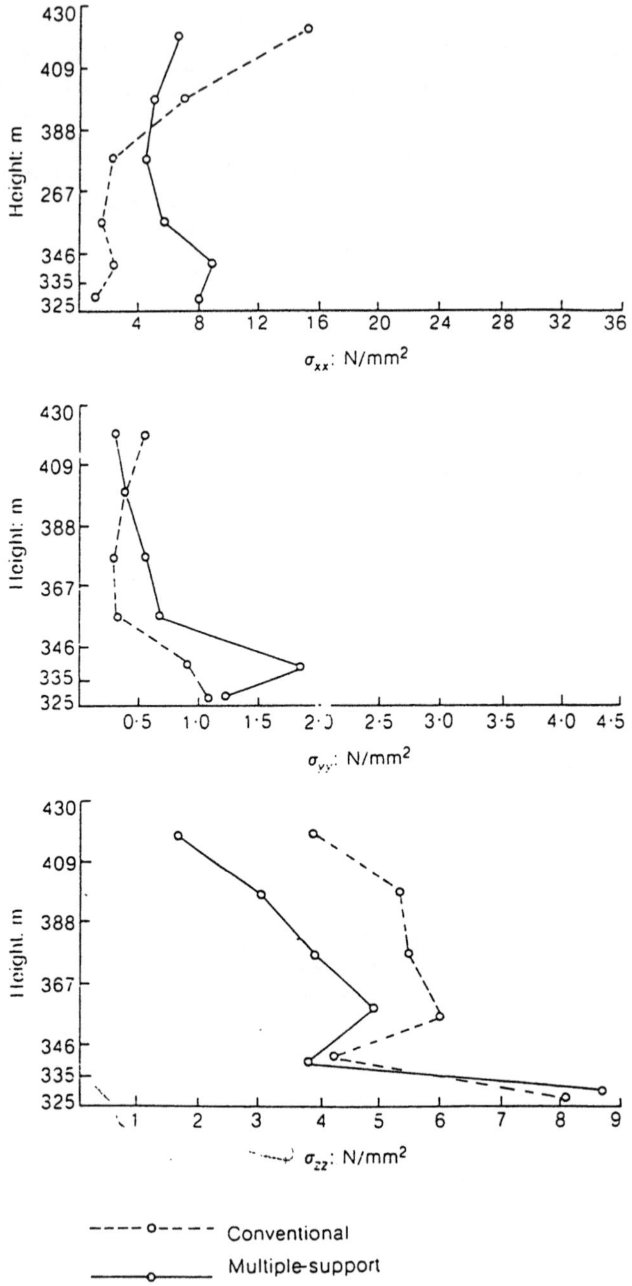

Fig. 9 Victoria Arch Dam. Stresses on section 1 on downstream face

REFERENCES

1	Altinisik D Severn R T	Multiple-Support and Asynchronous Base Excitation of Structures. Proc Instn Civ Engrs, Part 2, 1981, 71, June 407-426
2	Dumanoglu AA Severn R T	Dynamic Response of Dams and Other Structures to Differential Ground Motions. Proc Instn Civ Engs, Part 2, 1984, 77, Sept 333-352
3	Calciati F Castoldi A Ciacci R Fanelli M	Experience Gained During In-situ Artificial and Natural Dynamic Excitation of Large Concrete Dams in Italy; Analytical Interpretation of Results. Trans. ICOLD Congress, New Delhi 1979, 4, Q51, R32
4	Zienkiewicz O C Clough R W Seed H B	Earthquake Analysis Procedures for Dams - State of the Art ICOLD Bulletin No 52 1986
5	Altinisik D et al	A Seismic Design of Arch Dams Particularly the Contribution from the Reservoir and Multiple Support Excitation of the Base Paper 24, Dams and Earthquakes, ICE London 1981
6		Report on Observations of Kurobe Dam.Kansai Electric Power Company 1971

MIX
Papier aus verantwortungsvollen Quellen
Paper from responsible sources
FSC® C105338
FSC
www.fsc.org

If you have any concerns about our products,
you can contact us on
ProductSafety@springernature.com

In case Publisher is established outside the EU,
the EU authorized representative is:
Springer Nature Customer Service Center GmbH
Europaplatz 3, 69115 Heidelberg, Germany

Printed by Libri Plureos GmbH
in Hamburg, Germany

MIX
Papier aus verantwortungsvollen Quellen
Paper from responsible sources
FSC® C105338

If you have any concerns about our products,
you can contact us on
ProductSafety@springernature.com

In case Publisher is established outside the EU,
the EU authorized representative is:
Springer Nature Customer Service Center GmbH
Europaplatz 3, 69115 Heidelberg, Germany

Printed by Libri Plureos GmbH
in Hamburg, Germany